UNIVERSAL REALITY

The New

Theory of Everything

Edgar L. Owen

Copyright © 2016 by Edgar L. Owen

All rights reserved under International and Pan-American Copyright Conventions

First Edition, Version 1.0, September 1, 2016

Library of Congress Cataloging-in-Publication Data

Owen, Edgar L.
Universal Reality: The New Theory of Everything / Edgar L. Owen – first ed.
p. cm.
Includes biographical references.

ISBN-13: 978-1535596084 (Edgar L. Owen)

ISBN-10: 1535596082 (Pbk.)

EdgarLOwen.info

Self-published on CreateSpace.com

Printed in the United States of America

To my secret muse

PREFACE

Universal Reality is the sequel to the author's 2013 book Reality. In the intervening three years the theory has been extensively developed and clarified resulting in this entirely rewritten and expanded Theory of Everything with many new insights integrated in a much more coherent and easily understood overall structure.

This book assumes a general knowledge of modern physics, cosmology, cognitive science, and computer science, at least at the popular level, and some familiarity with the great perennial issues of philosophy will be helpful. It also assumes the ability to carefully and objectively examine the structures of mind and consciousness from the inside, in particular how the biological structures of language and mind construct our internal simulation of the reality in which we seem to exist.

But all that's really required is the desire to explore the deepest mysteries of reality from a completely new perspective with an open mind. Universal Reality begins with a general acceptance of modern science and the scientific method, but *reinterprets* what science and nature are really telling us in an entirely new light. By combining the fundamental insights from all fields of modern science it discovers an entirely new understanding of reality lying dormant and unrecognized at the core of all of them.

Whether or not you agree with the theories of Universal Reality, I think you will find them an extremely interesting and entertaining read, and a refreshingly new perspective on the universe. It's a completely new approach and certainly one you won't find anywhere else. The result is a theory that unifies all aspects of reality, including the core issues about which traditional science has nothing meaningful to say; namely consciousness, the present moment, existence, and realization.

The theory of Universal Reality explains all these in a manner completely consistent with modern science and carefully analyzed direct experience. The author is confident the reader will at least think the theories worthy of serious consideration as together they offer remarkably convincing solutions to many of the greatest problems of existence in a single unified Theory of Everything.

This is not another popular exposition of modern science, nor is it one of the innumerable New Age books replete with wildly irrational ideas and wishful thinking, though it does attempt to reveal the deeper meaning they both seek. Instead it's a completely new view of reality based solidly in science and logic, one that unifies all aspects of reality in a single logically consistent structure that is immensely compelling and satisfying.

This book was written primarily in an effort to clarify and further develop my own understanding of reality, but hopefully its publication will make it accessible to others as well and generate intelligent criticisms and suggestions for improvement. I personally believe it's the best, most accurate and complete view of reality that has so far been discovered, but reality itself is always full of mysteries and surprises and is always the final arbiter of truth.

To the extent Universal Reality is an accurate description of reality it is not something I have created, rather it is reality itself revealing itself to someone who has hopefully been able to observe and study it without projecting too much of his own personal programming onto it. Reality is continuously revealing itself to all of us in all its awesome glory, and I believe anyone willing to observe it carefully and open-mindedly will be able to personally verify and experience the truth of much of what this book contains.

I would like to thank everyone who has helped make this book possible and encouraged me while writing it. Thanks to all of you for putting up with my unusual hermetic life style. And a special thank you to all my wild visitors, including the occasional human, and to the beauty and profundity of nature, which always inspires me with joy and meaning. Thanks to reality itself for continuously revealing itself in all its glory to those who will only look with opened eyes, and thanks most of all to my secret muse. Thank you, thank you, thank you all!

And finally thanks to all those thinkers, scholars, scientists and visionaries throughout history without whose heroic efforts, genius and cumulative hard work this book could not have been written.

The author welcomes all comments and questions and can be contacted at Edgar@EdgarLOwen.com.

CONTENTS

INTRODUCTION 1
OUTLINE OF THE THEORY 5
EXISTENCE 12
    REALITY 12
    THE NATURE OF EXISTENCE 12
    THE AXIOM OF EXISTENCE 14
    THE PRESENT MOMENT 16
    TWO KINDS OF TIME 16
    HAPPENING 17
    THE UNIVERSE AS A LIVING SYSTEM 18
    ABSOLUTENESS 19
    IMMANENCE 20

CONSCIOUSNESS 23
    XPERIENCE 23
    REALITY AS XPERIENCE 25
    CONSCIOUSNESS 26
    NON-HUMAN CONSCIOUSNESS 35

COMPUTATIONAL REALITY 38
    THE COMPUTATIONAL UNIVERSE 38
    EXISTENCE & THE QUANTUM VACUUM 39
    EVIDENCE REALITY IS COMPUTATIONAL 40
    BIOLOGICAL PROGRAMS 46
    COMPUTATIONAL STRUCTURE 48
    PROGRAMS & DATA 51
    CONSISTENCY & COMPLETENESS 53
    SUPER CONSISTENCY 56
    IMPLICATIONS 56

FUNDAMENTAL PRINCIPLES 59
    THE EXISTENCE PRINCIPLE 59
    THE CONSISTENCY & COMPLETENESS PRINCIPLE 59
    THE PRINCIPLE OF DIFFERENTIATION 60
    THE STc PRINCIPLE 60
    THE CONSERVATION OF PARTICLE COMPONENTS 61
    THE MEv PRINCIPLE 62
    THE METc PRINCIPLE 65
    IMPLICATIONS OF THE METc PRINCIPLE 67
    THE SOURCE OF THE METc PRINCIPLE 72
    EXCLUSION PRINCIPLES 72
    THE PRINCIPLE OF CHOICE 74

- THE CONSERVATION OF INFORMATION 75
- SUMMARY 76

THE COMPLETE FINE-TUNING 77
- INTRODUCTION 77
- BRIEF REVIEW OF THE QUANTUM VACUUM 80
- COMPUTATIONAL SYSTEMS 81
- COMPUTATIONAL SPACE 83
- LOGICAL OPERATORS 86
- DATA TYPES 87
- ELEMENTAL DATA OF REALITY 89
- PARTICLE COMPONENTS 92
- FUNDAMENTAL CONSTANTS 96
- THE PLANCK CONSTANT 98
- ZERO-POINT ENERGY 98
- PARTICLE MASSES 100
- FOUR FORCES 103
- FOUR DIMENSIONS 105
- DEEP SYMMETRY 107
- THE REAL MICROCOSM & MACROCOSM 108
- THE ELEMENTAL PROGRAM 111
- HIGHER-LEVEL LANGUAGES 113

UNDERSTANDING TIME 116
- TWO KINDS OF TIME 116
- SOME THOUGHT EXPERIMENTS 118
- THE PRESENT MOMENT 121
- P-TIME 122
- CLOCK TIME 125
- THE STc PRINCIPLE 126
- TIME & GRAVITATION 129
- THE VELOCITY OF SPACETIME 132
- THE ARROW OF TIME 134
- CONFIRMING THE PRESENT MOMENT 135
- SPACE TRAVEL 136
- TIME TRAVEL 137
- OBSERVER SINGULARITIES 139
- SEEING ALL 4 DIMENSIONS 140
- ENTROPY 140
- TIME REVERSAL 142

COMPUTING SPACETIME 148
- MASS VIBRATIONS & GRAVITATION 148
- RETHINKING SPACETIME 152
- HAPPENING & THE P-TIME PROCESSOR 154
- THE UNIVERSAL REFERENCE BACKGROUND 156
- NEWTON'S BUCKET & MACH'S PRINCIPLE 159

    DIMENSIONAL DRIFT 161
    UNIFYING RELATIVITY & QUANTUM THEORY 165
    PROCESSOR CYCLES & THE STc PRINCIPLE 167
    PROCESSOR CYCLES & QUANTUM RANDOMNESS 168
    FOUR COMPUTATIONAL LEVELS 170
    THE EQUIVALENCE OF MASS-ENERGY & SPACE 174
    A NEW MODEL OF GRAVITATION 176
    GRAVITATIONAL ATTRACTION 179
    THE CLOCK POSTULATE 184
    MASS VIBRATIONS & THE HIGGS FIELD 186
    THE INCREASE OF MASS WITH VELOCITY 187
    DIMENSIONAL SPACETIME 191
    OBSERVER FRAMES 192

THE OTHER FORCES 196
    ELECTROMAGNETISM 196
    THE ELECTROMAGNETIC FIELD 200
    THE HELICAL FIELD MODEL 201
    PHOTONS 207
    ELECTRICITY 210
    THE STRONG & WEAK FORCES 212

QUANTUM REALITY 215
    EVENTS & THE ENTANGLEMENT NETWORK 215
    THE SOURCE OF QUANTUM INDETERMINACY 217
    WAVEFUNCTIONS & DECOHERENCE EVENTS 218
    QUANTUM GRAVITY 224
    A VISUAL MODEL OF THE UNIVERSE 226
    RESOLVING THE SPIN ENTANGLEMENT PARADOX 229
    FIELDS & VIRTUAL DECOHERENCE 232
    DIMENSIONAL FRAGMENTS 235
    THE NATURE OF PROBABILITY FUNCTIONS 240
    THE SOURCE OF QUANTUM RANDOMNESS 242
    RESOLVING QUANTUM PARADOX 243
    WAVEFUNCTIONS 245
    QUANTUM TUNNELING 246
    HALF-LIVES 248
    BOUND PARTICLES 249
    MASS-ENERGY STRUCTURES 254
    IMMANENCE & THE QUANTUM VACUUM 255

COSMOLOGY 257
    THE REAL VIRTUAL REALITY 257
    THE COSMOLOGICAL HYPERSPHERE 259
    CURVED LIGHT CONES 264
    COSMIC INFLATION 265
    THE HUBBLE EXPANSION 268

- REDSHIFTS ... 272
- A NEW DARK MATTER THEORY ... 274
- THE BIG BANG ... 276
- A BIG BOUNCE? ... 279
- TUNING THE FINE-TUNING ... 281

INFORMATION COSMOLOGY ... 283
- THE INFORMATION UNIVERSE ... 283
- INFORMATION DOMAINS & OBSERVERS ... 284
- THINGS ARE THEIR INFORMATION ... 287
- THINGS ARE THEIR COMPUTATIONAL HISTORY ... 289
- THE CONSERVATION OF INFORMATION ... 291
- THE SHERLOCK HOLMES PRINCIPLE ... 294
- FROM CAUSALITY TO CONSISTENCY ... 296
- HOW THE PRESENT DETERMINES THE PAST ... 297
- BEYOND THE ANTHROPIC PRINCIPLE ... 301
- THE PROBABILISTIC FUTURE ... 304

EMERGENCE ... 305
- THE EMERGENT UNIVERSE ... 305
- EMERGENT PROGRAMS ... 306
- EMERGENT LAWS ... 309
- THE GENERAL PRINCIPLE OF EVOLUTION ... 310
- INANIMATE PROGRAMS ... 312
- BIOLOGICAL PROGRAMS ... 314
- FREE WILL ... 318
- LARGE SCALE EMERGENCE ... 320
- CONVERGENT EMERGENCE ... 322
- OBSERVERS AND EMERGENCE ... 325
- THE INTELLIGENCE OF DESIGN ... 326

LIFE ... 328
- THE PRECURSORS OF LIFE ... 328
- THE ORIGINS OF LIFE ... 330
- NUCLEATED CELLS ... 333
- MULTICELLULAR ORGANISMS ... 335
- BRAINS & NERVOUS SYSTEMS ... 337
- REPRODUCTION ... 339
- FUNCTIONAL AUTONOMY ... 341
- FUNCTIONAL DESIGN ... 343
- LEARNING AND CULTURE ... 344
- TECHNOLOGY ... 346
- COMMUNICATION AND LANGUAGE ... 347
- ART AND WRITING ... 348
- FUNCTIONAL EVOLUTION ... 351
- THE BIOSPHERE ... 354
- THE FUNCTION OF DEATH ... 355

- HUMAN SELF DESTRUCTION 356
- PROBABLE FUTURES 360
- AN OPTIMAL FUTURE 361
- SUCCESSOR SPECIES 367

THE SIMULATION 369
- WE LIVE IN A SIMULATION 369
- EVOLUTIONARY ORIGIN 371
- SIMULATION STRUCTURE 373
- PERSONAL PROGRAMMING 376
- BIOLOGICAL PROGRAMING 379
- PERCEPTUAL ILLUSIONS 381
- THE ILLUSION OF INDIVIDUAL THINGS 381
- THE ILLUSION OF AN OBJECTIVE SELF 383
- OBSERVER ILLUSIONS 384
- QUALIA 387
- THE RETINAL SKY 388
- REALITY IS A RUNNING PROGRAM 390
- OURSELVES AS PROGRAMS 391
- ILLUSIONS OF TIME 394
- DREAMS 395
- OUT OF BODY EXPERIENCES 396
- ARTIFICIAL REALITIES 398

TRUTH AND KNOWLEDGE 400
- INTUITION 400
- KNOWLEDGE, INTELLIGENCE & WISDOM 401
- THE LOGIC OF THINGS 403
- SCIENTIFIC METHOD 405
- REALITY MATH & HUMAN MATHEMATICS 409

REALIZATION 412
- THE DIRECT EXPERIENCE OF REALITY 412
- APROACH 412
- THE FUNDAMENTAL REALIZATION 413
- REALIZATION OF TIME 414
- REALIZATION OF SPACE 417
- REALIZATION OF INFORMATION 418
- REALIZATION OF INFORMATION HISTORY 421
- REALIZATION OF IMMANENCE 423
- REALIZATION OF CONSCIOUSNESS 427
- REALIZATION OF TRUE NATURE 429
- REALIZATION OF CHI & ENERGY BODY 434
- DEFINING GOD 436
- BUDDHA NATURE 441
- REALIZATION OF LOVE 444
- ACCEPTANCE 446

PURPOSE AND ETHICAL PRINCIPLES	447
THE ENLIGHTENMENT EXPERIENCE	451
ZEN MIND	452
YOU ARE ALREADY ENLIGHTENED	453

EPILOGUE - TESTING THE THEORY	457
BIBLIOGRAPHY	460

INTRODUCTION

Universal Reality is a completely new science based Theory of Everything. It begins with the latest findings in modern physical, cognitive and information science and reveals the fundamental unity hidden within them. It then presents a model of reality that not only explains quantum theory and general relativity as parts of a unified theory but consistently incorporates other fundamental aspects of reality including existence, the present moment and consciousness about which modern science has had nothing meaningful to say.

The fundamental insight of Universal Reality is that understanding the universe as a computational system or program running in the substrate of existence that science calls the quantum vacuum leads to simple and elegant solutions of many of the most important problems of science and philosophy.

The quantum vacuum is the universal medium of existence in which the observable universe exists. The complete fine-tuning of the quantum vacuum determines the fundamental structure of the observable universe that exists within it, and the quantum vacuum also contains the elemental program that continually computes the evolution of the universe.

The apparent incompatibility of quantum theory and general relativity is due to their inconsistent models of spacetime. Both model spacetime as a pre-existing container in which events occur but their models are inconsistent. Universal Reality solves this problem by demonstrating how spacetime is computed by quantum events in the form of dimensional entanglements among particles in a manner consistent with both theories. In this manner both material structures and spacetime are computed together as the single unified structure of the observable universe.

This unified computational approach enables Universal Reality to discover literally scores of important new insights about reality that can't be found anywhere else.

For example it explains why everything in the universe is constantly moving at the speed of light through spacetime. This 'STc Principle' is a little known implication of relativity that scientists usually dismiss as a curiosity but which is actually of fundamental importance to the nature of the present moment, the arrow of time, and relativity, and it has important implications for the cosmological geometry of the universe.

Another important new discovery is the fact that there are two kinds of time, clock time and the time of the present moment. The very fact that space travelers always meet up in the *same* present moment with *different* elapsed clock times demonstrates there are two different kinds of time. This has been one of the most controversial parts of the theory but a number of new examples and proofs are included that clinch its validity.

The existence of two kinds of time immediately solves all sorts of important scientific and philosophical issues from the limits of time travel to the structure of the universe, and it confirms the most fundamental and obvious of all scientific observations, the undeniable existence of a present moment through which clock time flows.

Universal Reality envisions the observable universe as a single universal program that can be understood in terms of any number of individual programs running within it in interaction with each other. These programs include everything in the universe including you and I who are clearly immensely complex computational processes down through all the hierarchies of our structures to the processes of our individual cells and even the interactions of our elementary particles.

Though we are all running programs this doesn't diminish or alter us in the least. We are still living, free, purposeful, sentient, intelligent and conscious biological programs in the full sense of the words. We are entirely as we were before, we just gain the recognition that every aspect of our entire being is a computational process, and our computational process is an integral part of the universal computational process that includes all the other programs that exist and evolve together to create the observable universe.

A defining characteristic of biological programs is that they construct internal simulations of their environments that enable them to function more effectively within them. These simulations seem so real that they convince us they are reality, but nothing could be further from the truth. The actual nature of reality is completely different than our simulation portrays it as the observable world around us.

A major problem of philosophy is to understand that we all live within the simulated reality produced by our brains. Universal Reality takes this as a fundamental insight, and analyzes the nature of the illusions produced by our simulation to reveal the true reality hidden beyond their veils. By doing so Universal Reality puts realization on a firm scientific and rational basis by defining it as understanding and directly experiencing the true nature of reality, and everything this implies.

Universal Reality also provides the only truly convincing and logical theory of consciousness. Consciousness is clearly not anything physical so it cannot be the product of a physical universe in the usual sense of the word, and can only be explained by attributing the essential quality of consciousness, its actual self-manifesting immanent presence, to reality itself.

The self-manifestation of existence within all information forms imbues them with what can be called immanence. Everything in the universe, including us, is filled with the immanence of its existence. Thus all information forms glow with the internal light and life and reality of the existence in which they exist, and that manifests their being and gives them actual presence in reality. It is this immanence of things that is the key to consciousness. The internal glow of the immanence of things is not something visible to the eyes but it is visible to the mind as consciousness.

Thus consciousness is not something human minds generate and shine on things. Consciousness is simply the self-manifesting immanence of the information of things manifesting within our minds. Immanence is the invisible glow of being in all things. It's the presence of actual existence within things that lights them up with being. Immanence is invisible to our eyes but manifests as consciousness in mind.

Universal Reality is the best, most comprehensive and consistent Theory of Everything the author has been able to discover. From a few simple, understandable and quite reasonable and verifiable assumptions naturally emerges a unified and complete Theory of Everything that is completely consistent with modern science, though not its standard *interpretations*, and also opens the way towards important additional progress in our understanding of reality.

Most of Universal Reality is reasonably self-evident when we just look at what reality is actually telling us with open eyes and carefully analyze it in the context of the deep principles underlying established physical and cognitive science. What emerges are secrets that at once are incredibly profound but amazingly obvious when they are finally seen for what they are.

No doubt some points of the theory are speculative but it all fits together neatly and elegantly into a unified whole of great explanatory power that incorporates all aspects of reality. And the ultimate test of true knowledge is self-consistency over maximum scope.

The search for the Theory of Everything is the ultimate quest, and it promises discovery of the ultimate treasure. We hope to make this quest as simple, clear and enjoyable as possible while we explore the deepest secrets of the universe where the greatest most wonderful mysteries of both reality and of ourselves are waiting to be discovered.

This is a concise summary of the Theory of Universal Reality. The complete theory is presented in extensive detail in the body of the book. The author welcomes comments and questions, which may be directed to Edgar@EdgarLOwen.com.

OUTLINE OF THE THEORY

We begin with an outline of the core concepts of the theory of Universal Reality that explain how the observable universe emerges computationally from the underlying substrate or medium of existence. Complete details, convincing evidence, and the extensive implications of the theory are covered in the body of the book.

Existence is an originally formless substrate or medium within which all the individual **forms** of existence gain their individual existences. The formless substrate of existence and the forms of existence that exist within are all that exist. All the forms of existence consist only of **data,** which is the fundamental nature of all individual things that exist. Data as it's meaningful to observers is called **in*form*ation**. Data, forms, and information are different perspectives on the same thing, the actualized forms *of* existence that exist *within* the sea of existence.

All individual forms are forms *of* existence within an otherwise formless sea of existence in the same sense as waves are forms of water within an otherwise formless sea of water. Existence is the fundamental nature of the entire universe, and everything that exists is a data form *of* existence *within* the universal sea of existence.

The **quantum vacuum** is science's initial discovery of part of the nature of existence. Universal Reality considers existence to be the complete identity of the quantum vacuum. Thus existence and the quantum vacuum are identical and are all that exists.

1. **Intrinsic attributes** of existence:
    1.1. **Necessity**. Existence necessarily must exist and self-evidently does exist. And existence must exist because non-existence cannot exist. Thus existence has always existed and there is no need for a creator or creation event. (The big bang was an *actualization* event.)
    1.2. **Presence**. The presence of existence manifests as a universal current present moment within which all things exist and all events occur.
    1.3. **Happening**. Happening is the *universal processor* that continually recomputes the current data state of the universe in the current present moment. Happening computes both mass-energy structures and dimensionality including local relativistic

clock times running at different rates within the universal current present moment. Each complete computational cycle produces the next universal current present moment. Universal Reality calls this a *P-time* (present moment time) *tick*. Observers correlate P-time with their proper time (the time shown on their own comoving clocks).

- 1.4. **Logicality**. All the data of the observable universe is continually recomputed by the processor of happening according to logically consistent rules. Thus the universe is a logically consistent and logically complete computationally evolving data structure.
- 1.5. **Absoluteness**. Everything that exists is absolutely exactly as it is in the current present moment and once computed into existence could not be different in the slightest detail.
- 1.6. **Immanence**. Immanence is the self-manifesting immediate presence of existence in every one of its data forms. All things internally 'shine' with the immanence of their existence. The immanent self-manifestation of existence in things is their actual observable presence and being in reality.
   - 1.6.1. **Consciousness** is simply the manifestation of immanence within mind. The internal shining of immanence in things is not visible to the eyes but it manifests as consciousness in mind. The immanence of existence is what makes things actually *real* in the external world and makes things *conscious* in our mental simulations of the world.

2. **Data & programs**. The forms *of* existence that exist within the quantum vacuum, the sea of existence, consist entirely of data given actuality by the immanence of their existence. Both the actualized data of the observable universe and the virtual data of the complete fine-tuning exist within the quantum vacuum of existence. They are all data forms of existence within the universal sea of existence.
   - 2.1. **Virtual data**. The fixed virtual data of *the complete fine-tuning* that determines the possible structures of the evolving actualized data of the observable universe. This virtual data is observable only through its effects on observable data.
      - 2.1.1. **Logical operators**. The set of elemental operators executed by the processor of happening the elemental program uses to compute a logically consistent observable universe. By analogy the 'machine language' of reality.
      - 2.1.2. **Templates**. What particle components exist and which sets of particle components make valid elementary particles. The fact of 4 forces and 4 dimensions etc.
      - 2.1.3. **Fundamental constants**. (What science calls the fine-tuning) The precise values of the fundamental constants such

as the speed of light, the gravitational constant and the free constants of the standard model.
- 2.1.4. **The elemental program**. The simple set of subroutines that actually compute all particle events that make up the observable universe in accordance with the rest of the complete fine-tuning.
- 2.2. **Actualized data.** *The observable universe* of science and everything in it is actualized data in the quantum vacuum. Actualized data is observable through its interactions.
  - 2.2.1. **Particle components**. All elementary particles are composed of their particle components, which are the elemental data structures of the observable universe. These include lepton and baryon number, the charges, mass, energy, spin, spatial and temporal parity etc.
  - 2.2.2. **Particle events**. Particles interact in events that conserve each of their particle components separately.
  - 2.2.3. **The entanglement network**. Because all their particle components are conserved through events the particles emitted by every event become *entangled* on each type of their particle components. Chains of successive particle events form a universal entanglement network encoding the entanglement relationships among all the particles in the observable universe back to the original big bang event. The data that makes up the current slice of the entanglement network is the complete observable universe as it exists in the current present moment. It encodes the integrated mass-energy and spacetime structures of the entire observable universe as entanglement relationships among the particle components of individual particles.
  - 2.2.4. **Computational domains**. At the aggregate level the data of the entanglement network manifests overlapping hierarchical domains based on computational density, type and dimensional relationships. Observers tend to identify individual things on the basis of computational domains in their simulations of reality.
  - 2.2.5. **Emergent programs**. All emergent data structures act as emergent programs at the aggregate level because the interactions of all their data elements are continually being recomputed. Due to the complete fine-tuning aggregate data structures manifest as independent entities in the same sense as ordinary computer programs composed entirely of structured sets of machine language operations do.
  - 2.2.6. **Living programs**. Emergent programs that are purposeful and have internal simulations of their environments to

varying degrees. We humans are emergent programs being computed by happening.
- 2.2.7. **The universal program**. The single running program of the observable universe. All individual programs are integrated subroutines of a universal program whose interactions compute the universal program.
- 2.2.8. **Cosmology**. At the largest scale the dimensionality of the observable universe takes the form of a 4-dimensional hypersphere with the 3 dimensions of space its surface and past P-time its radial dimension back to the big bang at the center. The surface is the 3 dimensions of space in the current present moment and is the dimensional form of the entire observable universe.

3. **Principles**. The fundamental principles by which the observable universe is computed that determine its structure.
    - 3.1. **Differentiation Principle.** The single formless sea of existence has actualized into innumerable individual particle component forms according to the virtual data of the complete fine-tuning.
    - 3.2. **Computational Principle.** Everything in the universe is data that is simultaneously being computed by the processor of happening. Thus it manifests as a universal program made up of the interactions of innumerable individual running programs.
    - 3.3. **Particle Component Conservation Principle.** All particle components are conserved through all particle events. Thus particle components rather than elementary particles are the elemental components of reality.
    - 3.4. **Exclusion Principles.** All sets of particle components not allowed by the templates of the complete fine-tuning are excluded. This is why the particle component sets of colliding particles must split into allowed combinations in new particles. It also includes the Pauli exclusion principle that underlies all atomic and molecular structure. The interplay of conservation and exclusion determines the mass-energy structure of the observable universe.
    - 3.5. The **STc Principle**. The total space plus time velocity of everything in the observable universe is always equal to c, the speed of light.
    - 3.6. The **MEv Principle**. All forms of mass and energy are different forms of relative spatial velocity. Only if they are different forms of the same thing can they be interchangeably conserved by the conversion of one form of spatial velocity to another.
    - 3.7. The **METc Principle**. The total mass-energy spatial velocity and time velocity of all processes is always equal to c, the speed of

light. The METc Principle combines the STc and MEv principles. This principle underlies all the effects of general relativity.
- 3.8. The **STo Principle**. Space versus time oscillations of the processor of happening are the source of all quantum indeterminacy including the zero-point energy, the Uncertainty Principle, and the wavefunction descriptions of particles as explained in the next section.

4. **Computations**. The universal program and all its individual programs are computed at the particle event level by the processor of happening.
    - 4.1. The processor of happening is an intrinsic aspect of the quantum vacuum so all the data of the universe always exists in and is computed simultaneously by the processor.
    - 4.2. All processor computations occur in the current universal present moment in a non-dimensional computational space in the same sense as computer programs define computational spaces.
    - 4.3. Everything is computed at the particle and particle component level, by analogy the machine language of existence.
    - 4.4. Each separate *coherent process* is computed by an individual *application* of the processor executing the elemental program. Thus coherently entangled particles continue to be computed as a single process and the measurement of one immediately affects the others. This explains the spin entanglement 'paradox'. There is no non-locality in computational space.
    - 4.5. The processor allocates a fixed number of cycles per P-time tick to compute velocity in time and velocity in space so that the total velocity through space and time of all processes is always equal to c, the speed of light. This is the computational source of the STc Principle and all general relativity effects.
    - 4.6. At the quantum scale there is a continual random oscillation between processor cycles allocated to compute velocity in space versus velocity in time. This random oscillation is the source of all quantum indeterminacy (zero-point energy, Uncertainty Principle, wavefunctions) as the dimensionality of space versus time is conflated at the quantum scale.
    - 4.7. The processor of happening computes all aspects of mass-energy structures including their dimensional relationships. The dimensional entanglement relationships among observed events are what observers interpret as a physical spacetime. Thus spacetime isn't a preexisting container *for* events but is computed *by* quantum events.

4.8. Since dimensionality is computed by events as the dimensional relationships of particles the spacetime that emerges is compatible with both quantum theory and general relativity.

5. **Simulations**. All living organisms (living programs) construct internal mental models of themselves within their environments to varying degrees. These simulations are subroutines of the total programs of living organisms that have evolved to enable them to effectively compute their functioning within their environments.
    5.1. Every organism creates a unique individual simulation of itself within its reality that differs by individual and species.
    5.2. Every organism believes its simulation is actual reality though it's actually an illusion.
    5.3. While the *appearance* of reality in a simulation is an illusion, the *logic* of the simulation is a simplified mapping of the actual logic of reality. This logical correspondence enables organisms to function and survive within their environments on the basis of their simulations
    5.4. Simulations interpret the actual non-dimensional immanent data universe as a physical universe of material objects within a pre-existing spacetime. The simulation displays the logico-numeric data of the entanglement network as a bright dynamic physical world in the same manner as the logico-numeric data in a computer program can be displayed as a dynamic interactive environment in dimensional spacetime in a virtual reality headset.

6. **Realization**. Understanding and directly experiencing the true nature of reality insofar as possible in human form with no religious or metaphysical connotations.
    6.1. **Fundamental realization.** Our consciousness in the present moment is our direct experience of the fundamental process of the universe, the continual happening of immanent existence occurring within us as it recomputes the universe of data forms.
    6.2. **Realization of data.** We may directly realize that the true nature of things is only data forms by analyzing things into the individual data *associations* our simulations interpret as physical things.
    6.3. **Realization of data histories.** When we analyze the data forms of things we realize all things are the current computational results of all their past interactions. Thus things are their data forms, and their data forms are their computational histories.
    6.4. **Realization of knowledge.** Since all things are their data histories and their data histories encode the data of their

interactions with other forms the information of all forms is distributed through the other forms they have interacted with. This is the basis of the **Sherlock Holmes Principle** that enables information to be inferred from form to form and that underlies all knowledge.

6.5. **Realization of immanence.** Our consciousness of anything at all is the direct experience of the immanence of its existence. By emptying the mind of the forms of individual things we more directly experience the immanence of formless existence in our experience of consciousness itself. Realization is the direct experience of the immanence of existence in all things.

6.6. **Presence.** Realization is the recognition of the ever presence of the immanence of existence in all things, in all situations, in all places and at all times. There is nothing that is not part of immanent reality thus all is the ever-present immanence of existence.

6.7. **Illusion is reality.** Our simulations of reality are illusions that hide the true nature of reality, but since everything is part of reality so are our illusory simulations of it. Thus our simulation of reality is our only direct experience of reality and realization is just a matter of recognizing its true nature. Illusion taken for reality is illusion but illusion recognized as illusion is reality.

6.8. **You are already enlightened**. Because we always exist within reality and are part of the immanence of reality we are all already enlightened and always have been. It's just a matter of realizing it. Thus there is no necessity of a teacher or temple for realization since we already live within the temple of reality and it itself is the only teacher.

EXISTENCE

REALITY

We define reality as the 'true nature' of the totality of everything that actually exists in the present moment. It is clear that many things are not as they initially appear to us. We typically see only the surfaces of things rather than the hidden structures and processes that underlie them. The world has a deeper more fundamental reality that is often obscured by appearances and science certainly agrees. The illusory appearances of things are due partly to their representation in mind and partly due to the hierarchical complexity of reality itself. This book explores the hidden structure of reality and how it is simulated by mind to reveal the true nature of the whole system.

There are three fundamental questions with respect to reality. First why does something exist rather than nothing? Second why does what actually exists exist instead of something else? And third what does actually exist? We will attempt to answer all three of these questions.

THE NATURE OF EXISTENCE

All things that exist whatever their nature are said to have existence. Existence is what makes something real and actual and gives it being. Existence is most often considered a quality of individual things, but it makes much more sense to think of existence as a universal medium or substrate in which everything that exists actually does exist. In this view it's the presence of things within the universal medium of existence that gives them their individual existences. Thus existence is the common active ingredient of all things that exist that gives them their individual existences.

Since all things that exist have existence, it's quite reasonable to assume that all things that exist must share a common active ingredient of existence that makes them real. If there is not some common active ingredient then how can the myriads of disparate things in the universe all be said to exist in the single sense implied by the use of a single word?

What would saying something has existence even mean if there weren't something that gave it that existence? This is the obvious and logical conclusion.

So we can reasonably assume that the existence of all things that exist is the same existence. The forms of things differ widely but the fact they exist is the same for all. If we define reality as all that exists then the universal substrate of existence fills all of reality and the existence of any individual thing is its presence in this common substrate of reality. Without being present in the common substrate of existence, a thing would not be present in reality, would have no reality, and would not exist.

This becomes clearer when we consider the analogy of an ocean. Individual things gain existence as forms of existence within an underlying medium of existence just as individual ripples, waves, and currents become real by being different forms of water within a common substrate of water. Thus the big bang can be thought of as the appearance or actualization of various forms of existence in a previously formless sea of existence. All individual things are merely forms *of existence* within a universal sea of existence.

The insight of an underlying medium of existence common to all things is missing from traditional science. This is because there is no actual non-existence to contrast existence with. Thus things are just taken for granted as individual things and their underlying common nature doesn't tend to enter consciousness because there is no non-existence to compare their common existence against.

Individual things are recognized as individual things because they can be distinguished from all the other individual things that have different forms, but there is nothing different to contrast their existence with, and thus existence is rarely recognized as anything actual. Nevertheless existence is real and present in everything that exists in the universe, and is the largely unrecognized underlying presence of the reality of the universe. It is the underlying formless substrate of the universe in which all individual things gain their existence by their presence within it. If there were no underlying medium of existence, the universe and we within it would not exist.

THE AXIOM OF EXISTENCE

Something rather than nothing exists because only existence can exist. Non-existence or nothingness cannot exist because nothingness is non-existence and only existence can exist. Thus nothingness cannot exist and can never have existed. Only existence exists or has ever existed or can ever exist. Thus there is not and never was and never could or can be a nothingness out of which something came into being. There is and has always been only existence and whatever forms exist within it.

There is not even nothing outside of existence, or before or after or beyond existence. There is no outside or before or after or beyond existence. There is only existence and everything that exists is part of that existence.

Thus 'Existence exists', or more concisely just 'Existence!' which implies the necessary existence of existence, is the self-validating self-necessitating fundamental axiom of reality upon which all else depends. This is the ultimate turtle upon which all other turtles stand and the ultimate source of the entire logical structure of reality (Wikipedia, Turtles all the way down). Because the fact of existence is self-evident the axiom is self-evidently true. You would not be reading these words if existence didn't exist.

At first this may appear to be a mere sophism or tautology but it accurately expresses the actual logic of reality and is the only possible self-contained explanation for the fundamental fact of existence.

One might argue the axiom of existence is circular and of course it is but that is precisely the point since the fundamental axiom of reality must be circular; but it must also be self-evident and meaningfully so. A meaningful circular self-necessitating fundamental axiom is much preferable to a set of axioms that has no underlying logical foundation such as those of Euclidean geometry.

Because there never was a nothingness out of which something was created there is no need for a creator or creation event. All the interminable disputes about creators and the creation of the universe immediately become illogical and meaningless and must be abandoned. The axiom of existence immediately renders much of philosophy and religious doctrine moot and answers the first question of why something rather than nothing exists.

Thus the fundamental question becomes not why something exists, but why what exists *is* what exists. Through this proper self-consistent, self-necessitating definition reality becomes much simpler and illogical questions concerning non-existence disappear.

Reality is the existence of what exists, and existence is the manifestation of reality. Reality and existence are different perspectives on the same thing. Thus the only thing that can ever be real and actual is existence and the only thing that can exist is reality.

The question of how existence arose out of non-existence is nonsensical and meaningless and should not even be asked. It's based on a misapplication of the logic and language of everyday things where individual things do suddenly appear out of non-existence into existence. But whatever appears always actually appears out of something else, it's always a *transformation* of things rather than a creation out of nothingness. Nothing ever appears out of nothing at all or nothingness. The forms of reality often transform from one thing to another but since reality itself includes everything there is nothing for reality itself to transform from or appear out of.

Of course the physical universe as we know it originated in the big bang some 13.8 billion years ago but this was not the beginning of existence as the universe originated not from the absolute absence of anything but from the quantum vacuum which contained the unactualized virtual possibilities of all possible actualities. We must not mistake the apparent beginning of the physical universe and clock time at the big bang for the beginning of existence itself.

Thus existence has 'always' existed. By 'always' we mean here that there was never a *time* in which existence did not exist. In Universal Reality clock time is computed along with all the other processes of the universe and so clock time would only have begun with the big bang. Thus there was no clock time prior to the big bang and properly speaking no 'before'. Nevertheless there was a timeless present moment in which a prior virtual state of formless existence existed and there was never a time this was not true.

Existence must exist because non-existence cannot exist. The existence of non-existence is a logical contradiction, and logical contradictions cannot exist in a computational universe, since for the universe to be computational it must follow consistent logical rules that don't generate logical contradictions. Thus the axiom of existence provides a clear and convincing answer to the fundamental question of why something rather than nothing exists. It is simply impossible for nothingness to exist so there was always a something that existed and that was the originally formless substrate of existence itself.

THE PRESENT MOMENT

Existence has several intrinsic attributes. The first is presence. For existence to exist it must have presence and be present. The presence of existence manifests as a universal present moment in which everything exists. Since the present moment is the presence of existence there is no actual before or after or outside the present moment. Existence exists only in the present moment it creates by its presence.

The past is a non-existent logical projection inferred backwards from the present. It exists only in memories and its other computational results in the present moment. And the future doesn't exist because it has not yet been computed. Reality exists only in the present moment manifested by the presence of existence.

Many scientists deny the existence of a present moment because they believe it's inconsistent with relativity but this is based on a misunderstanding of relativity as explained in the chapter on Understanding Time. There is no doubt whatsoever that a present moment exists because it is the most fundamental and persistent of all observations both scientific and personal. The crux of scientific method is to develop theories that explain observations, never to deny them. Denying observations is the antithesis of science and the present moment is the most fundamental of observations.

TWO KINDS OF TIME

There are two kinds of time. There is the time of the present moment, which is universal and absolute and common to everything throughout the universe. And there is clock time, which flows through the universal present moment at different rates depending on local relativistic conditions. Clock time flows through the current present moment at different rates but the universal present moment is common to all observers throughout the entire universe. We will call the time associated with the current present moment *P-time* to distinguish it from clock time.

The fact there are two kinds of time is conclusively demonstrated by the established fact that relativistic observers always reunite in the exact same present moment even when their clocks read different clock times. The two kinds of time will be confirmed and explored in detail in the chapter on Understanding Time.

There is no doubt at all that there are two kinds of time and it is totally amazing that no one had recognized this obvious truth before I first pointed it out (Owen, 2007, 2009). This serves as an excellent example of the blindness of science to obvious facts that somehow don't register in the prevailing worldview or may not seem to have a mathematical basis.

HAPPENING

Happening, also called change or process, is another intrinsic aspect of existence. Because of happening change occurs and things happen. Existence continuously happens. Happening is the source of all the processes and change in the universe and the ultimate source of all the activity and life of living beings including us. It is also the source of experience and consciousness. Without happening there could be no experience, no consciousness and no life. Happening is the life force of the universe. Happening occurs and the universe comes alive, events occur, and the universe opens into an observable reality that is available to experience and knowledge.

Happening is the source of the flow of universal P-time and the local clock times computed within it. Clock time is the observational rate of happening at any relativistic location. Happening has nothing to do with clocks *per se*. Clocks are just standard physical processes that measure its local clock time rates.

Happening is the universal *processor* of existence that continually computes the evolving existence of the universe and all its individual data states. Because of happening time flows and the universe comes to life. Since at least the big bang clock time continually flows through the present moment.

THE UNIVERSE AS A LIVING SYSTEM

The universe can be considered a living system in the sense that it continuously happens and evolves of its own accord with nothing outside moving or causing it. In this sense, and only in this sense with nothing supernatural or biological read into it, the universe is alive. It's a living self-motivating system that continuously evolves all by itself without any external cause or force. It is after all everything that exists and it moves of itself. It's not a biological organism but it is a computational organism.

Thus everything within the universe can be said to share the life of the universe according to its own specific forms. Ultimately the life of the universe is the source of the lives of all biological organisms including us. For without the happening of the universe we wouldn't be alive or even exist. It is clear the reasonable conclusion is that the universe is much more than the blind clockwork system envisioned by traditional science.

We feel within ourselves the special feeling that we are alive, that we are living beings, that we have a life force that animates us and makes us different from inanimate objects. But what this actually is has long been a mystery. We are now in a position to provide an answer. Our personal life force is the same life force of happening that animates all of reality and us as well, each according to its forms.

The fact that reality continually happens in the present moment is the life force that we feel within us. Our life force is our participation in the real and actual presence of reality as it happens, as is our consciousness. We feel ourselves alive simply because we are part of the aliveness of the universe and because we are continually happening ourselves as the happening of the universe occurs within us. There is nothing esoteric or supernatural about this. It's simply the experience of the continual happening of the universe occurring within us.

Our life force is the direct experience of happening that animates all reality. What we are referring to here is not the biological definition of a living organism as an autonomous purposeful computational system. Rather it's this biological structure that is animated because it shares in the universal life force that animates all things each in their own way according to its form. Our feeling of being alive is simply what the life force of the universe feels like inside a biological organism. It's the feeling of happening within us flowing through our form.

So both our consciousness and the life force that animates us are not something unique to us but are due to our participation in the realness of reality because we are a part of reality. Of course the particular ways we express our lives and consciousnesses as biological organisms are due to our particular form structures but the fact that we are able to be animate and conscious, and our experience of that, is due to the self-manifesting happening of reality itself within us.

ABSOLUTENESS

Another intrinsic attribute of existence is that it's absolute. The universe and everything in it are absolute in the sense that everything is exactly as it is in the present moment with no possibility whatsoever of being anything else once it happens. And the intensity of its realness has no limits because everything is either absolutely real or doesn't exist at all.

Thus everything that exists is absolutely exactly what it is with not the slightest possibility of it being anything else or any different in any detail whatsoever. Things continually change but once they change they are exactly and absolutely as they are. In this sense all things are absolutely real and actual with no limits whatsoever to their reality.

Because what actually exists defines reality as it actually is there is and can be no alternative to existence as it actually exists. There of course can be theoretical (i.e. impossible) alternatives but there is no actually real alternative to exactly what is right now in the present moment of existence.

This is the solution to the second question of existence, why what exists is what exists rather than something else. The answer is simple though subtle. The existence of what actually does exist conclusively falsifies the existence of anything else at all, and this is true all the way back to the big bang and the fine-tuning of the universe. There simply is no possible alternative to any of it once it has happened.

The fact that we can imagine theoretical alternatives in no way implies an actual existence of those alternatives. None of these other alternatives have any existence at all because they just aren't part of existence except as ideas. Thus now that what exists is exactly what exists, and only what exists can exist, those alternatives have no reality at all, and certainly no possibility of existing either now or in the past whatsoever.

Thus the existence of what does exist conclusively falsifies all other alternative possibilities. There is zero probability they ever could have existed because they never did. When carefully analyzed and understood this logic is sound. We can legitimately imagine other possible outcomes for *individual events in time* because we can largely recreate the conditions that led to them, but the original event is entirely impossible to recreate as it has already occurred.so we must not fool ourselves into thinking it could ever have been any different than it actually was. The universe as a whole cannot be rerun thus there is no possibility whatsoever it could have been different in any last detail than it was.

It's important to understand that alternative possibilities meaningfully apply only to future states and never to past or present states. Alternative possibilities for the future are meaningful because the future is probabilistic because it has not yet been computed, and its computation is subject to quantum randomness but the actual present and past conclusively falsify any possible alternatives to the slightest detail and this includes the original complete fine-tuning which simply could not have been other than it actually was.

IMMANENCE

The final intrinsic attribute of existence is immanence. Immanence is the self-manifestation of existence in the actual presence and being of all things that exist. Immanence may initially seem to be a subtle or unfamiliar concept but it's actually what we experience all the time as the here now presence of things. Our experience of anything at all is simply our experience of the immanence of its existence.

In traditional philosophy immanence refers to the hidden presence of the divine within things (Wikipedia, Immanence). We use it in a very similar sense but without any religious or metaphysical connotation. In our usage immanence refers to the internal presence of existence within all things that makes them real and actual and gives them being. Take the information forms of all things, subtract their illusory physicality, and fill their information forms with the inner light of existence that brings them to life in the universe and you have their immanence.

The quantum vacuum is not just the invisible source of real particles, and thus of all things, but is what actually 'shines' within particles to make them real particles present in the observable universe. That 'shining' is what we perceive as their reality and their actuality. If the information forms of the world had no immanence of existence they would not exist and would never be experienced. We experience things only through their immanence, through the existence shining within them.

Nothing really mysterious, metaphysical or supernatural is implied here. It's just the idea that existence itself is more than just dark dead material things sitting in a material universe. Immanence is what brings the universe and everything in it to life in the real world of existence. Immanence is the discernable presence of existence in everything that exists.

The information of the things of the world is not just there like equations on a page in some dark traditional sense, it's filled with an inner light of being, that makes things really really real and actually there in a much profounder sense. Its being is absolutely real because it is absolutely what it is exactly as it is.

Immanence is analogous to a light bulb that is not just a light bulb in a box, but a light bulb that is plugged in and turned on with existence. Everything has a living essence of existence that internally illuminates its being into reality. Existence, the quantum vacuum, is the ubiquitous substrate of the universe. Thus the existence of an information form in the substrate or medium of existence is what makes it present in the universe and into a real actual thing and that existence is discernable within it as its presence in reality.

The old materialistic view of the universe is a blind clockwork universe of passive information forms that wait in the darkness to be observed, but our computational universe is a universe in which all forms actively glow with the inner light of their being, with an invisible inner light of presence, existence and happening. This makes them real and present in a much profounder sense. This is what is meant by the immanence of existence.

In this sense all the forms of the universe glow with life from within as existence continuously animates them into being. Existence is the life force of the universe that is actively present in the forms of all things making them real. Existence continuously manifests itself as immanence in all the forms of the universe as happening computes them into continuing existence and gives them actual reality.

This immanence of existence in all the forms of existence is the essential ingredient of things that makes them real and actual. It is also what makes things available to consciousness, and makes consciousness possible. Without the immanence of existence within things, forms might be blindly registered by brains in the same sense as a computer program registers a change in its data, but that registration would not be conscious. Immanence is of critical important in understanding reality because it's the key to understanding consciousness.

Reality is not just a lifeless mechanical, physical world composed of material bodies but something that *actively self-manifests* itself in the absolute realness and actual presence of its existence in the immanence of all the individual things that exist within it. All things that exist are different data forms of existence. All the forms of the world are empty of any self-substance other than that of their common existence.

Reality, existence, the present moment, happening and immanence are all names for different aspects of the single fundamental and only 'thing' that exists. Tao, properly understood, is an ancient approach to the concept of existence. The fundamental substance of reality is itself originally formless and nameless existence. We seek to describe it as accurately as possible in an English description of its aspects and characteristics since we must after all use language to speak of it. But as Lao Tzu rightly points out, "The Tao that can be named is not the Tao." (Legge, 2010).

CONSCIOUSNESS

XPERIENCE

All the programs of the universe effectively experience the other programs they interact with in the resulting changes to their own forms. All experience consists of internal changes to an observer's own forms. Our experience of external forms consists not of those forms themselves but of the changes our interactions with them make to the forms of our internal representations of them. Thus experience always consists of changes *to internal* forms even when it's *of external* forms. In all cases we experience both ourselves and other forms as changes to our own forms. Experience can only be of some aspect of our own form.

In a generic sense all programs, even inanimate ones, can be said to continually experience the existence of both their own forms and other forms they interact with as the continual recomputation of their own forms into existence. In this view experience is not something a form *has* but something a form *is*, of something that is part of it. All forms are in a continual state of recomputation, whether the form itself changes or is unchanged, and every recomputation of a form can be understood as an experience of something.

In this generic sense all the individual forms of the universe can be considered *generic observers*, and the process of their continual recomputation as experience. This is of course not experience in the sense that living organisms have experience but the fundamental nature of the process is exactly the same; the change of a form in response to its interaction with an external form or its own internal transformation. Thus we can coin a neologism and refer to the generic experience of forms as *xperience* in contrast with the *experience* of living programs.

Because all forms manifest the immanence of existence all xperience is actually real in the realest sense possible, however its expression is always that of its precise actual form. Xperience is the basic process underlying consciousness, and consciousness is simply xperience occurring in certain specialized forms of mind's simulation of reality.

Xperience is the view of the active manifestation of existence in all forms from the perspective of the forms themselves. *Immanence* is the perspective on the active manifestation of existence in all forms from the perspective of other forms.

Immanence and xperience are opposite perspectives on the same thing. Immanence is the active self-manifestation of existence in all forms, and xperience is the active self-manifestation of existence within a form's own forms. Immanence and xperience are opposite sides of the same coin of the active self-manifestation of existence in all forms.

Immanence and xperience manifest only through the actual forms they occur within. Thus unless the proper forms to register consciousness are present xperience remains unconscious. Thus most of the xperience of the universe is unconscious and not conscious experience in the human sense. Unless a program contains a specialized subroutine that monitors its xperiences and is able to register the fact it's experiencing them, its xperience remains unconscious.

So by and large almost all of the xperience of the universe is unconscious because it occurs in inanimate forms that are unable to record or report the fact that they are xperiencing. Nevertheless all this xperience is just as real as conscious experience, and xperience is the essential mechanism underlying conscious experience as well.

The fact we humans are able to experience the fact that we are xperiencing and report it to ourselves and others makes us believe that only we have xperience, but in fact all the programs of the universe also xperience in the exact same sense we do, they are just not able to consciously register or report the fact that they do because they lack the necessary forms.

A mountain xperiences its erosion in the computational changes to its forms in the same fundamental sense that humans experience things. It's just that the mountain has no specialized simulation model of its self and surroundings in which before and after states can be conceptualized and compared like we do. The mountain xperiences only its actual here now state in the present moment as it's computed. It doesn't experience comparative changes to its state nor does it experience the fact it's xperiencing as living beings do because it lacks the requisite forms and subroutines to do so.

REALITY AS XPERIENCE

Since everything in the universe consists of information forms and all recomputation of forms into continuing existence is xperience, there is a very real sense in which the universe itself consists only of xperience since xperience is simply the continual recomputation of all the forms of the universe into continuing existence. The universe wouldn't exist if it weren't continually being recomputed into continuing existence in the present moment and all computation of forms is xperience.

In this sense the observable universe becomes the xperience of itself, the xperience of its immanence self-manifesting in all its observable forms. Immanence and xperience are the essence or active ingredient of consciousness, thus the observable universe consists only of the active ingredient of consciousness, however most xperience is not conscious in the usual sense even though it contains the essential active ingredient of consciousness.

Thus the universe can be said to consist entirely of xperience, the xperience of all the programs of the universe continually xperiencing themselves and each other. In this sense the universe continually xperiences itself into existence, and xperiences itself in the xperiences of all its individual programs of the other programs they interact with. The universe must xperience itself to exist because xperience is the continual recomputation of forms and this is the manifestation of existence.

The universe is the xperience of itself, and all the individual programs running within it are the xperiences of themselves. In this sense if the universe didn't xperience itself it would cease to exist. The universe continually computes and xperiences itself into continuing existence. The observable universe is the largely unconscious xperience of itself.

In this sense every program in the universe is a generic observer, and the universe consists entirely of generic observers. All the programs of the universe are generic observers though almost all are unconsciously so. Almost all the xperience of the universe is unconscious xperience. In this sense the (generic) observer becomes an essential integrated aspect of the observable universe that automatically emerges from its fundamental nature. The fact of xperience is what makes the observable universe observable. Without generic observers to xperience it the observable universe would not exist.

The observable universe is the continual xperience of its own immanence. The entire universe xperiences its existence in the continual changes to all the forms that make it up. The universe xperiences its existence in its own computational happening. And since xperience is the active ingredient of consciousness the observable universe waits only the evolution of the proper forms to become fully conscious of itself, as consciousness is implicit in its fundamental design. Meanwhile the observable universe is conscious of itself through all the individual forms of consciousness it has evolved.

The usual perspective on the universe is of individual static things that undergo change, but this new view takes change itself as primary. Xperience is the fundamental nature of the universe, and the universe consists of constantly running processes always in the act of computational change. In this view static individual things don't exist in reality. They are only temporal snapshots of the instantaneous states of running programs at some moment in time snapped by some observer and recorded in its simulation.

The universe continually xperiences itself in its continual recomputation into existence. The experiences of living beings such as you and I are simply the participation of specialized neural structures in our brains in this same universal process.

CONSCIOUSNESS

Consciousness is the active presence of existence self-manifesting in things. Specifically biological consciousness is the active presence of existence in specialized forms in the simulation that monitor other forms in the simulation. Consciousness is simply the immanence or active presence of the actual reality of those forms. Consciousness is the *experience* that we are *xperiencing* representations of things in our simulation of them. Consciousness is the immanent reality of this process.

The phenomenon of consciousness *itself*, how consciousness can arise in a physical brain or more generally a physical universe, is the subject of Chalmers' 'Hard Problem' (Chalmers, 1995). This is opposed to the structure of the *contents* of consciousness, which Chalmers calls 'the Easy Problems' of cognitive science.

To fully understand consciousness we must clearly distinguish between consciousness itself and the contents of consciousness. The contents of consciousness are all the forms that become conscious within consciousness and their structural characteristics, and consciousness itself is the fact that these contents of consciousness are conscious.

This is immediately analogous to the distinction between existence itself and the contents of existence. Existence itself is the originally formless substrate of reality within which the individual contents of existence have existence. Likewise consciousness itself is the fundamentally formless presence of consciousness within which the individual content forms of consciousness become conscious. The analogy holds because the essence of consciousness itself is the immanence of existence itself.

In Universal Reality consciousness itself is easy to understand because it's simply our experience of the self-manifesting immanence of things that actually exist. The individual *contents* within our consciousness are the particular information forms generated by our minds that recursively encode the monitoring of the other information forms that we are conscious of.

The fact that this recursive process shares in the immanence of all processes is what we call consciousness. Immanence is the essence of consciousness because it explains both the popping into reality of thing forms in the world, and the popping into consciousness of thought forms in the mind. The popping into reality of forms in the world is their existence. The popping into reality of these recursive forms in the mind is their consciousness. These are both manifestations of the same fundamental process of the immanence of existence.

Our consciousness of individual things is not just the presence of representational information about things in our brains; it's the immanence of the information recursively encoding that information of xperience being there. This accounts for the moving focus of conscious attention around the information contents of the simulation. Our specialized neural structures encode the information contents of consciousness, but the fact that these contents are conscious is simply the immanence of their existence that pops them into reality.

Consciousness itself is difficult to define clearly. Everyone has the experience of consciousness but no one seems to be able to define it much less explain it. Basically consciousness of something is a mental representation of that thing somehow popping out of the mind's dark background into clarity as attention is focused on it to the temporary exclusion of most of the rest of the contents of mind.

Thus the salient feature of consciousness is the exceptional self-illuminating presence of something to attention. Stated this way it is clear that the self-illuminating presence of a thought to consciousness is very similar to the self-illuminating immanence of things into existence in the external world. In fact consciousness is exactly the same immanence of existence manifesting in specialized forms of the mind instead of the forms of things in the external world.

Immanence is the simple fact that everything in the world shines with the inner light of its existence. Thus our mind's simulation of the world also shines with the inner light of its existence and that is the essence and active ingredient of consciousness. So to explain the structural aspects of consciousness, we just need to determine what information structures in our minds filled with immanence would manifest as consciousness as we are familiar with it.

Our entire simulation of reality contains all the information that is potentially available to consciousness. It's a vast nexus of interconnected information encoding our mental model of all aspects of our self and the world around us being continually updated by sensory and cognitive inputs. When some particular information in our simulation is brought to the focus of attention that is our conscious *experience* of that information.

The main program that constructs and maintains the simulation is extremely sophisticated and extensive and is always running in the background. The huge volume of information that makes up the simulation continually shines with the immanence of its existence but it isn't yet conscious. Thus the immanence of the data in the simulation can't be all there is to consciousness because all the data in the simulation is present but it's clearly not all conscious at any given moment. Something is missing to explain why only small sets of the entire simulation are conscious at any given time.

The immanence of data representations in the simulation is not sufficient to make them conscious. They are real xperiences but not conscious experiences. All forms are the immanent realities of the exact things their forms are the form of. The immanent reality of the form of an external thing is the external thing. The immanent reality of the representation of a thing in the simulation is only the representation itself. It isn't the reality of the organism having the xperience of that representation. So for consciousness we need a form that encodes the fact that representation is being xperienced and it's the immanence of that form that will be consciousness because that is the reality of what that form is the data of. Consciousness is the immanence of forms that encode we are having xperiences of things in our simulation.

There is a specialized little program that roams the simulation focusing on particular areas of interest and bringing them to conscious attention. This is the moving focus of our conscious attention onto particular contents of our simulation. This is the key to consciousness.

This little program also generates information but this information is not information about the external world, but *the information that some information about the world is being xperienced in the simulation.* It's the immanence of *this* information that we experience as our consciousness of the original information about the world. So consciousness is the immanence of an internal information form representing the fact of our internal representation of the form of some actual thing in reality. Consciousness is the xperience of an xperience being xperienced. And the xperience of an xperience being xperienced is an *experience* and so becomes conscious due to the immanence common to all xperience.

The focus of attention program is essentially a specialized program that actively monitors things in the simulation and encodes specific information being represented and the fact it's being represented. This information of things being represented, of xperiences occurring, is our experiences of those things. And the immanence of these experiences is our consciousness of those things. The reality of these focus of attention forms is they are xperiences that things are being xperienced, and immanence makes this experience pop into reality and manifest as consciousness. Consciousness is the immanence of the specialized focus of attention program recording the fact that things are being xperienced in our simulations of them.

Everything in the universe has the common active ingredient of consciousness in the immanence of its information forms and its xperiences of them but everything is not fully conscious in the sense we are. Existence manifests immanence only through the actual forms it appears within. So whatever that form is, existence manifests it as the real actual thing it is the information of, and only that. The real actual thing the focus of attention program is is the fact a particular xperience is taking place, and the information something is being xperienced is the conscious experience of that thing.

Most information is the information of the running programs of the inanimate things of the world, and so the appearance of existence within their forms makes them the real inanimate and unconscious things of the world. However the appearance of immanence in the forms of our focus of attention on their representations in our simulation gives them the reality of conscious experience, because these are forms encoding not things but the experience of things being xperienced. Thus the reality of these recursive forms is things being xperienced, rather than just things, and immanence makes that xperiencing conscious because that is the reality of the focus of attention forms. Consciousness is the immanence of information forms encoding xperiences of other xperiences being xperienced. It's the immanence of a recursive process.

This little program of the focus of attention is what people tend to think of as their I or conscious self because it's a concentration of consciousness that can even be self-aware as it roams around the simulation experiencing things within it.

This focus of attention program operates like an adjustable spotlight. It has a strong central focus of attention but also considerable peripheral consciousness continually on the look out for subsequent areas of interest to turn its attention to. And the intensities and breadths of both areas are also adjustable to some degree. This program can also be turned off as in dreamless sleep. In this state the programs maintaining the simulation are still running and all the data is still there but is not brought into consciousness. The xperiences of all the myriad processes of our bodies and simulation of reality are still actively occurring but the focus of attention program that registers they are occurring is turned off and so they remain unconscious.

The focus of attention is guided by a number of complex adaptive rules that won't be examined in detail, but it tends to focus on updates taking place in the simulation, especially those judged of particular potential importance. Movement against a background is one example; recognition of faces or animals, imagined or real, in background forms is another that has evolved because it tends to promote survival.

We experience many aspects of our simulation as the conscious experiences of an external world but we are actually only conscious of our experience of our mind's simulation of the external world rather than the external world itself. However there is sufficient *logical correspondence* between the external world and our simulation of it that our mind can reasonably interpret our consciousness of our internal simulated world as consciousness of the external world it simulates.

So our simulation is not just an internal programmatic model of the *information* of the programs of the world. Our consciousness also experiences the immanence of the things of the world in the immanence of our simulations of them. The immanent existence of our simulation of the world that makes it conscious is the exact same immanence of existence that makes things in the world actually real. In this way we experience the same immanence of existence of the things of the world in the immanence of our conscious representations of those things in our simulations.

This is why our consciousness of representations of things in our simulation seems like the consciousness of real actual things in the world when it isn't. This is why our brain's simulation of the world seems like a real actual world when it isn't.

Immanence is the only reasonable explanation of consciousness, because consciousness is clearly not a physical phenomenon. That's why everyone has failed to explain how it could possibly arise in a physical brain in a physical world because it just can't. But the universe is not a physical world in the traditional sense assumed. It's actually an information universe that manifests its self-illuminating existence in the immanence of its forms. When we think carefully this is the only reasonable explanation of consciousness. When we recognize immanence as an intrinsic aspect of the universe, and as the active ingredient of consciousness, the Hard Problem is immediately solved.

Thus the very existence of consciousness is another conclusive demonstration that the universe can't be physical in the traditional sense because consciousness does exist and it isn't a physical phenomenon. All attempts to show how consciousness could be produced from a material universe have failed because it simply isn't possible.

Thus consciousness itself is not something mysteriously produced by human brains and shown like a spotlight onto things in the world. It's simply the intrinsic presence of the immanence of existence within all things manifesting within mind's internal representations of them as they are highlighted by the focus of attention.

This theory of consciousness can also be confirmed by the mental exercise of meditation in which the contents of consciousness are ignored and allowed to pass by freely until they largely subside. This is a simple mental exercise anyone can practice without any metaphysical or religious context.

When mind is emptied of forms in meditation consciousness itself shines brighter. Consciousness itself becomes the focus of attention. The focus of attention program focuses on nothing in particular and just experiences the pure immanence of its own existence. Though technically existence can only manifest through form as only form can be experienced, the experience of meditative immanence is simply the consciousness itself of the form of empty mind. It is as close as we can come to consciously experiencing formlessness while in human form. It's the direct experience of the immanence of reality itself, the direct experience of the universal substrate of existence.

In meditation the focus of attention can also explore deeper areas of the simulation not normally brought to attention such as the underlying energetic processes of the brain's primitive central areas some thinkers call the center of being (Wilhelm, 1931). Deep meditation is as close as we can come to consciousness of the formless immanence of the sea of existence that underlies and manifests in all the things of the world including ourselves.

Our consciousness of a thing is our detection and xperience of its immanence through its form. The particular forms of things are our mind's sampling of their data and fleshing the data out with qualia, but our consciousness of things is the immanence of their data representations in our minds.

The internal glow of existence, the immanence of things, is not something visible to the eyes but it is visible to the mind as consciousness. Consciousness is simply the direct experience of the living immanence of things configured by the structural forms of the mind that receives it. Consciousness itself is the internal glow of the immanence of things. The structural details of what is conscious, the forms of the contents of consciousness, depend on the structure of the receiving mind, but the xperience of immanence, the essence of consciousness itself is universal and all events participate in the immanent existence of things but only according to their forms.

Though our experiences may seem to be of external things, in reality all the experiences we have of things are experiences of our own internal representations of those things in our simulation. All our experiences, even those that seem to be of external things, are actually our experiences of computations in our own brains representing those things.

The reason humans have the conscious experiences we do is because we are participating in the same fundamental xperiential process of reality that all things do, namely the continual interactive recomputation of all forms into continuing existence as all forms manifest the immanence of their existence. Consciousness is our participation in the universal immanence of happening.

Existence manifests only through the actual forms it occurs within. Thus whether an xperience is fully conscious in the human sense or not depends on the structure of its forms. The form structures determine what the reality of its existence becomes because the forms are the information that thing actually is. Thus the immanence of every program in the universe is expressed only through the actual data forms of that program.

So when immanence manifests in inanimate programs those computations do pop into existence as the xperience of themselves. But since they have no additional monitoring forms encoding the fact they popped into existence they have no way to be fully conscious of the fact of their existence. They have the active ingredient of consciousness in immanence and xperience but no forms recording that they have it. They xperience their existence in reality in the immanence of their existence, but they are not conscious of this xperience because they have no secondary forms recording any context or knowledge of it. In the realest sense, they are the actual xperience of their existence, but they are unconscious that they are because they have no forms encoding they are. One could say they are unconscious that they are conscious.

As with inanimate forms, every biological form in our bodies even down to the cellular and particle level, and those of other species as well, has the same sort of proto-consciousness in this same unconscious sense, but only the specialized forms of the focus of attention xperience xperiencing in a knowable reportable sense. Consciousness is the *recursive experience* of the inner light of existence shining in our mind's internal model of reality.

One could say that all the programs of the universe have this sort of proto-conscious of their existence because they xperience the immanence of their existence. But inanimate programs lack information forms encoding the fact they are xperiencing. Thus they have the necessary ingredient of consciousness but no forms to experience that they have it and they aren't conscious of their xperience. They are not conscious of their xperiences; they just *are* their xperiences. All forms are their xperience, but the xperience of the forms that monitor other forms is the experience of consciousness of those other forms.

So for the immanence of existence to manifest as full

consciousness of things it must manifest in secondary forms representing the xperiences of those things rather than just representing the things. Since the secondary forms encode an xperience of the primary forms the immanence of their existence manifests as a conscious experience of the primary forms.

This explains why only programs with internal models of their worlds or selves actually experience their consciousness. Everything in the universe has consciousness in the generic sense of xperience, but this xperience doesn't become conscious experience without specialized forms xperiencing the xperience.

This is what misleads everyone, scientists and laymen alike, into believing that consciousness is something within human minds that is shown out onto things like a spotlight. Nothing could be further from the truth. In the same way that the ancient extramission theory explained vision in terms of eyes shining light on things (Cornford, 1997), so the theory that mind shines consciousness onto things is dead wrong. The active ingredient of consciousness is the universal immanence *within* all things. Human consciousness is simply this same universal immanence in specialized forms of mind.

The active ingredient of the generic consciousness of xperience is the same active ingredient that makes all forms actually real and present in the universe. The manifestation of the immanence of existence in external forms makes them real. The manifestation of the immanence of existence in specialized recursive forms representing the xperiencing of external things makes them real too, but their reality manifests as a consciousness of the primary forms representing external things.

Thus the recognition of the immanence of existence of all things in Universal Reality is the major paradigm shift that is necessary to fully explain consciousness and is the only possible rational explanation of consciousness. It's a simple and necessary addition to the Theory of Everything with great explanatory power. It's an integral part of an entirely new interpretation of science that is entirely consistent with logic and science.

No longer is the universe a mechanical clockwork system in the traditional sense. It has become a living happening computational system that actively self-manifests its existence in the immanence of all of the forms of existence that exist within it including ourselves and the forms of consciousness through which we experience its immanence.

NON-HUMAN CONSCIOUSNESS

Consciousness is not either or but a part of a vast spectrum of xperience across all the innumerable immanent forms of the observable universe. Every form in the observable universe continually xperiences its existence in the continual computations of every one of its subsidiary forms. To the extent a form has subsidiary forms that monitor the xperiences of any of its other forms it can be called conscious but consciousness must be understood to manifest only through the actual details of those specific monitoring forms and the forms they monitor.

Today many robots and other electronic systems from automobiles to industrial robots to the space station monitor ever more details of their functions and their environments, and self-monitoring is the basis of consciousness. Thus all such systems can be said to be conscious of themselves to the extent they self-monitor. The basic mechanism is the same as that of human consciousness, it's just a matter of how extensive the monitoring system is, and in what forms its results are encoded.

Thus there is no intrinsic reason that a robot with sufficiently complex self-monitoring systems and a focus of attention routine to xperience and report the fact of those experiences could not be considered reasonably conscious. Of course for a robot to reliably pass the Turing test of consciousness the robot would have to have a human-like simulation (Wikipedia, Turing test). This is why the Turing test is not a test of consciousness *per se*, but only of a human-like consciousness, and there are certainly all sorts of other variations of consciousness in other species in particular.

Biological organisms typically have very complex self and world monitoring systems in their internal and external sensory systems that transmit volumes of information through chemical gradients and nervous systems to continually refresh their simulations. Thus living beings of almost all species are clearly conscious though in all cases their consciousnesses are restricted to the actual forms of their simulations of self and environment. In all cases the details of consciousness are only those of the actual forms involved.

Humans are essentially biological robots. If we could build a human being from scratch by assembling exact duplicates of all the cellular components of a natural human being it would certainly be conscious in the exact same sense we are, even if initially lacking in memories and a trained learning system, though those could theoretically be uploaded into the brain.

Thus consciousness is not some mysterious separate component that must be added to a biological robot to make it conscious. It is the natural manifestation of the biological design of human beings out of natural organic chemical compounds. So an artificial being could certainly be conscious, though the structural details of that consciousness would depend completely on its design. Would it have electrical circuits that reported all the feelings and status of its body parts to a self-simulation it could monitor? If so it would be conscious of the feelings of those parts of its body just as biological organisms are though always according to the specific details of those particular forms.

Thus all sorts of robotic systems could potentially be constructed with consciousnesses each depending on their computational structures. This would result in all sorts of different varieties of consciousness depending on what they were conscious of and how they were conscious of it, and how they reported it to themselves and each other in what language with what terminology.

As a matter of fact there are already many artificial systems with all sorts of what can be considered specialized consciousnesses and the variety and complexity of these will certainly grow exponentially in the future. And of course there are the enormous varieties of biological organisms each with its own variety of consciousness. Even among humans there is great variation in the details and structure of consciousness that is largely unrecognized.

It is not clear exactly how much functionality consciousness adds to an organism. Massive automatic computations in the whole simulation at the unconscious level generate almost all actions independent of consciousness. Consciousness seems to function primarily as a high-level quality control system with only minimal capacity to modify unconsciously generated actions. However the sense of self provided by consciousness clearly conveys a sense of identity and self-worth to an organism and may well heighten its adaptiveness in that respect.

This self or I associated with consciousness is a subjective I. In humans in particular the simulation also constructs an objective self that is then identified with the subjective self. This objective self is essentially the structure of the self as an object among the other objects in the world rather than the direct experiencer of consciousness from the inside. When the objective self is added to the mix it's experienced objectively as a self that has consciousness rather than a self being conscious.

Self-consciousness is the consciousness of the focus of attention subroutine monitoring itself in the process of monitoring other areas of the simulation. It may also associate the objective self as having the experiences. Consciousness doesn't require self-consciousness to be conscious. All consciousness itself requires is the presence of information forms that encode experiences of things being illuminated by existence from within. This causes experience to shine with the immanence of being. This internal shining is invisible to the eye but visible to the mind as consciousness.

COMPUTATIONAL REALITY

THE COMPUTATIONAL UNIVERSE

There is a single elegant and parsimonious model of the universe that is consistent with and emerges naturally from our theory of existence. It's a completely new interpretation of science but is entirely consistent with science. There are a number of key components to this model that will be explained in detail further along:

1. We define the universe as the totality of everything that has reality, the totality of everything that exists and has existence.
2. The universe is a computational system. It consists entirely of information or data in a continual process of recomputation. The data that is computed is the total unified mass-energy and spacetime structures of the observable universe.
3. Thus the universe can be considered a running program. The entire universe is a single universal program that continually recomputes its evolving data state.
4. This single universal program can be understood in terms of innumerable individual programs that interactively compute all the evolving details of the universe.
5. The happening of existence is the processor that executes the universal program and all its individual programs.
6. Happening defines a single universal computational space within which the universal program runs and all the computations of the universe occur. This is a non-dimensional computational space in the same sense as computer programs define non-dimensional computational spaces.
7. All actual spacetime dimensionality is computed within this computational space and is relative to it.
8. The universal program that computes the current data state of the entire universe runs in the current universal present moment common to all processes.
9. All local clock rates are computed within the common universal present moment depending on local relativistic conditions.
10. The combined computational space and universal present moment in which everything is actually computed defines an absolute pre-dimensional reference background with respect to which actual spacetime dimensionality is relative.

11. Absolute rotation and actual world lines are relative to this computational space in which they are computed. All *actual relativistic effects* are with respect to the computational space in which they are computed and are those that are persistent and agreed by all observers.
12. On the other hand *observational relativistic effects* are due to relative motion between observers. They are observer dependent and cease as soon as the relative motion stops.
13. The data that makes up the universe is of two types, the *observable* data that makes up the observable universe and the *virtual* (non-observable) data that determines and computes the allowable structures of the observable data. The virtual data is observable only through its effect on the observable data.
14. The virtual data of the laws of nature is an essential real component of the universe and must exist somewhere. There is only one possible location that virtual data could exist known to science and that's the quantum vacuum.

EXISTENCE & THE QUANTUM VACUUM

The idea of a universal medium or substrate of existence as a common universal active ingredient of all things that exist is already nascent in science's concept of the quantum vacuum. In modern quantum theory the quantum vacuum is a virtual realm that fills all space and from which all actual particles emerge (Wikipedia, Quantum vacuum). Thus it's reasonable to assume that the quantum vacuum supports the existence of actual particles as well, since if it didn't exist actual particles could never emerge into reality.

Quantum theory implies the quantum vacuum is a universal substrate to the existence of all actualized elementary particles. Thus real actualized particles continue to exist within the quantum vacuum once they appear and its continuing presence is necessary to support their existence. So the quantum vacuum is very similar to our notion of existence as a universal substrate to the being of all things that exist and it's reasonable to identify the quantum vacuum with the substrate of existence.

In this view science has begun to discover a little of the nature of the substrate of existence in the quantum vacuum and we merely take this to its logical conclusion. In this view the quantum vacuum is the locus of both the actualized data of the observable universe and the virtual non-observable data of the laws of nature that determine the forms and computations of the observable data.

Thus the observable universe, the elemental program that computes it, and all the virtual data templates necessary to produce the observable universe reside within the quantum vacuum, which is the actively happening substrate of existence. Together we define all the virtual data of the quantum vacuum as the *complete fine-tuning*. See the chapter of that title below for a full discussion.

So our concept of existence is really just a new *interpretation* of the already widely accepted theory of the quantum vacuum. In our theory existence and the quantum vacuum are different names for the same universal substrate or medium of existence. They are identical and we use the terms synonymously as appropriate.

The quantum vacuum is experimentally confirmed by the Casimir effect (Wikipedia, Casimir effect), and is the basis of Hawking's accepted theory of evaporating black holes (Wikipedia, Hawking radiation). Thus identifying it with our notion of existence is a simple natural step that lends considerable weight to the notion of existence as a universal substrate of being. In our theory the quantum vacuum and the substrate of existence are different names for the same thing.

EVIDENCE REALITY IS COMPUTATIONAL

A computational model is by far the most reasonable and fruitful approach to reality. Our computational model appears both internally consistent and consistent with science and scientific method. This may initially seem counter intuitive but there all sorts of convincing reasons supporting it.

There is overwhelming evidence that everything in the universe is its information or data only and that the observable universe is a computational system:

1. To be comprehensible, which it self-evidently is, reality must be a logically consistent structure. To be logical and to continually happen it must be computable. To be computable it must consist

of data because only data is computable. Therefore the content of the observable universe must consist only of data being computed.

2. The laws of science which best describe reality are themselves logico-mathematical forms. Why would the equations of science be the best description of reality if reality itself did not consist of similar structures? This explains the so-called "unreasonable effectiveness of mathematics" in describing the universe (Wigner, 1960).

3. By recognizing that reality is a logico-mathematical structure the laws of nature immediately assume a natural place as an intrinsic part of reality. No longer do they somehow stand outside a physical world while mysteriously controlling it.

4. Physical mechanisms to produce effects become unnecessary in a purely computational world. It is enough to have a consistent logico-mathematical program that computes them in accordance with experimental evidence.

5. When everything that mind adds to our perception of reality is recognized and subtracted all that remains of reality is a computational data structure. This is explained in detail in the chapter on The Simulation and can be verified by carefully analyzed direct experience.

6. We know that our internal simulation of reality exists as neurochemical data in the circuits of our brain. Yet this world appears perfectly real to us. If our cognitive model of reality consists only of data and seems completely real then it's reasonable to assume that the actual external world could also consist only of data. How else could it be so effectively modeled as data in our brains?

7. This view of reality is tightly consistent with the other insights of Universal Reality, which are cross consistent with modern science. Total consistency across maximum scope is the test of validity, truth and knowledge.

8. This view of reality leads to simple elegant solutions of many of the perennial problems of science and the nature of reality and leads directly to many new insights. Specifically it enables a new understanding of spacetime that conceptually unifies quantum theory and general relativity and resolves the paradoxical nature of the quantum world.

9. These insights complete the progress of science itself in reducing everything to data by revealing how both mass-energy and spacetime, the last remaining bastions of physicality, can be reduced to data as explained below.

10. Viewing the universe as running programs computing its data changes nothing about the universe which continues exactly as

before. It merely completes the finer and finer analysis of all things including us into their elementary units. It's simply a new way of looking at what already exists in which the elementary particles themselves consist entirely of data while everything around us remains the same.

Thus there are many convincing reasons to believe that everything in the universe consists only of its data and that the apparent physicality of things is an illusory interpretation produced by our minds. All the apparently material things of the world around us are our experiences and interpretations of various types of information forms in our mental simulations of reality and by extension in the interpretations of science based on these human simulations of reality.

First, a computational universe immediately solves the vexing problem of how nonmaterial laws of nature could possibly control a material universe they were not a part of. This is a problem that was intractable in the traditional materialistic view of science (Penrose, 2005). However, if the universe and the laws that govern it are respectively actualized and virtual types of information then it's natural that both would be part of a single computational universe. The laws of nature are simply the programmatic structure of the programs that compute the information state of the universe.

Thus the laws of nature, being forms of information in a reality consisting only of data, are an integral part of nature as real as the data forms that encode actual things, and thus are as real as the things of the world. The laws of nature don't stand apart from nature in some mysterious metaphysical realm while controlling it as science mistakenly assumes. That the laws of nature find a natural place in our computational model of reality is strong evidence for its validity.

Second, it's quite clear that our experience of a seemingly physical universe, and everything in it, actually consists only of information in the neural circuits of our brains. While there is certainly a real universe external to our brains, the seemingly physical universe we experience our existence within is without any doubt an information construct in our brains. So if just information in our brains can produce such a very convincing illusion of a material universe, why couldn't the actual universe external to our brains also consist only of information?

That would immediately explain why neural computations within our brain's model of reality could enable us to function so effectively within actual external reality. How could our internal mental simulation of the universe so accurately map the actual workings of the universe if the universe itself were not also an information structure?

Third, all the laws of science consist only of mathematical equations imbedded in a logical framework, in other words they consist only of information. How could information structures accurately describe the universe if the universe itself didn't consist of information structures? This immediately solves the mystery of why mathematics works so incredibly well to describe the universe. Of course mathematics and logic would naturally provide the best description of a universe that was itself a logico-mathematical information structure.

Fourth, when we carefully analyze seemingly material things in our minds we find that they actually consist only of the information of what they are, and this is true of everything without exception. They all consist only of their information, the combined information of their colors, textures, forms, structures, chemical compositions and whatever else makes them up. These are all just different forms of information that in combination are interpreted by our brains as material objects. Our brain tells us these combinations make material object but even that interpretation is just more information.

This is also confirmed in the design of robotic control programs and pattern recognition (Wikipedia, Pattern recognition). In robots able to operate effectively within complex environments internal models of themselves within their environments must be laboriously constructed and continually updated from streams of raw sensory data. That data is then converted into simulations of purposeful action within the model, which are in turn tested, valuated and used to control appropriate motor activities. Internally it's all based on internal data models of the robots within their environments that work only due to the model's logical consistency with the actual data structure of external reality. All living organisms including us operate on similar principles though in much more complex systems.

The information that makes up even a simple physical object, not to mention that of a living being, is not a simple data string like the name or description of an object. It's an incredibly complex hierarchy of forms and multiple hierarchies of subprograms within subprograms, and their ongoing computational interactions and relationships with other forms and programs. Think of the hierarchies of total information content of anything down through its individual systems to its individual cells to the detail of every one of its elementary particles and their interactions, and that is the complete information structure that makes up that thing, and actually is that thing. These are the total running programs that things actually are.

Fifth, even modern science now has now reduced the entire materiality of the universe to just mass-energy, and spacetime. However, in the chapter on Fundamental Principles, we show how spacetime reduces to the information of dimensional relationships, and mass-energy reduces to the information of relative motion. So even the universe envisioned by modern science naturally reduces to pure abstract information.

Sixth, accepting a universe consisting only of information doesn't change the universe that we experience around us in the least. It still appears exactly as it did before, as a material universe. The only difference is that we now realize that its seeming physicality is an interpretation of its information structure produced by our minds, and that the underlying data structure of the seemingly physical world we live in is its actual fundamental structure.

Thus it's reasonable to conclude that the data structure of the universe is its actual fundamental nature and its seeming materiality is an illusion produced by our mind as it combines the information of things into the semblance of physicality.

Thus in our theory all the programs of things that make up the universe without exception consist only of their data in a continual process of recomputation. These programs have existence because they run in the substrate of existence, and thus they become the real actual things of the world, but the fundamental nature of all these things is information given being by its presence within the substrate of existence in the observable universe.

Thus at the most fundamental level the things that make up the universe are not material things, they are simply different information forms that arise in the originally formless sea of existence, as water waves, currents and ripples are different forms of water that arise in an ocean of water. And since the things of the universe are not physical, they have no individual self-substances that make them different things; the only difference among things is the differences in their forms, the different data that distinguishes them one from another.

Information takes innumerable different forms but the fundamental nature of all the data that makes up the universe is the same; it all consists of abstract data forms that are computationally evolving in a common non-material medium of existence. The only substance of all information forms is existence itself, just as the only substance of all water waves is water no matter how their different forms vary.

All things in the universe consist of information given actuality by existing in the universal medium of existence. It is their common existence, rather than any material substance, that makes them all real things. They become real things by appearing in the virtual medium of the quantum vacuum, just as water waves become real by appearing in water.

Thus all the seemingly physicality of the things of the world is actually interpretations of their information forms in our minds. The apparently physical world in which we seem to exist is our mind's internal simulation of an actual external reality consisting only of the information states of running programs. To this extent the physical world is completely an illusion, though certainly a very convincing illusion.

This is equally true even at the perceptual level. It's just a matter of becoming aware of what we actually see when we observe things. If we really take our perceptions of things apart into their individual components we find their every component reduces to the information of what it is, and that's all we experience because everything without exception is ultimately perceived in terms of its information. Only information is observable. There simply isn't any way to perceive anything except in terms of its information. Perception is information input and sensory information input is perception.

So everything is actually just its information or its data, but this is not data in the usual sense of data on a printout or even in a computer. The medium of this data, the data of actual things, is not marks on paper or electronic bits in a storage device. The data of reality exists in the medium of existence, and that's what makes it real and actual. It makes the data of existence into real actual living things. Its existence gives it immanence and being, it makes whatever form the data has into the real actual thing that has that form.

And since existence continually happens that data is continually recomputed and takes on a life of its own in interaction with the life of all the other data forms that make up the universe.

BIOLOGICAL PROGRAMS

If we take a human being and go down the hierarchy of its biological structure down to the particle level there isn't any doubt that each of us ultimately consists of the elementary particles that make us up. We lose nothing in this process we are just the same as we always were. It's just a more complete explanation of what we actually are.

Now just take it one more step down and think of those particles as being data or little bits of existence in a continual state of recomputation. This is the model that Universal Reality suggests underlies everything in the universe including us. It doesn't change us, or anything in the least; it merely explains us as ultimately computational processes, as running programs; the running programs of ourselves.

We are living, purposeful, free, sentient, emotive, intelligent programs that go about our business just as we did before. We just now have an explanation of how we function at the most basic level. We are composed of patterns of computations at the elemental level and that allows us to intelligently compute the living of our lives at the emergent level of being humans.

We can even confirm that everything is ultimately data or information at the classical level of our perception. The only way things can be perceived is in terms of their information. Whatever we look at in the seemingly physical world is actually associations of the information that makes it up. All we ever perceive of anything is its data and there is no evidence anything at all is anything other than its data, its data given an actual real happening existence in the medium of existence.

The apparent physicality and materiality of things is just an association of the different types of sensory information of those things. The information of color, of tactile feelings, of heft, of sounds, odors and tastes is all different kinds of information our minds combine and interpret as physical objects. The apparent physicality and materiality of things is an interpretation of their data in our mind's simulation of reality.

We can easily confirm this by identifying and subtracting each of the information types that makes up anything at all. We successively remove the information of the shape, color, feel, smell, location, orientation, function, meaning, etc. and what is left? Nothing at all is left after all the information is removed from something. Thus it always consisted only of its information, and its apparent physicality was an interpretation of the *combination* of that information in our mind's simulation of it. And this is true of everything that exists without exception.

Even with our closest friends and family the same is true. The actions, feelings etc. of people that makes us love and bond with them are all reducible to different types of data both in them and in our interactions with them. Everything without exception can be reduced to its information including the information of meanings and feelings we have about things and even people.

All is information only, complex associations of information forms that we interpret as a friend or loved one, a human, a dog a bird or a stone. Everything is the complete exact information of what it is, and nothing else. And not only that, we too are the complete data of ourselves, and only our data, in a continual process of recomputation by the program of ourselves which is our data continually being recomputed by the elemental program of the quantum vacuum at our particle level.

And this is further confirmed because it's exactly how we store our simulated worlds in our minds. Everything without exception is stored in our brains as the information those things are to us. As real as they seem, they all exist entirely as neural data in our minds.

Our brain's simulation of reality seems totally real even though it's just the data of things in our brains. Since our brain's data model of reality seems completely real, there is no reason whatsoever things can't also be their complete data states in external reality itself. How could our brain even model them as information so realistically if they weren't actually information themselves?

The individual programs of living beings compute actions intelligently on the basis of internal models of their environments. They are effectively intelligent, sentient, purposeful biological robots. You and I are the running programs of our biological robots. This description changes nothing about us and doesn't diminish us in the least; we are still fully human as before. It just describes what our humanness actually is in terms of its information processes.

Our internal mental model of us within our local environment is called our simulation. Our simulation interprets the information world around us as the familiar 'physical' world of our experience. We believe we are a physical being living in the physical world we appear to experience around us, but that physical world is largely an illusion. Its appearance is completely an illusion, and its logical structure is only a minute sample of the complete logical structures of the running programs around us.

Thus reality consists entirely of the information of the universal program running in the quantum vacuum substrate of existence. The actual universe has no physical nature; its apparent physicality is an illusory interpretation of an actual information reality produced by our simulation. The 'physical' world exists entirely in our own head. There is overwhelming evidence this is true.

COMPUTATIONAL STRUCTURE

The following overview of the basic computational structure of the observable universe explains how elemental quantum computations generate an entanglement network among all elementary particles. This entanglement network encodes and actually is the unified data structure of all aspects of the observable universe. It includes the data of all mass-energy structures and all their dimensional relationships we interpret as spacetime.

The internal logico-mathematical consistency of the dimensional relationships of mass-energy structures is what observers interpret as dimensional spacetime in their simulations of reality. All aspects of the observable universe, including spacetime, are computed by the elemental program from elemental particle interactions.

1. The observable universe consists of all actualized (non-virtual) observable data. The actualized data of the observable universe consists of particles, particle components, and their relationships. Particle relationships consist of the entanglements among particles on their particle components generated by particle events and interactions.
2. Particle components, such as spin, mass, energy and the charges are taken as the actual elemental data components of reality, which in valid combinations create elementary particles. This is reasonable since particle components are conserved through all particle interactions even when particles transform into other particles.

3. Every particle interaction creates entanglements among its particles on each of the particle component types involved in the interaction. The additive sum of all properly defined particle components is always conserved.
4. Together all the particle entanglements of the observable universe form an entanglement network that computationally connects all the particles in the universe and establishes relationships among their particle components.
5. The logico-mathematical consistency among all the dimensional data of the particle components is what we reify and interpret as an encompassing spacetime.
6. The entanglement network can be visualized as particle event nodes connected by the data trajectories of individual particles. It incorporates all the actualized particles in the universe and all their events.
7. The entanglement network extends back to the original event(s) of the big bang and represents the entire evolutionary *history* of the observable universe at the particle component data level.
8. Thus the entanglement network is the integrated mass-energy and spacetime structure of the entire observable universe throughout its entire history.
9. The actually existent current observable universe is the current present moment data state of the entire entanglement network.
10. All the actualized data of the entanglement network is continually recomputed by the elemental program in the current universal present moment of existence to produce the current data state of the observable universe. This includes local relativistic clock times, which are all, computed in the current present moment.
11. At the aggregate level the entanglement network of all particle data takes the form of hierarchies of overlapping computational *domains*. Domains are areas of computational density and similarity characterized by domain boundaries of lower computational density and similarity.
12. Individual observers tend to view the universal program in terms of individual things and programs on the basis of domains. For example surfers extract waves, smelt extract tides, and oceanographers extract currents from the overlapping computational domains of a single ocean.
13. Nevertheless the universal program is a single interconnected entanglement network ultimately computed on the basis of individual particle interactions by separate applications of a single elemental program.
14. This single elemental program consists of a small set of virtual subroutines that identify and compute all possible types of particle

interactions. Individual applications of this elemental program compute the evolution of the entire observable universe in the form of the entanglement network.

15. Domains and observer identified things and programs are *emergent* manifestations of particle aggregates. These emergent data structures act as independent programs due to the meaningful relationships among their particles. This is analogous to how computer programs designed to perform various high-level functions all consist of meaningful sequences of a small set of elemental machine language operations that perform the actual computations.
16. Computer programs consist of pre-written sequences of code executed sequentially. But the programs of reality consist only of the data of things in a continual state of recomputation. All the actualized data of the observable universe always exists in the current present moment in the quantum vacuum. Thus it all exists within the processor of happening where all the data of the observable universe is *simultaneously* recomputed by individual applications of the processor in the current present moment.
17. A simultaneous separate *application* of the elemental program and single processor of happening recomputes each individual particle interaction process.
18. Emergent programs and processes are meaningfully structured aggregates of elementary particle data. Emergence begins at the atomic and molecular level and extends up the hierarchy of aggregate structure to the highest classical levels and beyond. Nearly all of science other than particle physics is concerned with emergent processes and data.
19. Though emergent processes are meaningfully considered as programs the emergent laws of science only *describe* them rather than actually *computing* them since all actual computations occur only at the elemental level and are computed by separate applications of the single elemental program to each coherent particle process.
20. The biological programs of living organisms including us are all characterized by internal simulations of themselves within their environments to various extents. They are also characterized by purposeful actions based on instinctual imperatives that are part of their operational software passed from individual to individual and from generation to generation encoded in their DNA.

The quantum vacuum computes the observable universe at the particle and particle component level because it contains the actual elemental program that does the computing and is the processor that executes the program. However the observable universe itself acts like a universal program at the emergent level even though everything is actually computed at the elementary particle and particle component level.

Unlike a regular computer program the observable universe consists only of current data states rather than pre-programmed code sequences. Thus the quantum vacuum computes the interactions of current data states rather than emergent level sequential code strings.

The observable universe can't contain pre-programmed code strings because that would imply a pre-determined future, which doesn't exist. And if the universe consisted of multiple pre-programmed code strings that would imply multiple possible versions of the future that would inevitably lead to irreconcilable inconsistencies. So the present must be continually recomputed from the interactions of its current data states. This is consistent with science in which the evolution of the universe consists of the continual interactions of all its particles.

There is another important way the computational universe differs from the way ordinary computers work. In ordinary computers a single data computation occurs at every processor cycle. Though a computer can have multiple processors each processor can only perform one computation per cycle. This greatly limits the overall processing power of even supercomputers.

By contrast all the data of the observable universe exists in the quantum vacuum simultaneously, and the happening aspect of the quantum vacuum is the processor that computes reality. Thus all the data of the entire universe exists within the processor of happening and is recomputed simultaneously with every P-time tick to create the next current present moment. Thus the computational power of the quantum vacuum is limited only by its processor cycle rate, which we will see below has important implications.

PROGRAMS & DATA

Universal Reality proposes that reality consists only of information or data in the form of self-modifying programs running in a universal substrate of existence, which it identifies with the quantum vacuum.

There are a number of convincing reasons to accept this model of reality and it also leads to simple and elegant solutions of many of the fundamental problems of science and philosophy.

In this model the universe is a single universal running program that continually recomputes its current state, the current state of the universe, in the present moment.

All the individual things and processes of the universe are individual subprograms running interactively within the universal program. They are all computationally consistent parts of the single universal program. These individual programs computationally arise, transform, and fade as identifiable structures in continual interaction with other programs within the universal program. This overall process is the computational evolution of the universe.

The individual programs are all the actual processes of the world from the most elemental interaction of particles through human beings to cosmological processes on the grandest scale. Individual programs are processes identified on an *ad hoc* basis on the basis of the computational domains produced by the universal program and their personal meaningfulness to individual observers.

Even though the observable universe consists only of its current data state rather than code strings it acts like a single universal program that can be understood in terms of innumerable individual programs in continual interaction with each other. We just need to keep in mind that the actual computations occur only at the particle and particle component level and all the emergent level programs of the universe are aggregate manifestations of elemental computations.

This is analogous to silicon computer programs where the actual computations occur only at the level of individual machine language instructions, but structured aggregates of instructions form meaningful emergent level programs with specific higher-level functions. In a similar manner the data that makes up the universe consists of meaningful data structures, though not code sequences, at the aggregate level. One can identify code sequences but these are not stored pre-programmed sequences but simply post facto lists of sequences that have already occurred.

The aggregate data structures are those familiar to science as the compound particle structures that make up all the emergent level structures of the universe. Because these data structures are continually interacting at the particle level they act as programs and these running programs are the programs that appear to compute the universe at the emergent level.

Thus the observable universe can be considered a universal program composed of innumerable individual programs because its elemental data structures at the particle level are meaningful in aggregate at the emergent level due to the details of the complete fine-tuning which define them.

We know from information science that only data can be computed, and data can only be meaningfully computed on the basis of exact logico-mathematical operations embodying consistent rules (Wikipedia, Data (Computing)). Sequences of logico-mathematical operations on data are called programs. Information science also tells us that if a set of axioms and logical operations is consistent then all computational results of the system will be logically consistent and logically complete; that the results of all possible computations will produce a single consistent logico-mathematical system with no contradictions.

All data forms are forms of in-*form*-ation. Thus everything in the observable universe consists of forms of information or data that arise in the common medium of existence and thus gain reality as real actual things. The continual recomputation of all data forms generates the evolution of the observable universe.

CONSISTENCY & COMPLETENESS

Once we understand that the programs of the universe are simply the ongoing changes occurring in its information it's clear these changes must be computational because the only way information can change is computationally. Therefore all the processes of the universe must be running programs, and what we call the things of the universe are all the current information states produced by these programs.

Because it's computational the programmatic information structure of the universe must be a rule based logical structure. The structure and programs of the universe must follow logical rules contained virtually in the complete fine-tuning of the quantum vacuum. Thus every information state of the universe is computed according to these logical rules from its previous information state, and every information structure of the universe is logically consistent with every other.

If the universe is a computational system then it must be logically consistent and logically complete because computations operate only on the basis of logical rules and for the universe to exist those rules must be consistent and complete. Only computations produce one information state from another on the basis of consistent logical rules.

Logically complete means that every computation of the universal program and its individual subprograms will always be able to produce some result. And logically consistent means the universal program can never compute any result that is a logical contradiction to any other computational result it has produced or can produce.

There is a straightforward proof that the universe is logically consistent and logically complete. If a computational universe was not a logically consistent and logically complete system, it would tear itself apart at the inconsistencies and pause at the incompletenesses and simply could not exist. Thus the fact of its existence demonstrates the universe is in fact logically complete and logically consistent, which is in fact what we always observe.

In addition a computational universe must be logically consistent and complete to be meaningful and amenable to knowledge. Since the universe is a massively understandable system that also confirms it is logically consistent and complete.

For the universe to be logically complete and consistent the fundamental axioms of the complete fine-tuning must themselves be a complete and consistent set of rules, because it's these rules that govern the computations of the universal program. The fundamental logical rules of the laws of nature must be a consistent set of axioms that cannot generate inconsistent results. And this is necessary for the universe they compute to exist. This is one clue into the mystery of why the actual fine-tuning of our universe is likely the only fine-tuning that can exist, because it may be the only one that produces a logically consistent and logically complete universe.

Logical consistency and completeness of course refers to the computations that produce the actual information states of the universe rather than to the reasoning of individual actors within the universe. Within this overall consistency it is quite possible for human thinking based on false or incomplete premises or invalid logic to generate inconsistencies as we see far too often.

In particular the logic of human and other beings is based on the emergent logic of things and highly simplified models of reality, which ignore most of its actual computational details. This enables humans to quickly compute reasonably accurate descriptions of small fragments of reality and base fairly effective actions upon them.

But by computing on the basis of highly simplified individual details and events the complete logical consistency of reality is lost. Typically what human reasoning does is continually remap simplistic models of reality consisting of relevant sets of individual things, events and relationships, and replace them with others as needed, with no necessity of complete consistency among them. This leads to the general fuzziness and inconsistencies of human thinking, but it enables humans to reason fairly quickly and accurately from moment to moment on the basis of heuristic mental models of changing situations.

At the core of this process is employing the logic of things to model the world in terms of individual things, properties and relationships, rather than its actual enormously complex single network of computations. These individual concepts are redefined as needed to quickly model the relevant details of a current situation. This often results in dynamic overlapping identities of individual things, such as a forest when hiking or an individual tree when cutting wood. This enables humans and other species to quickly reason on the basis of minimally pertinent information sets. But to accomplish this humans lose the complete picture of the processes of reality in terms of their inherently contradictory heuristic models of reality from moment to moment.

It's important to note that the logico-mathematical system of reality isn't subject to Gödel's Incompleteness Theorem because every state is directly computed from its prior state. Gödel's theorem applies only to human mathematical systems in which it is possible to propose a well-formed statement that cannot be proven either true or false (Hofstadter, 1980). But reality doesn't make statements and then try to prove them like mathematicians do. It just computes the next results from the previous results and this can always be done, and is always consistent and complete.

Thus, unlike human mathematical systems, reality mathematics can be and must be logically complete and logically consistent. The universe produces every information state computationally and so only produces statements (data states) whose truth is automatically proven by their existence.

SUPER CONSISTENCY

The great mystery is that this process works at all because it requires a type of *super-consistency* which allows the simplistic inherently contradictory overlapping models humans have of reality to exhibit reasonable internal self-consistency sufficient to ensure effective action in an actual world that consists of enormously more complex information systems.

This super-consistency is necessarily inherent in the virtual information structure of the complete fine-tuning and is what allows us to effectively live our lives within an enormously more complex computational universe.

The logical consistency of the universe is what makes it understandable, and makes reason, knowledge and science possible. The ability to map the logic of reality in our simulations of reality makes this possible. The universal program is a completely self-consistent logical system and thus there seem to be no intrinsic limits to understanding it.

The universe acts as a running program in the medium or substrate of existence which compute it and gives it reality and the living vitality of its happening. And the immanence of the quantum vacuum gives all the information forms that make up the universe the immanence of their self-manifestation that makes consciousness of them possible.

At the heart of the reality of the universe is happening, the fact that things change, that its data is continually recomputed in the present moment in which everything exists. Happening is the life of the universe. It's what brings everything to life and gives it reality and being and immanence and makes consciousness possible.

IMPLICATIONS

Universal Reality is a science-based theory. The universe evolves according to the computational laws of nature, however it does so in a manner that logically integrates happening, the present moment, immanence and consciousness, the fundamental aspects of existence, into a single unified system.

The data of the observable universe is computed into a universal entanglement network of all particle and particle component relationships by particle events. This is the fundamental data structure of the observable universe. It consists of the computed relationships of all the particles and particle components of the universe including their dimensional relationships. This entanglement network is the unified data structure of all mass-energy relationships including their dimensional relationships and is a single integrated structure.

Because spacetime is computed as the dimensional relationships of mass-energy structures both atomic and molecular structures and the spacetime they exist within are computed together as a single unified structure consistent with both quantum theory and general relativity and it's the computational emergence of spacetime that is the key to their unification as explained in upcoming chapters.

The universe actually is its computed data. Its current data state is the observable universe in the present moment. Thus the observable universe is not the physical or material structure it appears to be but the complete data of what we humans and other species *interpret* as a physical universe, each in our own way.

Thus the true nature of every individual thing in the universe is the complete data and programs of what it is. Everything is its running program, and its snapshot at any point in time is its instantaneous data state. This includes us as well. We are the programs of ourselves, and the data of ourselves in the continual process of recomputation by our programs. While at first this may seem counter intuitive and even crazy, we can actually confirm this through careful analysis.

Science is a wonderfully accurate and comprehensive model of reality and is extensively confirmed by experiment. Universal Reality generally accepts all experimentally confirmed science but fundamentally reinterprets it. In Universal Reality what science interprets as material things in a physical spacetime container actually consists of the integrated data structures of those things and this data evolves computationally rather than causally.

The only way anything can happen is for it to be computed, and the only way something can be computed is for it to consist of data. This view has long been implicit in science itself since science consists of the same logico-mathematical structures that underlie computation. However the clear implications of this view have been suppressed by the archaic belief that reality is material and physical.

However it's easy to demonstrate that the apparent physicality of reality is an interpretation generated by our mind's simulation of reality to help us make sense of the world. How this comes about is explored in detail in the chapter on The Simulation, but there is plenty of other evidence.

Science itself has progressively reduced the apparent physicality of the universe to fewer and fewer elements until currently only mass-energy and spacetime remain as the last bastions of the old material world.

However as we will discover below mass-energy and spacetime also reduce entirely to data, thus demonstrating in one more way that the fundamental nature of the universe is only data. Mass-energy reduces to the data of relative motion and spacetime emerges computationally from mass-energy interactions. This makes it clear that a computational approach to reality is quite reasonable, and by adopting this view we lose nothing essential to the workings of science.

This paradigm shift leads to significant new insights about our universe, and a completely new interpretation of the reality in which we exist. It also provides conceptual solutions to many of the important unresolved problems of science such as the nature of quantum paradox, the apparent inconsistency of quantum theory and general relativity, and the source of quantum randomness.

We will begin to explore how this all works with a discussion of fundamental principles in the next chapter, and the details of the complete fine-tuning in the subsequent one. At this point the details of how particle events and atomic and molecular structures are computed together with dimensional spacetime in a manner consistent with both quantum theory and general relativity can be clearly explained.

FUNDAMENTAL PRINCIPLES

THE EXISTENCE PRINCIPLE

Existence! Existence exists! The very fact of existence, the fact that something rather than nothing exists. In Universal Reality the quantum vacuum is identified as the medium or substrate of existence from which all that exists gains its individual existence. The quantum vacuum includes the observable universe and the virtual data of the complete fine-tuning that computes and structures the observable universe.

THE CONSISTENCY & COMPLETENESS PRINCIPLE

In Universal Reality the universe is a computational structure consisting entirely of data computed by applications of an elemental program according to logico-mathematical rules embedded in the quantum vacuum.

Thus the observable universe must be logically consistent and logically complete. If it weren't logically consistent and logically complete a computational universe would tear itself apart at the contradictions generated by inconsistencies and pause at the incompletenesses and thus could not exist. Incompletenesses would manifest as elements of a computational universe suddenly vanishing.

Thus the logico-mathematical structure of the entire universe can contain no contradictions and its computational processes can never halt. They will always continue to compute so long as there is observable data to compute.

It is this logico-mathematical consistency of the universe that makes it meaningful in terms of logic and mathematics and which is the basis of all knowledge.

THE PRINCIPLE OF DIFFERENTIATION

This refers to the fact that different individual forms of existence appear within the originally formless sea of existence; that there are separate identifiable forms rather than a single formless thing. This refers specifically to all the irreducible different data, virtual and observable, within the quantum vacuum. It includes all the individually identifiable things that exist within the observable universe, as these are all manifestations of the differentiated virtual data of the quantum vacuum.

This principle is certainly involved with the actualization of the observable universe, which consists of myriads of differentiated entities, but it also applies to the quantum vacuum itself, which has a differentiated virtual data structure.

THE STc PRINCIPLE

The STc Principle states that the combined vector velocity through space and time of everything in the universe is always equal to the speed of light, c. This is a consequence of relativity that is well known to scientists though usually viewed as a mere curiosity (Greene, 1999, 2005). However it's actually a fundamental principle with profound consequences. It means that every clock in the universe runs slower in time proportional to its velocity through space so that the combined spacetime velocity through time and space always remains equal to c.

This is one of the fundamental conservation principles of reality. It means that the total velocity of everything in the universe is always c, and that velocity is distributed between velocity in space and velocity in time. This is a critically important principle that underlies relativity as well as the fundamental nature of mass and energy.

The STc Principle is a direct consequence of a fixed number of processor cycles used to compute the happening of the universe. Processor cycles go first into computing the spatial velocity of processes. The remaining cycles compute the internal evolution of processes, which manifests as their clock time rates. In this manner the processor cycles that compute each process are distributed among computing velocity in space and velocity in time so that the vector sum of velocity through space and time is always c. This is explained in more detail in the chapter on Computing Spacetime.

By definition observers don't move relative to themselves, thus all the c spacetime velocity of every observer is completely through time on its own clock. Thus c is actually the speed of clock time and light moves at that velocity through space because it has no internal processes to compute and thus no velocity through time on its own internal clock. Thus c is the speed of time on all observer's own clocks and the baseline velocity of all processes in the universe.

The particular value of c in our universe provides enough time for things to happen and enough space for things to happen in. If the speed of time was zero nothing could ever happen, and if it was infinite the whole history of the universe would be over before it began, so a viable universe requires a reasonable finite non-zero value of c, which is encoded in the complete fine-tuning.

The value of c must also be quite large, as it is, relative to typical velocities through space. If it weren't the spacetime dilations of mass-energy would produce gravitation so intense as to crush all possible material structures, and routine spacetime distortions so great as to make ordinary processes unintelligible.

THE CONSERVATION OF PARTICLE COMPONENTS

The particle components that make up elementary particles are the basic computational data elements of reality. The particle components combine in valid sets to form the elementary particles that make up everything in the observable universe. Every elementary particle is composed entirely of its unique set of particle components. The particle components are the basic constituents of the observable universe.

The total values though not numbers of all particle components are conserved through all particle computations. The subroutine that computes the conservation of particle components must exactly redistribute all the incoming particle component values in any particle interaction among the resulting particles or an interaction cannot occur.

Particle component conservation in particle interactions produces the network of entanglement relationships among elementary particles that observers interpret as mass-energy structures in spacetime. Thus the conservation subroutine computationally creates spacetime in the form of entanglement networks from the interactions of elementary particles and their particle components.

There are a few rare exceptions where currently defined particle components are conserved only in combinations. But it should be possible to define a proper set of particle components such that they are always conserved.

We can identify two general classes of particle components, those intrinsic particle components that uniquely identify *types* of particles, and dimensional particle components such as spin, mass and energy. Mass is a form of energy and total mass-energy is conserved in every particle interaction. This is key to understanding how processor cycles compute quantum events as discussed in the chapter on Quantum Reality.

THE MEv PRINCIPLE

Total mass-energy is conserved in every particle interaction. All forms of mass-energy are inter-convertible in particle events. The conservation of mass-energy is a basic principle of science.

For something like mass-energy to be conserved all its different forms must be forms of the same underlying thing. In the case of mass-energy all its forms are different forms of relative motion whose values are their spatial velocities. Thus the conservation of mass-energy is always the conversion of equivalent amounts of relative velocity from one form to another, and all forms of relative velocity are equivalent to some type of mass or energy. This MEv Principle is a fundamental principle with profound consequences.

Physics already understands some forms of energy, such as kinetic energy and heat, as relative motion. Kinetic energy is half the square of the linear velocity of a mass relative to an observer. And heat energy (temperature) is the average kinetic energy of the velocities of atoms and molecules within a substance.

The electromagnetic energy carried by photon is also a form of relative motion. It's the relative motion of its vibrational frequency, the vibrational velocity of electromagnetic waves. The amount of energy a photon carries is directly proportional to its frequency, which is the velocity of its wave cycles.

With $E=mc^2$ Einstein showed that mass is a sort of frozen energy. This is easy to understand if we model masses as forms of very fine very high frequency in place vibrations. By vibrating rapidly in place they have the same amount of motion relative to all stationary observers and thus particles have fixed rest masses.

This is compatible with the observed increase of mass as velocity approaches the speed limit of light. Near the speed of light it becomes more and more difficult for linear velocity to increase so some of the linear velocity is converted to internal vibrational velocity and this increases the mass as predicted by relativity (Wikipedia, Special relativity). It is worth noting that String Theory also models particles as rapidly vibrating strings (Susskind, 2006, p. 199), (Wikipedia, String theory).

Chemical energy is the binding energy of electrons in their orbitals around atomic nuclei. Binding energy is the energy that holds all forms of atomic and molecular matter together and is responsible for the different chemical compounds. It is due to the electromagnetic attraction between negatively charged orbital electrons and the positively charged protons of nuclei.

Different molecular configurations will have different binding energies and the release of chemical energy in explosions is due to the transition from molecules with higher binding energies to molecules with lower binding energies. An explosion converts the difference in binding energies to heat energy and the kinetic energy of rapidly expelled particles. The internal velocities of the binding energies are converted into the linear and wave velocities of emitted particles.

Binding energy is due to the rapid oscillatory velocities of electrons within atomic orbitals. Like the vibrations of mass, the relative motion is locally constrained so it manifests as fixed energies to stationary observers.

When particles combine in molecules there is actually a very minute increase in total mass over that of all the combined particles. In other words some of the binding energy is converted to, or manifests as, mass due to the orbital velocities.

Nuclear energy is similar to chemical energy but involves binding energies of the quarks and gluons in nuclei that hold protons and neutrons together. Nuclear energy is due to the strong force rather than the electromagnetic force that holds electrons in their orbits. Because there is much more total velocity converted in nuclear explosions they are enormously more powerful than chemical explosions. Nuclear explosions are transitions from nuclei with higher binding energies to nuclei with lower binding energies thus resulting in the conversion of the lost binding energy to other forms of energy. The internal velocities of the binding energies are converted into the linear and wave velocities of emitted particles.

This is confirmed by the modern theory of the quarks and gluons in nuclei (Wikipedia, Quark#Mass). The total mass of these particles is just ~1% of the total observable mass of nuclei so most of the mass of nuclei is actually due to the equivalent amount of relative velocities of their quarks and gluons.

As for potential energy it's sort of an accounting trick rather than an actual form of energy. When we say that an object in a system has potential energy what we are usually saying is that there is some energy in an external system that is blocking an equivalent energy in the system under consideration. The notion of potential energy just makes it easier to isolate systems for analysis. Potential energy is the amount of energy in a system that can be released by removal of the equivalent blocking energy. A compressed spring or a weight suspended by a wire are examples.

The potential energy of a charge due to its position in a force field, including a mass in a gravitational field, is not an actual amount of existing energy but a measure of predicted future energy that doesn't yet exist. It's the potential for some form of relative motion. This is another accounting trick because the conservation of mass-energy properly applies only to conversions of actual forms of relative motion as they occur.

Universal Reality models force fields as fields of velocity density of various forms specific to the force as explained later. Thus the velocity gradient of the field generates the linear velocity of kinetic energy and total velocity is conserved.

Gravitational energy is another kind of relative velocity. A gravitational field is very effectively modeled as an area of vibrational density in space itself produced by the vibrational motion of mass or energy. This dimensional vibration effectively produces the curved spacetime of general relativity as explained in the chapter on Computing Spacetime. It's because of the spatial velocity density of fields that the clocks of objects in gravitational fields run slower in accordance with the

STc Principle. This is also explained in the chapter on Computing Spacetime.

When we understand that all forms of mass and energy are just different forms of relative motion the reason for the conservation of energy among its different forms becomes clear. The conservation of energy is just the transformation of one form of relative motion to an equivalent amount of another form of relative motion. For example the conversion of mass to energy in a nuclear explosion is just the conversion of some of the mass vibrations of particles of fissionable material into equivalent amounts of wave frequencies and linear velocities of ejected particles.

Thus all forms of mass and energy are different forms of relative velocity computed by the elemental program of the universe. Since all forms of mass and energy are relative velocities they are fundamentally data forms rather than anything physical. They are part of the data of dimensional relationships computed by the elemental program of the quantum vacuum.

Thus Universal Reality demonstrates that both spacetime and mass-energy, the two remaining fundamental components of 'physical' reality, reduce naturally to computational data without diminishing the explanatory power of science. Universal Reality leaves the enormous explanatory power of science intact, but it gives us a completely new interpretation of science.

THE METc PRINCIPLE

The STc Principle states that the vector sum of space velocity and time velocity always equals the speed of light c. But from the MEv Principle we see that mass-energy is equivalent to spatial velocity. By combining these two principles we arrive at an even more fundamental principle of the conservation of mass-energy and time. This METc Principle states that the total mass-energy spatial velocity and time velocity of everything in the universe always equals c, the speed of light.

To be precise the total spacetime velocity of everything is c, and the spatial velocity component of anything is always some form of mass-energy. Thus all forms of mass-energy slow the velocity of time because they are all forms of spatial velocity. The total vector mass-energy velocity and time velocity of all processes is always c.

Thus the fundamental fabric of spacetime is composed of spacetime velocity, which has the same c value everywhere apportioned between spatial velocity, which is equivalent to some form of mass-energy, and time velocity. This is a deep fundamental principle of reality that underlies the computational unity of mass-energy and spacetime and is key to really understanding both quantum theory and general relativity.

The METc Principle is a direct consequence of the fixed cycle rate of the quantum vacuum processor that computes the observable universe. It's a direct consequence of the fact that the observable universe is a computational structure. The processor cycle rates that compute the happening of everything in the universe are allocated between calculating the internal changes of processes and their spatial motion. Motion in space is their mass or energy and the rate of change of their internal processes is their velocity through time.

Thus we have the fundamental principle of the conserved equivalence of mass-energy and time that underlies general relativity and nearly every aspect of reality. The total relative velocity of mass-energy plus the total velocity of clock time for all processes is always equal to c, the speed of light, which is the cycle rate of the processor that computes everything in the universe.

This principle demonstrates that mass-energy and clock time are two aspects of a single fundamental entity and are inter-convertible so long as the total vector velocity of both is always equal to c.

The total spacetime velocity is always equal to c. The spatial velocity manifests as either the intrinsic velocity density of a field or the linear or wave velocities of a particle or object, and the remaining velocity manifests as the velocity of time, the local clock time rate. This confirms that all forms of mass-energy are just excitations of space, and space itself is a ubiquitous field of energy as Quantum Field Theory proposes (Wikipedia, Quantum field theory).

For mass-energy to exist in an individual form it has to be packaged in a valid set of particle components. It then pops out of the virtual quantum vacuum as an actual particle so there is something able to move relative to the background energy field of space itself. Thus charges are quanta of the velocity field of empty space packaged in particles.

Mass-energy fields are velocity density gradients in space, and flat space itself is the uniform velocity density of the zero-point energy. Charged particles are velocity density fields centered on velocity concentrations packaged in valid particle component sets as particles. Particles and their fields are concentrations in particle component packages of the universal field of velocity density of space. Thus everything that exists, all particulate structures, consists of arrangements of little elemental bits of space in time. All mass-energy structures are composed of little quanta of spatial velocity packaged as particles.

Spacetime velocity is what holds the observable universe open so events have room to occur. Every point in spacetime is a c valued combination of space and time velocities. The total amount of velocity in the universe holds it open and gives it the volume that it currently has which is the volume of the observable universe in the present moment.

The velocity that creates and opens the observable universe is produced by the processor of happening that injects life into the universe. The fixed number of processor cycles in every P-time tick for each process goes first to computing velocity in space and the remainder computes velocity in time. The total fixed processor cycles always compute a total velocity of c for every process in the universe and at every point in the universe.

IMPLICATIONS OF THE METc PRINCIPLE

Universal Reality's METc Principle is already implicit in general relativity and certainly consistent with it. Einstein's special theory of relativity demonstrated that time and space were both aspects of a single 4-dimensional spacetime, and that mass and energy are equivalent forms of a more fundamental mass-energy. And in Einstein's theory of general relativity mass-energy tells spacetime how to curve, and spacetime curvature tells mass-energy how to move. This clearly suggests an actual equivalence of spacetime and mass-energy that has not been previously acknowledged (Wikipedia, General relativity).

Universal Reality reveals this previously unrecognized connection between mass-energy and spacetime. In Universal Reality mass-energy and spacetime are essentially two aspects of the same thing. Mass-energy is the velocity shape of spacetime, and spacetime is the distribution of mass-energy.

The entire spacetime universe is simply the direct manifestation of the distribution of all the mass-energy, all the relative spatial velocity, in the universe. Both mass-energy and spacetime are ubiquitous and coterminous, each is an aspect of the other, and together they are the single observable structure of the universe. We can say that mass-energy manifests as spacetime to have room to exist.

Even flat space is the manifestation of the presence of the zero-point energy of the quantum vacuum. The existence of the zero-point energy of the quantum vacuum creates a universe of flat space for it to have room to exist. And the value of the zero-point energy defines the flatness of spacetime in terms of the value of c, which is the fixed velocity of all processes in spacetime. It is the observational value of the combined space and time velocities of everything that exists.

All other forms of mass and energy are additional distortions or relative motions in flat spacetime; they are all local shapes of spacetime in addition to the flatness of its zero-point energy. Thus spacetime itself, with all its dilations and relative motions is the observable distribution of all the mass-energy of the universe. Spacetime and mass-energy are two aspects of the single computational nexus that is the observable universe.

Any deviation from the flatness of the zero-point energy space is some additional form of mass-energy, and any velocity relative to the benchmark flatness of the entanglement network that encodes spacetime is an equivalent additional form of energy. This includes the stretch warping of space around galaxies and galactic clusters due to the uneven expansion of space around cosmic scale mass-energy distributions and leads to a new theory of Dark Matter as explained in that section.

For all forms of mass and energy to be conserved they must all be different forms of the same underlying thing, and that can only be relative motion. Mass and energy are conserved through all transformations only because equivalent amounts of relative motion are being converted from one form to another. Since the underlying nature of all forms of mass and energy is relative motion, mass-energy must actually be relative motion; it must be an aspect of spacetime itself.

This explains how particles can pop into existence out of the flat spacetime of the quantum vacuum and vanish back into it. The modern theory of the quantum vacuum is that it's composed of virtual particles that pop in and out of existence so quickly they can't be observed (Wikipedia, Vacuum state).

The METc Principle is also compatible with a big bang in which all the particles of the universe could appear out of the nothingness of the quantum vacuum presumably reducing its zero-point energy by the amount of particle energy that was actualized. Thus spacetime can crystallize into particles around valid particle component sets in specific cases, and all particles may have originated as velocity eddies in computational space. There are several experimentally confirmed effects in which particles spontaneously appear out of spacetime at very high energies or relative velocities (Wikipedia, Unruh effect).

A completely flat entirely virtual zero-point energy universe likely existed prior to the big bang in a non-observable state. Only distortions or eddies within flat spacetime are potentially observable against the background and able to make observations as well. There must be some non-uniform mass-energy distribution, some spacetime structures, for observations to take place. Without some non-uniform structure nothing can happen and there will be no computable clock time. It would be completely uniform in a state of maximum entropy and no energy could be exchanged to make anything happen.

The total mass-energy content of the universe including the zero-point energy is the source of the hyperspherical geometry of the universe as explained in the chapter on Cosmology. This energy content of the observable universe on the largest scale is the attractive force that curves overall spacetime in on itself into a hypersphere. Since there seems to be no other possible viable geometry from the perspective of P-time this very fact itself requires the existence of a zero-point energy just as the existence of the zero-point energy requires a hyperspherical universe.

Thus what is called flat spacetime is actually the minutely curved surface of the cosmic hypersphere at the largest scale. Gravitation is equivalent to a spacetime curvature. Since curved spacetime is energy, the zero-point energy itself manifests as a very slightly curved spacetime. The cosmic hypersphere is the direct manifestation of the total energy of the universe; it's the inward attractive effect of the total mass-energy of the universe.

If mass-energy is relative motion in space then if space itself begins to vibrate in the right mode that vibration can effectively crystallize spacetime into particles that pop out of the quantum vacuum. In this view particle charges are little crystals of spacetime each of which has a particular mode of vibration, a particular crystal structure, corresponding to the type of charge.

Thus the force charge particle components including mass nucleate combine around other particle components to form particles that are crystals of spacetime. For example mass will be one sort of vibrating crystal, and electric charge another. Only valid combinations of these little crystals can combine to form actual particles. Thus actual particles will be combinations of particle component crystals that pack together in allowable combinations.

Thus particles can be thought of as *phase changes* of spacetime where specific forms of relative motion crystallize into particular types of particles. Each particle is a little standing vibration of spacetime; one of a small set of possible forms of persistent localized relative velocity.

In this view the charge particle components are the possible crystal structures that can be produced from spacetime. They are localized phase changes in spacetime that are conserved in all particle interactions. And they are centered in surrounding fields of velocity density or equivalently spacetime dilation each with a specific form resulting from the vibrational mode of the charge producing it.

Only the particle components carrying the charges of the four forces, including the mass charge of gravitation, produce spacetime dilations or the equivalent velocity density fields. Other particle components like particle identity or spacetime parity don't carry vibrational energy or produce fields.

Particles are composed of valid combinations of multiple particle components. Thus some carry charges of more than one force. For example electrons carry both mass and electric charge and are combinations of two kinds of spacetime structures each with its own vibrational mode, and each with its own surrounding spacetime velocity density dilation field.

There are limits to the crystal metaphor, but by extension the entire structure of the universe can be thought of as complex combinations of a few fundamental types of spacetime crystals and how they pack together into atoms, molecules and emergent material structures according to computational rules and interactions based on their basic crystalline structures or alternately their vibrational modes and forms.

As we recall from the previous chapter, what we interpret as spacetime is the underlying entanglement network produced by the conservation of particle components in all particle interactions. It is these conservation laws that generate the relative scales of spacetime dimensionalization produced in the resulting entanglement network. So from this computational perspective as well, spacetime and mass-energy are two aspects of a single entanglement network. Both mass-energy and spacetime are aspects of the same data structure computed by the conservation of particle components in particle interactions.

In addition to the flat spacetime of zero-point energy, there are three categories of spacetime effects corresponding to categories of mass-energy. These are the vibrational relative motion of charges of the four fundamental forces, the vibrational frequencies of photons, and the linear relative motion of charges and their fields with respect to the absolute background of the underlying entanglement network.

Thus all distortions in the flatness of spacetime are forms of mass-energy in addition to the zero-point energy. There are only four possible intrinsic forms of vibration and associated spacetime dilation corresponding to the four forces, though the actual shapes of the fields of the associated spacetime dilation fields in aggregate can be quite complex.

All spacetime dilation or velocity density fields produce gravitational effects since they curve or densify spacetime. However the fields of the electromagnetic, strong and weak forces primarily affect particles carrying the same type of charge since the fields of these charges couple to damp or reinforce. However there will be some residual effect on other particles with disparate charges due to the general gravitational effect of spacetime dilations of any form.

Thus electric charges mainly attract or repel other electric charges, but their fields do produce some gravitational effects as well. This is consistent with the Einstein field equations where the presence of any form of energy curves spacetime (Wikipedia, Einstein field equations).

Though electromagnetic velocity density fields do produce measurable gravitational effects the dilation fields of the strong and weak forces are constrained to nuclear scales and are too limited in range to produce measureable gravitational effects.

Thus the four forces of nature can be understood as four types of velocity density or equivalently spacetime dilation fields produced by the particles that carry their charges. Each of these velocity density fields, in particular those produced by mass, tilts the c balance of space and time velocities so that some of the usual c velocity through time becomes velocity through space. Thus the spacetime at every point in the field is distorted to slow time and increase distance proportional to the strength of the field.

THE SOURCE OF THE METc PRINCIPLE

The happening of the quantum vacuum is the processor that computes all processes in the universe. The processor has a fixed cycle rate that sets the combined space and time velocities of all processes to c, which is the fixed velocity of all happening in the universe. This is the fixed rate at which everything in the universe is computed.

The fixed processor cycles are used to compute both the internal changes of data states and their relative motion. The relative spatial motion is computed first and any processor cycles left over compute changes to internal data states. The computational rate of change of internal data states manifests as the clock rate of the process.

The processor cycle rate is the source of the METc Principle, which states that the combined velocity of everything is always equal to c, the speed of light. The METc Principle is the source of most relativistic effects. A universe that is computed on the basis of the METc Principle automatically incorporates general relativity as explained in the chapter on Computing Spacetime.

EXCLUSION PRINCIPLES

Exclusion principles are the inverse of conservation principles. They appear to be based on templates in the complete fine-tuning of the quantum vacuum though there are likely more fundamental principles at work as yet undiscovered.

The first exclusion principle determines what sets of particle components produce valid elementary particles. Most combinations are excluded for unknown reasons. This principle is responsible for the redistribution of particle components in particle interaction events.

When multiple particles collide in the same location the total particle component set doesn't produce a valid particle. This is forbidden by the exclusion principle so the particle components must be redistributed in valid sets to create new particles, which are then ejected in different directions so the exclusion principle is satisfied. When the Exclusion Principle is violated the Exclusion subroutine makes a call on the Conservation subroutine to conserve and redistribute the particle components involved into valid particle sets.

The Pauli exclusion principle states that no two fermions (matter forming particles as opposed to force carrying bosons) can occupy the same quantum state (Wikipedia, Pauli exclusion principle). This principle is responsible for the structure of atoms and molecules as it determines how electrons can fill atomic orbitals. It is largely responsible for the fact that atoms and molecules are stable and occupy dimensional volume, and that two material objects cannot be in the same place at the same time.

However the Pauli exclusion principle is more accurately just a result of the computational rules that govern the interactions of individual particles and the balance of forces as explained in the section on Bound Entanglement. This principle is responsible for much of the structure of the matter in the universe. For example it's responsible for the fact that ordinary bulk matter is stable and occupies volume since electrons can't all collapse to the lowest orbital in atoms.

The filling of orbitals by electrons in turn determines the chemical structures of atoms and gives rise to the Periodic Table of the Elements and thus determines the emergent structure of the observable universe.

Another example is the Uncertainty Principle, which asserts a fundamental limit to the precision with which pairs of observable properties of a particle known as complementary variables, such as position and momentum, can be simultaneously known (Wikipedia, Uncertainty principle). However Universal Reality interprets this as uncertainties in the structure of spacetime rather than in particles with respect to spacetime as explained in the chapter on Quantum Reality.

The Exclusion principle is involved in all particle identities and interactions, and thus in all the interactions of particulate matter in the universe, and the computational evolution of the universe.

THE PRINCIPLE OF CHOICE

The Principle of Choice, aka the Randomness Principle, states that whenever the Conservation Principle could be satisfied in multiple ways nature has no rules to decide and so either makes the decision randomly or postpones it by encoding it in the form of a probability distribution called a dimensional fragment to be decided probabilistically by subsequent events. Subsequent events decohere the probability distributions of dimensional fragments to exact dimensional values. Thus in these cases subsequent events make the postponed decisions.

Importantly this occurs in dimensional particle component interactions where there are often ranges of possible exit velocities (speeds and directions) that conserve energy and momentum particle components. It also occurs where the novel spin orientations of two entangled particles can potentially have any possible orientation with respect to a measuring device so long as they are equal and opposite so total spin is conserved.

The conservation of most particle components results in only a single possible deterministic choice, but the conservation of dimensional particle components may necessitate probabilistic choices among a distribution of possibilities.

This principle is the source of all quantum randomness and thus of all the randomness and freedom in the universe, including ultimately the free will of living beings. It's the sole reason our universe is not completely deterministic and completely predictable to any future time. One could say it's the reason the future doesn't already exist.

The operation of this principle generates entanglement networks that incorporate constrained randomness into the structure of spacetime as it's created from the interaction of elementary particles. Thus there isn't a fixed pre-existing spacetime that particles are probabilistic with respect to, but a spacetime that itself incorporates a degree of randomness in its structure as it is built. This is explained in detail in the chapter on Quantum Reality.

Understanding how this principle works resolves all the apparent paradoxical nature of quantum processes. Quantum processes are paradoxical only with respect to an assumed fixed pre-existing spacetime background that doesn't actually exist. When it's understood how spacetime emerges computationally in the form of entanglement networks generated by the Conservation subroutine the quantum world is no longer paradoxical.

THE CONSERVATION OF INFORMATION

The particle components are the basic data elements that make up the universe. The total values of each type of particle component are conserved through all interactions though not their absolute numbers since new particle components can be actualized out of the quantum vacuum by the addition of energy in high-energy collisions, and destroyed in particle anti-particle collisions.

If we assume that particle components are ultimately digital in nature then the quantum vacuum acts as a reservoir from which new instances of digital data types can be actualized by the relative motion of energy so long as their numeric values by type sum to zero. Effectively nothing can split into opposite values of something so long as their sums add to the nothing they came from.

This all has to do with the addition of relative motion to nothing (the quantum vacuum) to produce opposite pairs of somethings (actualized particles). This provides a clue to the nature of the big bang that is explored in the chapter on Cosmology.

Data itself doesn't seem to be strictly conserved as events continually create the data encoding new particle component entanglement relationships. Though events reduce the complexity of previous particle relationships and gradually submerge it below the level of retrievability in the data of new events it's not clear there is any conservation of particle relationship data through events as there is with particle component data.

Instead data is constrained by its internal logico-mathematical consistency so there is a sense in which its forms are conserved in that they are strictly limited by the rules of logic.

SUMMARY

The computational structure of the universe is a balance between the active principles of conservation and exclusion, guided by the Principle of Stochastic choice at the quantum level and the METc Principle at the relativistic level. The computational interplay of these principles driven by the processor cycles of happening produce the entanglement network, which is the combined data structure of the observable universe of mass-energy structures and their dimensional relationships that we interpret as material objects within a physical spacetime.

All the emergent laws of physics that describe the processes and structure of the observable universe derive directly from the computational interplay of these few fundamental principles.

THE COMPLETE FINE-TUNING

INTRODUCTION

In modern physics the fine-tuning is the set of irreducible fundamental constants for which there are no known source or cause. It includes the specific values of the gravitational constant, the speed of light, and the values of the constants that determine atomic structure, the free parameters of the Standard Model (Wikipedia, Fine-tuned Universe).

It's clear that if the values of these fundamental constants varied by only a small fraction the universe as we know it and in which we were able to evolve wouldn't exist. Thus it seems these constants are somehow fine tuned to values that enable our existence, or at least the evolution of intelligent life.

The fine-tuning constants are essential so far as they go but in a computational universe there are clearly a number of other fundamental types of data necessary for our universe to exist in the form that it does that are not included in the standard fine-tuning.

Universal Reality proposes a *complete fine-tuning* that includes all the data necessary to produce and sustain our universe that are not reducible to, or derived from anything more fundamental. The complete fine-tuning is the complete set of fundamental irreducible information necessary and sufficient to compute the observable universe that actually exists.

This complete fine-tuning includes the values of the fundamental constants, but also several other types of essential information including an elemental set of programmatic routines that actually compute reality, the even more fundamental set of logical operators by which they operate, the basic data types of the universe, the elemental particle components, the four forces and four dimensions, and the fundamental templates that determine the structures of particles and atomic and molecular matter.

This chapter describes each of these fundamental components of the complete fine-tuning in detail. The complete fine-tuning is basically the virtual data structure of the quantum vacuum that encodes and implements the fundamental principles described in the previous chapter.

Since standard physics doesn't understand the universe as a computational system it doesn't provide a location or mechanism for its fine-tuning and relegates it along with the laws of nature to a metaphysical realm for which it has no explanation and so generally ignores. The inability of modern science to understand that the universe has to be computational for anything to happen, and that the laws of nature and the fine-tuning have to exist as virtual data to facilitate this, and that there is already an accepted place where virtual data actually exists called the quantum vacuum is truly one of the fundamental failings of modern science.

But in a computational universe it is natural that the laws of nature and the complete fine-tuning exist as real virtual data in the quantum vacuum. The quantum vacuum is the only reasonable locus for the complete fine-tuning and there isn't anywhere else that it could exist. So it's only natural to assume the complete fine-tuning and the elemental laws of nature exist as virtual data in the quantum vacuum and that the quantum vacuum also contains the processor that actually computes the universe on this basis. This fundamental model is simple, elegant, straightforward and highly explanatory.

The only difference between the virtual data of the elemental program and complete fine-tuning, and the actualized data of the elementary particles that make up the observable universe is that the observable universe evolves computationally and is directly observable through its interactions while the virtual data is fixed and observable only through its effects on the actualized data. Both the virtual and actual data of the universe exist entirely as data within the quantum vacuum of existence.

An ocean of water is a useful analogy. The ocean of water is initially formless but specific forms can arise within it. It's the innate or virtual nature of formless water that determines the type of forms and dynamic interactions of the water waves, ripples and currents that can appear within it. Likewise the complete fine-tuning is the innate nature of the quantum vacuum that determines all the specific types of actualized data and computational processes that can appear within it to produce the observable universe. Thus the complete fine-tuning of the quantum vacuum determines the specifics of the observable universe that appears within it.

Just as an ocean is observable only in terms of the forms that appear within it, so the quantum vacuum is observable only in terms of the observable data and observers that appear within it.

Thus the complete fine-tuning completes the picture of all the information necessary to account for the actual universe, some of which exists as actualized emergent running programs as the observable universe, and some of which exists as the virtual nature of the substrate of existence in which they run that determines their natures and the natures of their interactions. In our theory the quantum vacuum is not just the virtual reservoir from which real particles emerge but also contains the complete fine-tuning rules that support the existence of real particles and govern their behavior.

This is fairly obvious since the fine-tuning couldn't generate particles with specific properties if it was not also the source of the rules that govern the natures of those particles. There would be no way to specify the forms of actual particles as they emerged if the quantum vacuum was not the reservoir of the templates for those forms. The complete fine-tuning has to exist as data in some form somewhere, and the quantum vacuum of existence is its only possible locus.

The complete fine-tuning determined the structure of everything that came into being at the big bang, and continues to determine the fundamental structure and evolution of the universe to this day. The basic routines that govern the information of particles and their interactions include everything necessary to compute the big bang and the entire evolution of the observable universe since the big bang.

The great mystery of the complete fine-tuning is how its precise details automatically produce all the immensely complex myriads of running programs that emerge from it. It's clear that the programs of reality compute not just data but effectively compute new programs that are naturally self-organizing and self-modifying, as we see so clearly in our own programs.

The other great mystery is to what extent the complete fine-tuning is the only one possible, and if so why is it as it is. Hidden within this question is the deepest secret of the universe, and the nature of existence itself. Universal Reality provides at least a partial answer but first things first.

BRIEF REVIEW OF THE QUANTUM VACUUM

The quantum vacuum, the medium or substrate of existence, is all that exists. It's an active computational system that computes the existence and evolution of the observable universe, which exists as an evolving data structure within it.

The data of the quantum vacuum consists of the virtual data of the complete fine-tuning and the actualized data of the observable universe. The elemental program computes the actualized data of the observable universe on the basis of the complete fine-tuning.

In particular the quantum vacuum computes the observable universe according to the fundamental principles outlined in the previous chapter that are the fundamental manifestations of the complete fine-tuning.

The actualized data of the observable universe consists of the data of particles, particle components and their relationships. This universal structure is computed as an entanglement network of particle components that encodes the data of both particle structures and spacetime as a single unified structure.

The quantum vacuum also has several qualitative non-data aspects. These include the most fundamental aspects of reality namely existence, happening, the present moment and immanence.

Existence, which Universal Reality identifies with the quantum vacuum, is the medium or substrate within which all the data of reality exists and so gains its individual existence and reality.

Existence exists, and for existence to exist is must be present. The presence of existence manifests as the current universal present moment in which everything that exists does exist, and in which all the computations of the universe are simultaneously occurring. The current present moment is the manifestation of the presence of existence and is the only moment in which anything actually exists or happens.

Happening is the aspect of existence that brings existence to life. It's the processor that simultaneously computes the continual evolution of all the data of the observable universe within the quantum vacuum in the current present moment. Happening computes all the data states of the observable universe simultaneously because it's intrinsic to the nature of the ubiquitous quantum vacuum in which they all exist.

Existence, the quantum vacuum, is a living system in the sense it's continually self-activating. Existence continually happens, and the happening of the quantum vacuum is the ubiquitous processor that executes the code that continually computes the current data state of the entire observable universe in all its individual interactions.

Immanence is the self-manifestation of existence and the essential ingredient of consciousness. The universe no longer exists just as material objects in a physical spacetime container. Everything in the universe glows with the inner light of its existence because it's real and actual, and absolutely what it is, and it has an actual here-now presence beyond just its material configuration. Everything that exists is really really there. Things exist in an immanent sense and are infused with their own living reality as part of the living reality of existence.

Immanence is the source of consciousness. Consciousness is simply opening a properly configured mind to the innate immanence of things. Consciousness is not something generated by human or other minds and shown out onto things like a spotlight. It's simply the living immanent presence of things within mind. Just as seeing things depends on light coming from them so consciousness depends on the unseen glow of immanence in the existence of things. The internal glow of immanence is not something visible to the eyes but it's visible to the mind as consciousness.

COMPUTATIONAL SYSTEMS

To understand the computational system that computes the universe we need to look to computer systems to see how they work because their general applicability and computational power clearly derive from their being modeled on the computational system of the universe. There are clear differences but the general design must be very similar.

When programming computer programs one declares variables and constants with label names and assigns them values. Variables can be declared as different data types; typically Booleans, integers, floating point numbers, arrays and strings. One then refers to declared variables by their label names in the code of the program where they are logically and mathematically manipulated to produce desired results.

Then either at run time or before, this code is interpreted or compiled into the sequences of machine language operators in terms of which the program actually computes. The machine language operators perform elemental logical, mathematical, and data manipulation (fetches, stores, and branches) operations on data elements. In general only a small set of machine language operators are necessary to compute all higher-level programs on any particular computer.

Programs are typically programed in higher-level languages because each higher-level code instruction is equivalent to a whole sequence of machine language operations. This makes coding much more efficient and easier to follow since higher-level variables and subroutines can be assigned meaningful names and programmed in terms of those names.

The various declared data types are all actually stored as binary data bytes with their declared names and types referencing the type of data they represent. Program code is also stored as sequences of binary bytes representing machine language operations with its storage locations providing context that it's code. Storage locations are accessed by pointers to addresses that can contain either data or code routines.

Thus all the bytes of a program are loaded into the fixed memory locations of the computer with pointers and contexts indicating whether the bytes at any particular location are to be interpreted as code or one of the declared data types.

Programs are executed by a processor(s) executing individual machine level operations in a stepwise manner. Conditional branches or calls and returns to and from other code segments are based on push and pop stacks of stored memory locations.

It is almost certainly true that the elemental program that computes the observable universe consists only of the analogue of machine level code. However it seems likely that the emergent programs of the universe can be best explained in terms of higher-level languages programmed not by any programmer but by evolution itself that make calls on the elemental program to compute the processes they encode. DNA is perhaps the clearest example of a higher-level language code programmed by the evolution of the elemental program. But the actions of living organisms also function in terms of calls on a whole hierarchy of lower level routines consisting of multiple elemental operations as for example running involves the coordination of vast numbers of individual cellular operations and muscle coordinations initiated when an animal just decides to run.

COMPUTATIONAL SPACE

Computational space is the logico-mathematical space defined by the computations of the observable universe in the same sense as computer programs define computational spaces. This is the data space in which all the data of the observable universe exists and is computed. It's the data space in which the current present moment entanglement network of all particles, particle components, and their relationships exist. It contains the data of the entire observable universe and is the actual fundamental level of the observable universe.

All the emergent dimensionality of the universe is computed with respect to this computational space and thus actual world lines and rotations are with respect to its parameters so that all actual motions and orientations are relative to it. This computational space exists in the current present moment where the processor of happening continually recomputes all the data that makes up the observable universe. Thus computational space provides an absolute space and time reference with respect to which all the relativistic dimensionality of the observable universe is relative. The overall distribution of mass-energy structure of the galaxies will be roughly aligned with this array since all mass-energy structures are computed within it.

Computational space can be simply modeled as a 3-dimensional data array whose contents are updated at every P-time tick by interactions of their individual cells though it's not certain this is the actual model used. This array has only numeric rather than physical dimensions in the same sense as an array defined in a computer program does, but it's the background within which the observable dimensionality of the universe is computed.

An important point is this array can be Cartesian (flat) because the curved spacetime of general relativity can be easily modeled with an equivalent flat space where every point has a velocity density. This greatly simplifies the encoding and computations of general relativity in our computational space.

Within this computational space the presence of particles is encoded by valid sets of the particle component data that forms the particle. The particle component values are data in the cell of the array where the particle is located.

Every point of this array has an intrinsic c valued combined spacetime velocity. The intrinsic nature of this computational space is combined spacetime velocity with a total value of c at every point. Space is equivalent to a velocity field, and it's the presence of the c valued velocity at each of its points that manifests as computational space.

Each point in the computational space array has an intrinsic c spacetime velocity that initially is completely through time. We can visualize this by an imaginary velocity meter in which the vertical axis is velocity in time and the horizontal axis velocity in space. Thus initially in empty flat zero-point energy space the velocity meter of every point will point straight up indicating all the c spacetime velocity is only through clock time at that point.

The presence of a charge, for example a mass, near a point tilts the velocity meter arrow towards the horizontal axis proportional to the velocity density field produced by the mass at that point. Because masses and other charges are forms of energy they are equivalent to fields of spatial velocity density by the METc Principle and tilt the STc Principle meter of adjacent points from velocity all in time towards more velocity in space. For the vector totals of space and time velocity at the point to remain equal to c, the velocity of time must decrease. This is indicated by the tilt of the meter.

A mass charge is not just a point but also a spherical field of spatial velocity density that falls off by the square of the distance from the mass. Thus a gravitational field is equivalent to a field of spatial velocity that increases the spatial velocity density of space. This velocity density is what slows the velocity of time in gravitational fields in accordance with general relativity.

And because this velocity density field is modeled as fine vibrational waves or pulses in the computational space the actual distance across points in the field is greater because the waves of velocity density must be traversed. This produces a model equivalent to the curved space of general relativity where the greater distance is modeled as traversing the longer curves of space near gravitating masses.

This Universal Reality velocity density model is equivalent to the curved space model of general relativity but much easier to visualize, understand, and compute because it can be represented in terms of a flat space Cartesian array.

This computational space is the underlying stuff of particles because it's a field of spacetime velocity. Charged particles are local excitations of the field of computational space associated with valid particle component sets. The presence of a particle charge is modeled as a tilt in the intrinsic c valued spacetime velocity of points in computational space from all in time to some in space.

Thus dimensional spacetime is a universal field of velocity density with vectors tilted towards spatial velocity when mass-energy is present and mass-energy velocity density fields produce tilts in the spacetime velocity meters of the background computational space.

The correct combined relativistic effects of gravitation and linear motion of particles or objects traveling through computational space can now be computed simply by adding the intrinsic spatial velocity density of the points traversed to their own linear spatial velocity.

And because a gravitational field is a velocity density field the field will contain a spatial velocity gradient that produces a resulting intrinsic spatial velocity vector towards the gravitating mass at every point in the field. This velocity vector is the source of gravitational attraction because it indicates the natural direction and strength of inertial motion in a velocity density field.

If computational space is an array it will have a minimum resolution and background spacetime will be granular at some minimal level far below the scale of particles.

The computational space array can also be hyperspherical simply by curving it over and connecting opposite edges. This eliminates any edges or infinities in dimensional spacetime and neatly models the hyperspherical geometry of our universe as explained in the chapter on Cosmology.

This flat array model appears to be a promising approach to representing the computational space in which the observable universe is actually computed though it needs to be simulated on computers to explore its limits. We have already made considerable progress towards this end.

LOGICAL OPERATORS

The first category of virtual data in the quantum vacuum is the set of fundamental logical operators happening uses to compute all the programs of the universe. These logical operators are by analogy the 'machine language' of the universe. They encode the most fundamental logico-mathematical, and data manipulation operations upon which all the computations of the universe are based. They are largely equivalent to the operators of Boolean algebra, and the other basic logical, numeric and functional operators of silicon computers.

These fundamental operators most likely include all the essential rules of logic and computation of Turing computers since the reason silicon computer programs are so universal is they are directly modeled on the computational logic of the universe. We can except equivalent concepts to is, implies, and, or, negation, equals, greater than, less than, stores, fetches, the basic arithmetic operations, etc. to be part of the virtual data of the complete fine-tuning and the computational basis of the elemental program that computes the particle interactions that make up the observable universe.

Sequences of just these few operations are sufficient to generate all possible computations that can occur in the universe. One can get a good general idea of what these might be by examining the basic operator set of silicon computers though there are some important differences as explained in the next section.

These operators necessarily encode an internally consistent system of logic. This means that is they can never generate a logical contradiction. This is critically important to the existence of the universe. If the computations of the universe were not logically self-consistent then contradictions would arise and a computational universe would literally tear itself apart at such inconsistencies and would cease to exist.

The set of fundamental operators must also be logically complete in the sense that they must always produce a result. If they didn't then a computational universe would halt at the incompleteness and also couldn't exist.

Thus the set of fundamental logical operators that is used to compute the evolution of the entire universe must be logically consistent and logically complete. It is likely there is only a single set of viable fundamental operators that satisfies this necessary criterion. This greatly limits the number of possible complete fine-tunings that produce viable universes.

DATA TYPES

The virtual data of the quantum vacuum must also contain some analogue to the basic data types of computer programs. However there will be important differences between the data types of human computational mathematics and those actually used to compute the universe.

For example it's unlikely the string data type would be part of the complete fine-tuning since it's primarily used to store text in computer programs. Another obviously unnecessary example would be the currency data type.

It's not entirely clear to what extent if any abstract numbers divorced from actual instances of things are necessary to compute the observable universe. It could be that abstract numbers may be a human invention extracted from multiple instances of actual things and not necessary to compute the universe. It is possible to perform most arithmetic operations on the basis of comparisons of actual instances of things without using numbers.

For example it appears that the conservation of particle components could be computed on the basis of comparing instances of actual entities to templates rather than counting them, assigning numbers, and then performing arithmetic on those numbers, and applying the results back to the actual entities, a less parsimonious process.

There is also the question of how probability distributions and random choices among them might be made in the absence of abstract numbers, so whether any abstract numbers separate from instances of things are necessary is still uncertain. It may be that the probabilistic relationships among particles might require a specific data type to encode them, or more likely they could be inherent to the operation of the processor rather than stored as data as we suggest in the chapter on Quantum Reality.

For example it's possible to simply assume a degree of randomness to the space versus time allocations of the fixed number of processor cycles used to compute elemental processes as we do to explain quantum indeterminacy below. In this case probability distributions need not be stored as data.

In any case if abstract numbers are used in the computations of reality they are certainly not identical to those used in human mathematics. The number types of human mathematics are clearly generalized simplifications of instances of things encoded in the complete fine-tuning.

For example in neither nature nor a computational reality can there be infinities or infinitesimals. Infinity is not properly a number but a continual unending *process* of adding one forever. This is obviously nothing that actually occurs in reality. Thus infinite or infinitesimal numbers will not be part of the complete fine-tuning data types. If floating-point numbers exist they would have a limited resolution in which case properly scaled integers might be sufficient. Thus none of the irrational numbers used in science to describe phenomena such as the value of p need exist in nature. Such numbers seem be part of the emergent laws of nature useful to *describe* reality at the aggregate scale but are not necessary to *compute* it since all the actual computations of reality occur at the elemental level and have a minimum resolution.

Possibly not even zeros need to exist though it's not entirely clear whether they could be sufficiently coded as absences of things. And likely no reason for imaginary numbers even though they are used to encode wavefunctions since Universal Reality shows how wavefunctions are simply emergent *descriptions* of quantum phenomena and don't actually compute them as explained in the chapter on Quantum Reality.

For anything to be consistently computed it must be exact and therefore digital. Thus the actual data of reality must be stored as bits, numbers or in some digital form perhaps just as the data of the particle components. In any case it must include the data of all particle components, particles and their interactions necessary to explain the observable universe, and must also apply to even the virtual data of the quantum vacuum.

Based on the computer analogy, there is possibly only a single fundamental binary data type consisting of contextually meaningful sets of on and off binary bits. Both the data and code of computer programs are stored as binary bits and just interpreted differently by the processor on the basis of their context. Whether the fundamental subroutines of reality also use binary code is not clear but certainly possible. If so what the fundamental data bits actually consist of with respect to the quantum vacuum is a fundamental question.

It seems that numbers may be necessary to store dimensional values. However integers may be sufficient if their units reflect a minimum scale of granularity. Both positive and negative integers including zero may be necessary to encode dimensional values though this is unclear since all the data of the universe exists simultaneously within the quantum vacuum and thus its processor. Thus array indices may be unnecessary.

Clearly the fundamental data types of the computational universe is a area that needs further work to see what works best in simulations on silicon computers.

ELEMENTAL DATA OF REALITY

Modern physics interprets the *elementary particles* that make up the matter and energy of the universe as the basic components of reality. However it makes a lot more sense to take what science calls the particle properties or quantum states as the true elemental components of reality. Thus Universal Reality refers to them as *particle components*. The particle components include particle number, mass and the other force charges, spin, and space and time parity.

The reason that particle components rather than particles should be considered elemental is that they are conserved though all particle interactions while the elementary particles they compose can be broken apart into their particle components and those reassembled into new particles. This occurs frequently when particles interact. Whenever it happens, the total additive amounts of all the particle components remains the same, they are just reassembled into new particles.

Thus it's clear that the particle components are actually the little components of reality necessary to make something real in this universe.

They are not so much *properties* or quantum *states* of elementary particles but the actual *components* that make them up and thus make up all the mass-energy structures in the observable universe.

The view that the elementary *particles* are the basic building blocks of reality is an artifact of the outmoded belief in a material universe that remains even though the modern view of particles is far from material in the usual sense. Even standard quantum theory implies that the wavefunctions used to represent elementary particles can't be physical entities since they must be represented by probability waves in a non-physical imaginary space.

Thus Universal Reality considers the particle components the actual components that make up all particles in the universe rather than mere qualities or descriptive attributes. These particle components are the true elemental data structures of the observable universe. Certain valid sets of particle components make up all elementary particles, and thus all the mass-energy structures of the observable universe.

Though this is a completely new interpretation it's a natural extension of the progress of science that led first to the reduction of all material objects to their chemical constituents; then the reduction of chemicals to elements; and finally elements to atoms and elementary particles.

We now take this to its logical conclusion and propose that all elementary particles are composed of an even more fundamental set of particle components that are conserved through all the computations of the universe. These particle components are the basic indivisible data types of the universe because it's only they that remain unchanged through all particle interactions. This is precisely the same criterion previously used to reduce all forms of matter to atoms and then elementary particles.

Though the particle components are very nearly equivalent to the particle components, there are a few rare cases in which a few of the currently defined particle components are not conserved individually but only in combination. For example color charges can be switched in weak force interactions (Wikipedia, Quantum chromodynamics) and there are rare cases in which electric charge and spacetime parity are only conserved in combination (Wikipedia, CPT symmetry).

These rare exceptions require an explanation. The exact particle component set needs to be defined as whatever fundamental components are necessary to make up all actual particles, and be conserved through all particle interactions. These will then be the elemental data structures of the universe out of which everything is made.

But even taking these rare cases of conservation violation into consideration particle components are much more strongly conserved than elementary particles and should still be considered the elemental units of reality. The few exceptions are most likely hinting there is an even more elemental unit of data even particle components are made up of, perhaps just binary data bits, that is conserved in all cases without exception.

Or it may be just a matter of properly identifying a set of particle components that is always conserved. In either scenario it will be the elemental set of data units that is always conserved and necessary and sufficient to uniquely identify individual particle types that are the true elemental data units of reality that make up the entire observable universe. So it's always possible there is an even finer binary digital structure that makes up the particle components but for now we assume properly defined particle components are fundamental.

Thus the elemental program of the quantum vacuum that computes the observable universe, is actually computing particle component relationships when it computes elementary particle events.

In Universal Reality the particle components are the fundamental data structures possible in 4-dimensional spacetime, and elementary particles are composed of valid combinations of particle components. A small set of unique possible combinations of particle components forms all known particles. These sets are the basic templates of reality. The particle components are the little bits of the different components of reality necessary to make something real and actual in the form of particulate matter and energy in our universe.

The particle components are specific actualizations of the quantum vacuum that combine according to complete fine-tuning temples in valid sets to form elementary particles. Thus the particle components are the possible forms that can precipitate or crystallize out of the quantum vacuum to form particles under proper conditions.

The observable universe consists of the data of all particle components and their relationships. Their actualized data appears to be all that is necessary for the elementary program to compute the evolution of the entire observable universe.

Why the particle components that exist are the ones that exist, and why the valid particles they form are the ones they form, are both fundamental questions, but it's apparent that this small set and their allowed combinations are required to build a viable universe such as ours, and it's quite possible the set that exists is the only possible set that could exist.

Presumably the particle components that exist are the only possible forms the quantum vacuum is able to produce due to the internal structure of its virtual nature. So the particle components that exist and the particles they form provide a window into the hidden internal structure of the complete fine-tuning.

Contemporary physics interprets the particle components as some of the conserved *quantum numbers* of elementary particles. However the actual necessary and sufficient intrinsic particle components are often lumped together with other quantum numbers, some of which are derivative (hypercharge) or some like flavor perhaps superfluous (Wikipedia, Hypercharge) (Wikipedia, Flavor (particle physics)).

These other quantum numbers are useful in predicting and interpreting particle interactions but aren't necessary to uniquely specify particles. The intrinsic particle components that actually compose the elementary particles will be those sufficient to uniquely and fully determine the identity of particles.

PARTICLE COMPONENTS

It is useful to distinguish between dimensional (spacetime related) and internal particle components. Particle components related to dimensionality are mass-energy and the other force charges, spin (related to rotational symmetry), and space and time parity (handedness in space and time). Internal particle components include lepton and quark numbers.

The particle components include the charges of the four fundamental forces: mass, electric charge, weak isospin charge, and the color charge of the strong force. However mass and the other force charges are all forms of energy which produce the various vibrations and velocity density fields of their forces in spacetime (Wikipedia, Einstein field equations).

The charge particle components of the four forces including mass can be modeled as vibrations of specific forms whose relative motion generates velocity density fields in the surrounding space. These velocity density fields are equivalent to the spacetime dilation or curvature of general relativity as discussed below in the chapter on Computing Spacetime. The specific form of the velocity density field corresponds to the type of charge. These spacetime dilation or velocity density fields are the charge *fields* of the four forces. Thus the charge particle components are not points but velocity density fields in the computational background.

Spacetime is the total field of all velocity density distribution. It's produced by the presence of the virtual particles of the zero-point energy and the velocity density fields of all actualized charged particles. The velocity density field of spacetime is a dimensional projection of the numeric dimensionality of the entanglement network. The resulting spacetime consists of specific forms of velocity density fields in the otherwise flat spacetime created by the presence of zero-point energy.

The particle components also include spin, which is intrinsic angular momentum relative to the entanglement network of the universe. Spin is the source of the magnetic force and the poles of the axis of spin are effectively the North and South poles of a little magnet. When the spins of the particles in a material are oriented in the same direction the material becomes a magnet. The intrinsic spin orientation of the computational background is also the source reference for absolute rotation that solves the Newton's bucket problem as explained below.

Space and time parity are also particle components. They define the particle's handedness relative to the dimensionality of the entanglement network in which the particles are computed. Handedness is reversed in antiparticles, along with the signs of charges. Handedness may be necessary so the spacetime of the universe can turn itself inside out through big bounces though this is speculative. If so the imbalance of particles and antiparticles may be a relict of the last big bounce.

Identity, which physics currently calls number as in lepton and quark numbers is another possible particle component. The identity particle component identifies the type of particle the particle component set manifests. Only matter-forming particles have non-zero identity particle components; the force carrying bosons have null identity values.

Leptons have lepton number +1, and antileptons have lepton number -1; quarks have quark number +1 and antiquarks have quark number -1.

There is also flavor, which is a quantum number that identifies the 6 leptons and 6 quarks of their three generations of two types individually. However flavor is not conserved under the weak force, which allows leptons and quarks to be converted among generations, so in this sense it probably isn't an intrinsic particle component.

It's also necessary to extend the rest mass particle component to total the mass-energy of the particle since total mass-energy, rather than just mass, that is conserved in all particle interactions. The particle components of the other three force charges also carry energy, but the amounts of those charges are independently conserved. In any case total mass-energy is conserved through all interactions, which means that even if forms of mass and energy are converted the total amount of relative velocity is conserved.

Momentum is also conserved through all particle interactions and in this sense is similar to a particle component since the momentums of individual particles can change through events just as total mass-energies can. How velocities (momentums) are used to express the conservation of total mass-energy is explained in the chapter on Computing Spacetime.

It should also be noted that linear kinetic energy and momentum are frame dependent and conserved only in the same frame. When observer frames are switched energy and momentum are not conserved because the relative motion of the new frame constitutes energy itself and changes the total energy of the system by adding the energy of its own spatial velocity to that of the process under consideration.

If we also consider the spacetime 4-position relative to the aggregate entanglement network, the background array of computational space, as a particle component then we have a unique identification of all separate particles in the universe and a way to tell different particles of the same type apart. Otherwise all particles of the same type are identical and cannot be distinguished. This may be necessary to identify and track individual particles in a computational universe though it's not clear there is any need to track individual particles from P-time tick to tick.

Particle components presumably exist as free virtual data elements in the reservoir of the quantum vacuum in an unassociated state. Enough energy, enough relative motion, can knock them in valid sets out of the quantum vacuum to form actualized elementary particles. These particles actualize out of the virtual flux of particle components and become part of the observable universe carrying the relative motion that created them.

Particle component values may be positive, negative, or zero, except for mass-energy, which can only be positive. And all particle components other than mass can have only a few small proportional values.

The total additive sums but not numbers of particle components are conserved. Thus it's their additive sums rather than their numbers that are elemental so in this sense particle components behave more like numbers than physical entities. They seem to be arithmetic data rather than physical entities, which supports our theory that everything is data at the elemental level.

For example very high-energy particle collisions generally create large showers of many new particles made up of even greater numbers of new particle components. Such high-energy events can create many new pairs of positive and negative particle components out of the quantum vacuum so long as their additive totals are conserved through the event.

This creation of many new particles rather than one or a few very high-energy particles allows the programs that compute them to limit the energies of individual particles as necessary so as not to challenge the c relative motion speed limit of individual particles. It becomes more efficient to distribute the large amounts of energy involved among more new particles instead. Nature tends to distribute high energies rather than concentrate them. This appears to be due to the principle that entropy tends to distribute energy (relative motion) uniformly throughout the observable universe as opposed to concentrating it.

Note also that the basic particle component values (other than mass) should probably be redefined as integral whole units. Physics currently notates some particle component values as fractional but this is mainly a historical convention due to the fact that integer combinations of these fractional units were originally thought to be their base units. An example is the fractional electromagnetic charges of quarks, which were only discovered after whole units of electromagnetic charge had been assigned to the protons and neutrons that are actually three-quark combinations (Wikipedia, Elementary charge).

FUNDAMENTAL CONSTANTS

Another aspect of the complete fine-tuning is the nature of the fundamental constants of nature, the constants not known to be consequences of anything more fundamental. Why does the specific set of fundamental constants exist and why do they have the particular values they do? Are they completely independent or are they interrelated in any way, and can they be derived from anything even more fundamental? And must these constants be exactly as they are to produce a viable universe or could there have been a different set with different values?

It's just these apparently independent fundamental constants that scientists generally refer to as the *fine-tuning* of the universe, but as we have just seen there is much more virtual data necessary to compute a functional universe. This is why Universal Reality prefers the term *complete fine-tuning* to include all aspects of the quantum vacuum irreducibly necessary to the existence of the observable universe as it is, and this obviously includes much more than the values of the fundamental constants.

In fact there is not even a scientific consensus as to the agreed list of what the fundamental constants are. They are often taken as the 26 or so free parameters of the Lagrangian equation of the standard model but this is clearly not the complete list as the standard model is known to be incomplete, and the number of its free parameters may be reducible as well (Wikipedia, Fine-tuned Universe).

In Universal Reality the unified data structure of mass-energy and spacetime is the complete observable universe, and we must clearly include the fundamental constants necessary to explain all aspects of this unity. These seem to fall into at least two general groups, dimensional and non-dimensional constants.

It's also important to note that the actual values of the constants used to compute the universe will most certainly not be those science uses to describe them. It depends on the system of units used to compute the observable universe and there's no reason to believe reality uses the metric system. The speed of light has a different value in meters per second than it does in miles per second. So we must think in terms of the intrinsic strength of dimensional constants such as the gravitational force as opposed to the numeric value of G in any particular set of units and look towards the possibility of a more natural fundamental system of units.

At least some of the fundamental constants are expressible in terms of dimensionless ratios that might be those actually used by reality. The values of dimensionless constants are independent of the choice of units and in this sense more fundamentally meaningful. One can also choose combinations of constants to arrive at dimensionless constants, or play with the units as is done with the Planck units.

Planck was able to derive a set of units of measurements by setting the values of 5 basic constants of nature: the speed of light, the gravitational constant, the reduced Planck constant, the Coulomb constant, and the Boltzmann constant all to 1 (Wikipedia, Planck units).

Using these values a basic Planck unit for length, mass, time, charge, and temperature can be derived. Because the Planck length and time are very small physicists have developed the unfortunate habit of assuming the so called 'Planck scale' is the minimum scale of granularity of the universe but there is no reason to believe this is true as the Planck mass is larger than the mass of elementary particles.

The importance of the Planck constants is they establish a set of natural units seemingly intrinsic to the universe, which simplify many of the equations of nature by eliminating constants from them. So it's likely that computational reality does use some sort of similar set though likely not the Planck set.

In any case the fundamental constants are integral to computing the structure of the entanglement network, which is the data structure of the history of the observable universe. So whatever set of fundamental units works best to compute the universe will likely be used by reality.

The ultimate goal is to reduce the number of fundamental constants of nature as much as possible. Universal Reality takes steps in this direction by redefining gravitational force as fields of intrinsic spatial velocity. Thus to determine the relativistic effect on a particle moving through a gravitational field we can simply sum the total amount of relative motion a particle experiences and take gravitation and the gravitational constant out of the calculation.

Likewise Universal Reality's suggestion of a possible relationship of c to the zero-point energy and the total mass of the universe promises to take us further towards this goal.

THE PLANCK CONSTANT

The Planck constant h is the quantum (minimal unit) of action, which is energy over time, or equivalently momentum over distance. Almost every quantum equation contains the Planck constant to express the minimal values energy or other variables may take (Wikipedia, Planck constant).

In Universal Reality energy is simply spatial velocity, and space itself is the presence of intrinsic c valued velocity so that the Planck constant defines a minimal unit of spacetime, which expresses the granularity of the observable universe and of dimensional spacetime itself. This is related to how the processor computes quantum events, as explained in the chapter on Quantum Reality.

ZERO-POINT ENERGY

The value of c, misleadingly called the speed of light, is the fundamental fixed rate of the velocity of everything in the universe through spacetime. In any particular relativistic situation this velocity is distributed between velocity in space and velocity in time so that their vector sum always equals c.

The particular value of c can be thought of as a function of the fundamental density or resistance to relative velocity of spacetime, its intrinsic resistance to motion through it. This resistance to motion is what restricts all velocity through spacetime to a finite non-zero value.

Flat space is likely a direct manifestation of the presence and strength of the quantum vacuum zero-point energy. Since flat space is the basis in which all spacetime velocity is through time, it's reasonable to suspect the value of c may be related to the value of the zero-point energy if that determines the intrinsic density or resistance to motion through flat spacetime. Thus the zero-point energy may be a measure of the maximum allowable velocity in our universe just as c is.

The zero-point energy can be thought of as the residual observable energy of the quantum vacuum out of which all the mass-energy of the big bang actualized thus the current value of the zero-point energy may also be related to the total mass-energy content of the universe. Thus the zero-point energy value may also be a measure of the curvature of the universe, which is a function of its total mass-energy density. This also suggests a possible relationship of the value of c to the total mass-energy of the universe.

The zero-point energy is also possibly related to the cosmological constant that determines the rate of Hubble expansion of the universe since that is most likely due to repulsive gravitation due to the energy density of empty intergalactic space (Wikipedia, Cosmological constant).

Thus it seems likely there may well be a relationship among the values of the zero-point energy, the speed of light, the cosmological constant, and the total mass-energy content of the universe. Thus all these four fundamental constants could be expressible in terms of a single even more fundamental constant, most likely the total mass-energy content of the universe.

The total mass-energy content was presumably produced by the big bang, likely setting the zero-point energy value as its residual energy. This would in turn set the value of the speed of light in spacetime and the allocation of processor cycles computing it. And in turn the value of the cosmological constant would be a function of these processes.

And since the four fundamental forces can be modeled as curvatures or velocity densities in spacetime, there could also be a relationship to the strengths of the four forces and the particle masses as well as these are also integrated aspects of a single spacetime mass-energy equivalence.

One of the outstanding problems of the complete fine-tuning is the disagreement of over 100 orders of magnitude between measured values of the zero-point energy, which are consistent with general relativity, and those predicted by quantum field theory under various assumptions in accord with the standard model. This discrepancy, called the vacuum catastrophe, clearly indicates a major problem with the standard model whose resolution may well lead to additional insights into the values and relationships among the fundamental constants (Wikipedia, Cosmological constant problem).

It's also important to note that the zero-point energy is the fundamental source of the Heisenberg uncertainty principle (Wikipedia, Zero-point energy). The fact that complementary variables of a system, for example its position and momentum, cannot be simultaneously specified with unlimited precision is due to the fact that all quantum systems exist in the non-zero energy of the quantum vacuum. For the complementary variables of a system to be specified to unlimited precision requires the system must have a ground state energy of zero, but since all systems exist in the quantum vacuum this is never true.

Thus there should be a relationship between the value of the zero-point energy and the Uncertainty Principle's Planck constant minimum precision of the complementary variables of all physical systems that is computed into the entanglement network as it's generated.

PARTICLE MASSES

The distribution of particle masses is one of the great mysteries of the complete fine-tuning. It is precisely because particle masses don't come in exact multiples that particle interactions create spacetime. For particle transformation events to occur some of the vibrational velocities of mass must be converted into linear relative and/or wave frequency velocities, which require a dimensional spacetime to occur within, and so spacetime has to be computationally created to conserve total mass-energy in events. Spacetime has to exist and must be created to allow particle events to occur due to the non-proportionality of particle masses. Thus the non-proportional values of particle masses are the ultimate source of spacetime and the observable universe in which we exist.

The fact that there are precisely three generations of two types of both leptons and quarks, the two categories of matter forming particles, has to be more than a mere coincidence. It has to reflect some deep fundamental symmetry. It's also significant that leptons appear to be free fundamental particles on their own but quarks can only exist in combination in protons, neutrons and mesons.

The fact that the only difference between particles of different generations of all types is in their masses makes this correspondence even more significant.

Amazingly the generational differences in mass of both leptons and quarks are in accord to great accuracy with a simple but little known equation called the Koide formula (Wikipedia, Koide formula) and its extensions (Goffinet, 2008), which virtually clinches their significance. (The masses of the neutrinos and some quarks are not yet known with sufficient accuracy to confirm their compliance but they are within the correct ranges.)

Also the basic three generations of two types template immediately suggests a possible relationship with the three dimensions of space and one of time. It suggests there might be one generation of particles for each dimension of space, and possibly two types corresponding to the two aspects of time, clock time and P-time. However this is currently speculative with little if any supporting evidence.

However the Koide relationship among generational masses does recall the Pythagorean formula of the STc Principle relation of the squares and square roots of space and time variables and adds circumstantial evidence to this assumption.

The Koide relationship states that the ratio of the sum of the masses of the 3 generations divided by the square of the sum of the square roots of those masses is equal to 2/3. This certainly suggests a sort of reverse or inside out connection to the structure of spacetime as expressed in the STc Pythagorean equation. The STc equation involves the square root of the sum of the squares of *space and time* variables, while the Koide relationship involves the square of the sum of the square roots of *mass* variables.

Recalling that mass-energy is equivalent to spatial velocity, and mass-energy and spacetime are two aspects of the same fundamental computational structure as Universal Reality proposes then these two equations may be somehow describing aspects of an underlying symmetry from the opposite perspectives of mass-energy and spacetime.

It is also worth noting that the square root of masses prominent in the Koide formula is how the gravitational effects of mass fall off with distance in 3-dimensional space. Due to the simple geometry of 3-space, the strength of gravitational fields fall off by the square of the distance or equivalently by the square root of the mass which gives the effective gravitational mass at any distance and is thus a basic statement about how mass manifests in 3-space.

This could well be meaningful especially in determining how mass affects spacetime during the dimensionalization of distance. The square root of mass simply tells the entanglement network computations how its effects are to fall off with distance as distance is computed. It's telling the conservation subroutine what form to use as it constructs the velocity density field around a mass.

Thus it seems the mass particle component comes only in values reflecting the way its field strength falls off in 3-space and this restricts the values it can take for the 3 generations of massive particles allowed in a 3-space universe in some manner.

While this idea is certainly speculative it may shed some light as to why there are precisely three generations of all types of massive particles whose masses are related by the manner the effects of mass fall off in the three spatial dimensions that characterize our universe. This is certainly possible when we recall that gravitation is a field of velocity density surrounding massive particles.

The Koide relationship doesn't uniquely determine what the masses of particles within a generation are, only what their relationship must be. The relationship holds for both leptons and bosons but their masses differ. So there is obviously some other as yet undiscovered rule necessary to specify exactly why the four leptons and bosons have different masses but still obey the Koide relationship. We need something else that determines at least the base mass in each generational series and the relationships among the four series.

This is likely related to the fundamental question of why only mass doesn't come in a small set of proportional values like the other force charges, and why there are no negative masses.

For example the existence of the universe is extremely sensitive to the ratio of electron to proton mass. If it were only slightly different atoms, molecules and all the structures of the universe could not exist. They would either fly apart or collapse in on themselves and never form in the first place.

The explanation of the relative masses of the four series of matter forming particles is not clear, but is undoubtedly related to the fundamental constants of spacetime given that mass is a form of relative motion that creates a spacetime dilation or velocity density field. If mass and spacetime are both aspects of the same single entity then there must be a deeper connection between the fundamental constants of both.

It should be noted there is another layer to the 4x3 matrix of lepton and quark generations that contains their antiparticles. The antiparticles in this layer have the same masses but their charges and parities are reversed. As a result they are facing backwards in time and have an opposite handedness in space, and the rotational directions of the helical spacetime dilations or velocity densities of their electromagnetic fields are reversed as explained in the chapter on The Other Forces.

FOUR FORCES

There are four fundamental forces in the universe each with a class of bosons thought to carry the force and act upon its charges. Universal Reality also models these forces as four distinct types of spacetime vibrations and their associated spacetime dilation, or velocity density fields.

Why there are four and only four forces in our universe is possibly related to the manner in which the quantum vacuum must manifest as a 4-dimensional spacetime, and the distinct types of vibration-dilation structures that are possible in 3-space though this is speculative.

There are several different ways to represent the elementary particles and their particle components in charts. One is to add the four forces to a 4x3 matrix of leptons and quarks across the bottom to obtain a 4 x 4 matrix (Wikipedia, Elementary particles).

From left to right we have the recently discovered Higgs boson thought to give all particles their masses. Next the photon that carries the electromagnetic force, then the weak bosons that carry the weak force, and last the gluons that carry the strong force.

This last row is a little different from what is seen in the usual charts, but they are somewhat arbitrary in this respect as there isn't any natural association with the force carrying cells of the standard charts and the particle classes above them. Thus it seems more natural just to add the four forces in separate cells of the chart rather than to first omit the graviton, and then put the two W and Z weak force bosons in separate cells, but lump all gluons of the strong force together in the last cell as the chart illustrated does. That doesn't make much sense.

Each of the four forces has a different number of charges so the actual number of particles in each position in this bottom row increases from left to right. The gravitational force has only one charge corresponding to the existence of positive mass. The electromagnetic force has the two plus and minus electric charges. The weak force has the three isospin charges of 0, ½ and – ½. And the strong force has 6 color charges (±red, ±green, ±blue).

Particle physics considers the four forces to be mediated by exchanges of particles called bosons. The strong interaction is mediated by exchanges of 8 types of gluons, the weak interaction mediated by exchanges of Z and ±W bosons, the electromagnetic force by exchanges of photons, and gravitation by exchanges of the hypothetical graviton.

However general relativity models gravitation as fields of spacetime dilations, and Universal Reality propose the forces are actually different forms of vibrational velocity density equivalent to different types of spacetime dilation. This is a good conceptual model but since Universal Reality considers spacetime to emerge from particle component interactions the two models must have an equivalence that arises *in the relative scales and forms* of the entanglement networks surrounding charges of the various forces.

As the entanglement network is generated by particle component interactions it's given a relative scale and dimensional form by the force interaction involved. This can be interpreted as vibrations and spacetime dilations of forms corresponding to the four forces as outlined in the chapters on Computing Spacetime and The Other Forces.

Thus our chart takes the four forces as primary, and the numbers of their charges and force carrying bosons vary increasing from left to right just as the masses of the quarks and leptons increase from generation to generation.

Our matrix also associates the recently discovered Higgs boson with the gravitational force, rather than the hypothetical graviton. The graviton is not even part of the standard model and doesn't appear on the usual particle charts but the existence of the Higgs is confirmed and part of the standard model. The Higgs is the particle thought to give all particles their masses in the standard model so it is the closest thing the standard model has to a graviton. And it is imagined as a field that pervades all of space as part of the quantum vacuum so it's clearly related to how the constants of spacetime and mass-energy interrelate.

In Universal Reality spacetime is actually a field of mass-energy including the ubiquitous zero-point energy of the quantum vacuum. And mass-energy is excitations in this field including the ubiquitous zero-point energy excitation of the whole field. In particular all the particle charges are localized excitations in the mass-energy field of spacetime.

This is the equivalence of mass-energy and spacetime in a single computational structure. And this is why the fundamental constants of both mass-energy and spacetime must necessarily be related.

Thus the values of the zero-point energy, c, G (the strength of gravitation), the strength of the other forces, and the masses of all particles due to the Higgs field are almost certainly related since they are all constants of a single mass-energy-spacetime equivalence.

And more generally it is quite likely that the three generations of massive particles correspond in some subtle manner to the three dimensions of space, and perhaps bosons have some symmetry connection to the time dimension. Perhaps the matter forming leptons and quarks are more a manifestation of spatial excitations or symmetries, and the force carrying bosons more of symmetries in time?

In Universal Reality all forms of mass-energy are forms of relative velocity or motion in space. So the values of the constants of the strengths of the four fundamental forces, must be proportional to the amounts of relative motion each carries. For all forms of mass and energy to be convertible they must all be forms of the same underlying thing, relative motion in space. Thus the values of the constants of the force strengths must be proportional to the amounts of relative motion those forces carry and measures of that relative motion.

FOUR DIMENSIONS

It is fairly clear that there can only be 4 dimensions in a viable universe such as ours. There must be three space dimensions because it takes at least three for meaningful structures to exist, and one and only one dimension of clock time for things to happen, and to happen in a consistent manner.

For example in a two dimensional space beings with internal digestive tracts couldn't exist because having one would split the being into two separate parts. For the same reason most types of compound structures with internal components couldn't exist and no complex life could exist. More fundamentally there couldn't even be any atoms or molecules in a 2-dimensional space since these are all 3-dimensional structures and there are no 2-dimensional equivalents. Basically there is simply not enough room in two dimensions for any meaningful structures to exist or evolve. Entities would always be bumping into each other and interfering with the movement and structures of other entities as there would be no third dimension for anything to go around anything else.

And in a universe with more than three spatial dimensions other types of meaningful structures cannot exist. Knots for example can only exist in 3-dimensional space. Stable planetary orbits can't exist because the strength of gravity would vary much more rapidly with distance and small perturbations would cause planets to quickly fall into their stars or escape their orbits. It's unlikely the basic laws of physics and chemistry produce viable mass-energy structures in more than 3 dimensions.

There must be at least one dimension of clock time for events to be able to occur at all. However two or more dimensions of clock time would allow physical processes to run at different rates or towards different futures in the same location and that would immediately lead to logical contradictions that would tear the fabric of a computational universe apart.

However in addition to the 4-dimensions of spacetime a separate over arching dimension of P-time is necessary in a viable universe to reconcile, contain and logically relate all the relativistic clock times that occur at different locations in a four dimensional universe with a finite speed of light.

Without an STc Principle with some finite speed of light clock time would either not pass at all and nothing would ever happen, or clock time would pass instantaneously and the entire history of the universe would immediately be over before it even began.

Thus only a universe with 3 spatial and 1 clock time dimensions such as ours seems viable, and there must also be a separate present moment P-time, and a single underlying computational space for computational processes to evolve in a consistent relativistic manner, and to exist in the same universal present moment of existence in the same observable universe.

Universal Reality models the charges of the four forces as either different forms of relative vibrational motion and dilation fields in 4-dimensional spacetime or as equivalent particle exchanges of bosons so there's no need for the hypothetical but unverified additional compacted dimensions proposed by String Theory (Wikipedia, String theory).

String Theory proposes vibrations in multiple compacted dimensions to account for different types of particles, but in Universal Reality it's the data of particle components that are the fundamental elements of reality and the force carrying particle component charges are modeled as different forms of vibration velocity density fields in standard 4-dimensional spacetime. And these are produced by the conservation of particle components in particle interactions.

Thus it seems reasonably certain that the four familiar spacetime dimensions of the universe along with P-time are precisely what is required to construct a viable universe, and this aspect of the complete fine-tuning of the quantum vacuum can only be exactly as it is. There are likely no viable universes possible with either more or fewer dimensions than our own.

DEEP SYMMETRY

The fact that there are precisely four types of massive particles that make up all matter, four forces that govern the interactions of this matter, and four dimensions in which this matter exists has to be much more than a mere coincidence especially considering the deep equivalence of mass-energy and spacetime as two aspects of the unified computational structure of the quantum vacuum. So if we add the four dimensions of spacetime to our matter and force table we get a beautiful 4 x 4 x 4 unity that summarizes essentially the entire observable universe in a nutshell. Mass-energy and spacetime constitute the entire observable universe, and this 4 x 4 x 4 unity boils it down to its essence.

Clearly we have the glimmer of a much deeper hidden secret here. The fact all these fundamentals come in sets of four certainly hints at a deeper unity. The unity of mass-energy and space and the fact both are composed of four fundamental units must certainly be more than a coincidence.

The force charges of the elementary particles are actually excitations of spacetime itself. The shape of spacetime is the distribution of mass-energy in the universe. Mass-energy is spacetime, and spacetime is mass-energy, and mass, energy, space and time are all forms of the single fundamental element of relative motion or velocity. Everything in the observable universe is part of a single unified data structure computed by quantum events.

Everything takes place in the abstract computational space of the quantum vacuum. It is there that the fundamental subroutines compute the interactions of the data of the particle components that make up the universe, and in doing so they generate the entanglement network that we observers interpret as our familiar physical spacetime universe.

Since mass-energy and spacetime are both aspects of the single conserved element of velocity it's clear that the fundamental constants of mass-energy and spacetime must be related at a deeper level. We have seen hints of this but there is certainly much more to be revealed.

Thus it seems quite likely there is an undiscovered relationship among the values of the zero-point energy, the speed of light, the cosmological constant, the total mass-energy content of the universe, the nature and charges of the four forces, and the specific particles and particle components that exist. All these fundamental constants are almost certainly expressible in terms of a single elegant set of even more fundamental constants, or perhaps ultimately a single constant or principle that uniquely defines our universe and the computational structure of its mass-energy spacetime equivalence.

It is unclear if, or to what extent, the complete fine-tuning might be altered and still produce a viable consistent universe though probably not much if at all. Many aspects of it, such as the basic rules of logic and the four spacetime dimensions, are clearly necessary exactly as they are and it's quite likely that the complete fine-tuning of our universe is the only one possible. Important additional evidence for this will be explored in the chapter on Information Cosmology.

THE REAL MICROCOSM & MACROCOSM

The templates of all the actual information structures that exist in the observable universe exist in a virtual form in the complete fine-tuning of the quantum vacuum. The computational space of the quantum vacuum is not a dimensional structure but it contains the virtual template for dimensionality and dimensionality is computed within it on the basis of that template. This is true of all aspects of the observable information structure of the universe. They are all implicit in the virtual information structure of the quantum vacuum in which they are computed.

Thus the fact that for something to be real and actual in the universe it must be composed of elemental bits of identity, charge, mass, spin, space and time handedness and so forth indicates that the quantum vacuum itself consists of all these aspects of reality in a virtual form.

The fact that dimensionality has characteristics of position, relative velocity, scale, orientation, etc. indicates that these fundamental elements of reality also exist as virtual templates in the quantum vacuum. They are there in the virtual data of the complete fine-tuning and whenever anything becomes actual it must crystallize into actuality according to these virtual templates.

Thus when particles crystalize out of the computational spacetime background they take on the forms implicit in the computational spacetime from which they crystallize. Particles and dimensional spacetime and their myriad relationships are all crystallizations of the absolute background computational spacetime in which they are computed. They are actualized manifestations of the underlying structure of the complete fine-tuning that exists all around us in a hidden virtual form.

We and all the particulate structures and programs of the observable universe including our dimensional structures, are computed within and exist within the quantum vacuum that supports our existence with its own. We are the direct manifestations of the underlying template forms of the complete fine-tuning that exist within us in virtual form and that buttress our observable structures and processes. Without the actual existence of the templates of the complete fine-tuning within our own being and the being of the observable universe neither we nor the observable universe would exist.

The traditional concept of microcosm and macrocosm is certainly incorrect but this is perhaps the true meaning of the microcosm and macrocosm being reflections of each other. There is only a single kind of thing that exists within the sea of existence of quantum vacuum and that is data given existence and being by its presence.

This data takes an elemental set of highly related virtual template forms in the complete fine-tuning. The quantum vacuum that pervades the entire universe and exists within all things includes the presence of these forms as virtual templates. Thus they exist within all individual things and when things appear they appear according to these forms. Everything that exists is an actualized instance of the virtual template forms that exist within them, and only the continuing presence of these virtual forms in the actualized forms maintains their existence as their data is continually recomputed.

So spacetime itself with its positions, velocities, orientations and clock time rates is a direct manifestation of the continuing hidden presence of the quantum vacuum, which defines what dimensionality is and what characteristics it must have to become dimensional.

The same is true for all particulate mass-energy structures. They are all observable forms constructed around the hidden blueprints that determine what particle components they must have to become actual. Every particle in the universe consists of its particle components built around a hidden instance of the complete fine-tuning template that determines what these actual particle components must be.

Thus in everything that exists there is combined the actual and the virtual, the microcosm and the macrocosm. All the elementals that exist are all manifestations of the universal. The universal exists within every elemental and all the elementals all exist within the universal.

Also all the charges that exist are little elemental crystals of spacetime that crystalize out of the computational background around valid sets of the other particle components necessary to make real and actual particles in our universe.

It is this crystallization process that creates the observable universe of particulate dimensional structures we humans interpret as a material world within spacetime. These particle crystals presumably appear when little units of spatial velocity move relative to their background and seed the process. When that happens the quantum vacuum gives up elemental quanta of the particle components it takes to make an actual particle so there is something to manifest an individual velocity.

We have the minimum units of everything on the one hand and the maximum speed limits etc. of all those things and their universal templates on the other. The myriads of minimum units of the tiny fill in the maximum scale and size of the universe. It's as if there was an original pulling apart or separation of the fabric of existence into the overarching single units of the maximum and the innumerable individual units of the minimum. From this perspective the original differentiation of the formless was not into male and female but into the large and the small, a differentiation of innumerable individual actualized units of the virtual templates implicit in the formless sea of virtual existence.

One can imagine the fundamental forms of existence being pulled out of the quantum vacuum into little actualized examples of those principles. The observable universe consists of innumerable specific instances of the universal principles of the complete fine-tuning and then begins to evolve via the interactions possible to their nature.

Though clearly the specific instances are created from the general principles one wonders if there isn't some possible feedback mechanism in which the probabilistic evolution of the specific instances towards convergent evolutionary ends implicit in the complete fine-tuning somehow feeds back into influencing the general principles of the virtual complete fine-tuning?

Perhaps this could somehow occur when and if the universe turns inside out in a big bounce and entropy and time reverse and the small somehow determine the principles for the new bounce that might possibly be created as explained in the chapter on Cosmology. In this way could the universe perhaps continually evolve its complete fine-tuning through each big bounce towards some ultimate goal of universal self-awareness? There is certainly much to be discovered…

THE ELEMENTAL PROGRAM

We now come to the last category of the complete fine-tuning. This is the elemental program that actually computes everything in the universe. This elementary program consists of subroutines that implement the fundamental laws of nature. It implements the fundamental principles outlined in the previous chapter.

The elemental program that computes everything in the universe consist of a small set of fixed routines that operate on particle component data. These are the elemental laws of nature of the universe that compute all the data interactions of the observable universe using sequences of the basic logical operators to implement the fundamental principles and complete fine-tuning. The elemental program computes the existence of particles, their interactions in particle events, and their relationships in terms of their particle components. These computations are performed on the basis of the principle of conservation of particle components and compute only events that distribute particle components in allowed sets among emitted particles and exclude any combinations that don't form actual particles.

The basic subroutines compute the STc Principle, the Exclusion Principle, the Particle Component Conservation Principle, and the Principle of Constrained Randomness. For completeness the two fundamental principles of existence and logical consistency give consistent existence to the data of all computational results.

There are also subroutines that compute the bound entanglements of the mass-energy structures of atomic and molecular matter. These are based on balances of the fundamental forces and their effects on individual particle interactions in bound states.

There are also routines that compute particle charge velocity density fields surrounding the charges. This includes the gravitational fields produced by mass-energy and the electromagnetic fields produced by electric changes. It is not clear whether the forces carried by the strong and weak forces are best modeled by fields or boson exchanges but these are likely equivalent models as charges are modeled as velocity excitations of space.

Particle interaction events can be modeled as all the particle components of multiple particles being collocated at the same point. This triggers the exclusion principle, which disallows particle component combinations that don't form valid particles. As a result the conservation principle is triggered which distributes all particle components among valid new particles that are then ejected with the left over energy conserved in the event.

Thus events are not data entities but computations. So they don't properly have frames of their own and are all computed within the background frame of the quantum vacuum.

This single elemental program is the only program that exists in fixed virtual form in the complete fine-tuning and it can be said to make up all emergent programs and compute the entire universe. All the emergent programs of the observable universe compute by making calls on the elemental program and its subroutines to compute the interactions of their constituent particles. By analogy the elemental program is the fixed firmware of the universe.

All the emergent level laws of nature that are the subject of most of science *describe* but don't actually *compute* the structures and behaviors of emergent level processes. All emergent processes are actually computed by organized mass calls on the elemental program to simultaneously compute every one of their particle interactions at each P-time tick.

Thus all the emergent programs running in the universe from atoms and molecules on up are aggregate manifestations of these fundamental logico-mathematical routines. All emergent programs consist only of ordered data structures produced by the elemental programs, which in turn are executed in terms of a small set of logico-mathematical operations by analogy the machine language of the universe.

These fundamental subroutines appear to be rather simple, much simpler than the plethora of emergent laws that describe but don't directly compute aggregate behavior. This small set of relatively simple computational routines seems to be all that is necessary to compute the entire evolution of the universe.

More work needs to be done to nail down the precise details of the complete fine-tuning and the elemental program. Ultimately this is probably best done with computer simulations to see what most effectively models the fundamental computational processes of the observable universe. We have made some progress in doing this on the XOJO programming platform.

HIGHER-LEVEL LANGUAGES

Because they operate by making calls on the elemental program the emergent programs that compute the vast majority of processes in the observable universe can be thought of as examples of higher level programming languages evolved by reality itself.

For example all the processes of chemistry and biology, though all actually computed in terms of elemental logico-mathematical operators do consist of more or less fixed calls on sequences or sets of the elemental operators. Thus each of those calls could be considered a higher-level language operation calling subroutines of machine language operations.

The programs of mind's construction and operation of the simulation are another clear example of an evolved higher-level language. Organisms' simulations of their environment have evolved through the progression of species from the beginnings of life. Basically these all operate in terms of a higher-level language we can call *the logic of things*, the basic rules of how individual things and processes are extracted from the perceptual background and work at the classical level. Organisms function within their environments on the basis of the logic of things, which is effectively, a higher-level language ultimately implemented via calls on the elemental program.

Thus the emergent laws of science that describe but don't actually compute the processes of the observable universe are effectively descriptions of elemental processes in terms of the higher-level language we call science. Though most of the programs that compute the processes described by science are just emergent manifestations of elemental level computations, it's also clear that these elemental processes have evolved higher-level programs in the form of living organisms, which operate intentionally to foster their function and survival. These programs operate in terms of the higher-level programming language of their simulation to direct their actions, which are all implemented by calls to the elemental programs that actually compute reality.

These purposeful programs have evolved because by optimizing their actions in support of their functioning they tend to be selected and perpetuated over those less successful. Thus, based on the mysterious rules of the complete fine-tuning, the elemental program itself naturally evolves higher-level purposeful programs in the form of living organisms.

Perhaps the clearest example of a higher-level language evolved by the universe is DNA. DNA clearly is implementing a higher-level language, ultimately based in calls to the elemental program, which codes for the creation, growth, and functioning of biological beings and consists of coded instructions in that language. Many known emergent processes can also be modeled as higher-level languages and we should certainly look for others as well.

All these emergent programs and the higher-level languages they implement are programmed by the general principle of evolution and reused over and over because they are adaptive results of the elemental program. They can be thought of as separate programs, though actually manifestations of the elemental program in the same sense that emergent patterns automatically arise from the elemental operations of cellular automata (Wikipedia, Cellular automaton).

UNDERSTANDING TIME

TWO KINDS OF TIME

There are two kinds of time, the time of the present moment and clock time which runs at different relativistic rates within a single universal present moment. But amazingly this fundamental fact of reality was completely unrecognized until first pointed out by the author in 2007 in his paper 'Spacetime and Consciousness' (Owen, 2007), and again in his 2013 book 'Reality' (Owen, 2013).

Obviously this proposal is controversial and requires good evidence to be taken seriously. However it's pretty straightforward to demonstrate. First, to prove clock time and the present moment are two separate kinds of time we need only demonstrate there is a single universal present moment within which clock times vary since it's already an experimentally proven and widely applied fact that clock times do run at different rates according to relativistic conditions. Thus if we can demonstrate this occurs in a common universal present moment, then clock time and the present moment must indeed be two different kinds of time.

The existence of a common present moment throughout the universe is not a new or strange idea. It was the standard accepted view of time throughout history until the advent of relativity. Clock time was thought to be flowing at the same rate throughout the universe and the present moment was thought to be the common universal present moment of *clock time* rather than a separate kind of time. The present moment was the current reading of a universal clock, which ran at the same rate throughout the universe. Thus there was only a single kind of time, clock time, and the present moment was the present moment of this universal clock time.

But with the advent of relativity it became clear that clock time didn't flow at the same rate everywhere so there couldn't be a universal clock time that was the same everywhere. So because time was still considered a single entity the newly variable clock time was still considered the only kind of time and the very notion of a universal present moment inconsistent with this view was wrongly discarded. The notion of two separate kinds of time to reconcile clock time with the present moment doesn't seem to have occurred to anyone until now.

Even though the present moment is a central experience of our existence scientists post relativity couldn't come up with any explanation of what the present moment was, and it was either ignored or even denied and replaced with truly outlandish theories such as block time in which all times exist simultaneously and there is no special present moment.

The main reason for the current ignorance of the present moment in modern physics is most likely that it has no obvious measure and physicists have an unfortunate tendency of ignoring or denying the existence of anything without measure even though neither consciousness, nor existence, nor the present moment have measure and they are our three most important and fundamental experiences of reality.

So the important insight of Universal Reality is to retain a universal present moment and recognize that relativistic clock times run at different rates within this common present moment. This is a very simple and reasonable insight in accord with our direct experience of reality and it has profound consequences. And it turns out this concept of two kinds of time is even an implicit though totally unrecognized principle of relativity itself without which relativity doesn't even make sense.

Every comparison of different clock times in relativity only makes sense if there is a common present moment in which the comparison takes place. There must be a common present moment that serves as a common background reference. Thus a common universal present moment in which *relativistic* comparisons can be made and shared is a hidden and completely unrecognized assumption of relativity used by physicists all the time but which they actively deny! This is one of the great blind spots of modern science.

For example if two space travelers with different clock times were really in each other's pasts and futures they would be completely unable to compare their clocks. They can compare the different readings of their clocks only because they and their clocks are both in the same present moment. Their present moment is the same but their clock times are different, thus it's clear there must be two different kinds of time.

When one of two twins embarks on a relativistic space journey they part in a common present moment. They then each continuously experience their separate existences in a present moment throughout the duration of their separation. And when they meet again they always meet only in a present moment common to both even though their clocks now read different times. So there is no reason whatsoever to think the present moment they both experienced during the entire duration of the separation was not the same present moment for both. Their ages and clock times are now much different but their present moments never lapsed and must have always been the same even though their clocks were running at different rates.

Thus it's reasonable to assume, without any evidence to the contrary, that there is a single common universal present moment throughout the entire universe, and to assume that every observer in the universe is always in the same current universal present moment as everything else. Thus all that exists, the entire universe, exists in the single common current universal present moment.

This is really quite obvious, but it has been by far the most contentious aspect of Universal Reality with all sorts of arguments being raised against it. Many of these arguments have been based on a specific misunderstanding of the theory, that it violates the relativity of simultaneity in which different observers can have different valid observations of whether two events occur at the same *clock time* or not (Wikipedia, Relativity of simultaneity).

But Universal Reality accepts and incorporates all the equations of relativity including those of the relativity of simultaneity. The relativity of simultaneity correctly describes the behavior of *clock times,* but says nothing about *present moment times*. Non-simultaneous clock times often occur in the present moment so this argument has no bearing on whether there are two kinds of time and can be disregarded.

For some very strange reason the existence of two separate kinds of time is beyond the comprehension of many otherwise intelligent people. Even though it's really a quite simple and straightforward idea most people have great difficulty wrapping their heads around it.

SOME THOUGHT EXPERIMENTS

There are a number of useful thought experiments that more clearly demonstrate a universal present moment. To prove our point we must demonstrate both that there is a common universal present moment for all observers stationary with respect to each other and in motion or acceleration with respect to each other.

We already know that whenever any two observers are spatially collocated they both experience the same common present moment because they can communicate more or less instantaneously to confirm this. And this is true whether they are moving or stationary with respect to each other.

Consider first a universe completely filled with stationary observers packed together like sardines. We already know every one of them is continuously in the same present moment as the adjacent observers on all sides and this is true for all observers in the universe even if all their clocks are running at different rates due to different gravitational potentials. Therefore every observer across the entire universe must be in the same common present moment and this present moment is universal.

This is a simple proof that there must be a single common present moment throughout the universe that holds whether the observers are stationary or in relative motion because it also holds if the observers are moving relative to each other and just happen to assume the packed sardine configuration when the experiment is done.

This proof also holds whether or not there are gravitational fields involved. The clocks of observers in gravitational fields will be running slower but across the entire universe all clocks will be running at varying rates in the same universal present moment and this will be agreed by all observers as each confirms their existence in the same current present moment with all those adjacent.

Now the counter argument might be raised that there is some difference in present moment across the entire universe too small to be noticed at the slight differences in location among adjacent observers. But what could a difference in present moment even mean? The present moment most certainly doesn't correspond to differences in clock times because we already know different clock time rates exist in the same present moment. So what would such a difference in present moments amount to? What would cause it and why would that difference be magically erased when space travelers meet with different clock times? If their present moment times were different during their separation how could they become the same again when they met? What would be the mysterious mechanism involved? It wouldn't make sense.

Now consider another thought experiment involving two space travelers who part with synchronized clocks, accelerate at different velocities in different directions and periodically redirect to cross paths. We know they are in the same present moment whenever they cross paths so take this example to the extreme and assume they accelerate with enormous and varying velocities but turn and cross paths more and more frequently until the interval between meetings approaches zero. Now they are crossing paths every minute fraction of a second. And every time they cross paths they both confirm they share the same present moment, and no matter how they change their accelerations and how much their clock times vary this is always true.

So it seems outlandish to assume that somehow in the minute fractions of a second they are separated their present moments somehow become different. Their clocks can run at very different rates and be read differently every time they cross paths but this is always happening in the same shared present moment. And in fact they both will confirm to each other their clocks continue to run at different rates every time they cross paths in the same present moment.

This is also confirmed by ground communications with the International Space Station. Time aboard the ISS progresses at a measurably slower rate than on earth because of the velocity it's traveling along its world line (Wikipedia, Time dilation). Yet the ISS is in continual contact with the earth at all times. The ISS and earth are both in the same shared present moment at all times even as their clock times run at different rates. True there is a slight communication delay but this delay occurs in the common present moment.

Now imagine the ISS accelerating enormously but maintaining the same circular path around the earth. Clock time and all physical processes aboard the ISS will begin running perceptibly slower from the point of view of earth observers and vice versa, and this can be confirmed by continual mutual communication. The slight time delay in communications remains the same confirming it's irrelevant. Observers on both earth and the ISS would now see each other moving in slow motion but they would both continually communicate this fact back and forth in the same shared present moment.

Their times would certainly seem very strange to each other but since they remain relatively close throughout they both would be able to continually observe this mutual strangeness in the same present moment. They would both continually observe each other's clock times run at different rates within the same shared present moment.

There is a similar situation observed in particle accelerators where particles travel at near the speed of light around the accelerator and their decay rates slow as scientists observe them due to their enormous spatial velocities. Thus the decays and the observations are obviously occurring in the same present moment, as they must be to be continually observed.

So it's clearly possible for two observers to actually watch each other's clocks run at different rates in the same present moment in some cases. This can only be true if the present moment and clock time are two different kinds of time. Science must be based on observation and the present moment and two different kinds of time are clearly observable facts.

In light of these thought experiments it seems undeniable that there is a single present moment common to the entire universe that is completely different than clock time, and therefore there are two distinct kinds of time.

THE PRESENT MOMENT

It's important to clearly understand what is meant by the *same* present moment. There isn't a single present moment that stays the same over all time as clock time flows through it at different rates. Instead there's a separate present moment kind of time that also progresses but at the same rate throughout the universe. The current present moment right now isn't the same as the previous present moment. For things to happen there must be a current present moment that is not the same as the previous current present moment.

The present moment time that Universal Reality proposes is the successive present moments of the active process of happening in the observable universe, which continually manifests as a universal present moment in which everything exists, and in which all the computations of the universe are occurring. This is a simple and elegant theory that is completely consistent with science including relativity and with basic personal experience and informed common sense.

The happening of existence is the processor that continually computes the current data state of the entire universe. Since all the data of the universe exists simultaneously within the processor of happening the *presence* of existence manifests as a universal current present moment in which everything happens. This current present moment is universal and common to all observers no matter what rate their clocks are running. Thus happening is the source of the fixed flow rate of present moment time throughout the entire universe. The time of the present moment flows at the same universal rate throughout the universe, but within that universal flow the different relativistic flow rates of clock time are locally computed.

If the happening of existence is the universal processor that computes the current state of the universe in the present moment that existence manifests by its presence, then there is a flow of present moment time that corresponds to the successive processor ticks of universal happening. However it's not easy to pin this down because present moment time has no intrinsic metric since all clock time metrics are computed within it. Thus present moment time is prior to the computation of any dimensionality of time. Happening computes all dimensionality including clock time metrics so it has no observable measure of its own.

Thus there is a present moment time that progresses with happening but no intrinsic associated metric other than the various clock time rates which happening locally computes. So two distinct kinds of time do exist and both progress, but only clock time has a measurable metric associated with it. Thus present moment time can only be measured in terms of the clock times it produces.

There is also another way to get a sense of the dimensionality of present moment time in terms of its effect on cosmological geometry as explained in the chapter on Cosmology. In fact the concept of a separate present moment time is doubly important because it immediately nails down the previously uncertain geometry of the observable universe.

P-TIME

If we define P-time as the time of the present moment, then every actually occurring event takes place simultaneously for all observers at the same P-time throughout the entire universe. In other words if an event takes place in the present moment for one observer, it also takes place at the exact same P-time for all observers. This will always be true no matter how different the clock times of various relativistic observers are or how fast they are running. This universal simultaneity of P-time is what allows relativistic observers to compare their different clock times at the same present moment.

This P-time simultaneity has nothing to do with the relativity of simultaneity of *clock time*. Different observers can still have different views of the clock time simultaneity of events due to the finite speed of light between observers and events.

There is another good argument for a universal P-time, but we first need a couple of basic definitions from relativity. Relativity defines *proper time* as the clock time reading of an observer's own comoving clock, the clock on his wrist or wall. *Coordinate time* is the time an observer sees on another observer's clock and is a measure of the intrinsic rate of processes associated with that other observer from the perspective of the first observer. An observer's measurement of another observer's coordinate time is always by comparison to his own proper clock time, and that measurement, like all observations, is always made in the first observer's present moment.

The proper times of all clocks are continuous; there are never any gaps in the flow of their P-time. There is never any time that proper time is discontinuous during the separation of observers or at any time for that matter. Thus there must always be a one-to-one correspondence of proper clock times for any two observers. There must always be some proper time reading on one observer's clock for every proper time reading on the other's clock whether they are different or not.

This gives us an additional means of testing the universal present moment theory. Every observer can specify his progressive P-times in terms of his own proper clock time readings. There is always a proper time clock reading for every current present moment since both clock time and P-time are continuous. For example, observer A can say he was tying his shoes in his present moment when his proper time clock read 12:00 AM. Every observer's proper time clock reading serves to uniquely identify his own current present moment P-time at that time, and this is true of all observers. For every current present moment of every process in the universe there was always a corresponding proper time reading that can be used to identify it.

Therefore if there is a single universal P-time that flows at the same rate for all observers so they all remain in the same common universal present moment, then there must also be a unique one-to-one relationship between the proper times of all observers even when they differ. There must be one and only one proper time on every observer's own clock that corresponds to every single P-time they all shared when their clocks read those proper times. If there isn't then the universal present moment theory is falsified, but if there is it's confirmed.

Simply stated, for every present moment proper time of any one observer every other observer must have been doing something at that exact same present moment at their own corresponding proper times, no matter how differently their own proper times may have been flowing relative to each other.

In many cases this proper time correspondence can be calculated and confirmed. If the twins exchange flight plans before they separate the one-to-one correspondence of their proper times throughout their separation can be calculated. Each will know the complete relativistic history of the other and thus know exactly how much proper time has elapsed on the other's clock for any proper time on his own clock. Each will know what the other's clock is reading at every moment on his own clock. To be absolutely clear this *calculation* of the current *proper time* of the twin is not the *coordinate time* that would be *observed* on the other twin's clock. It's the *proper time* of the other twin's clock, which is not generally observable but which can be calculated.

Thus it's always possible for any observer who has knowledge of the relativistic circumstances of any other observer to calculate what proper time reading of that other observer correlates to each of his own proper time readings. There is always a one-to-one correspondence of proper time readings between any two observers that tells them what proper time of one corresponded to the proper time of the other in their common present moment even when they are separated in different relativistic circumstances with different clock time rates.

However in the general case of any two observers in the universe, where an initial proper time correspondence can't be determined or the relativistic history of the other observer is not known, it can be impossible to calculate the current proper time correspondence even though it's certain one must exist. However if any observer in the universe can determine the relativistic variables of any other observer he can calculate the proper time rate of that observer relative to his own proper time rate and confirm the existence of a common shared P-time.

Thus it's easy to show there will always be a one-to-one proper time correspondence between any two observers in the universe, and this is all that's necessary to demonstrate a universal P-time present moment common to all observers. It's sufficient to note that time is continuous for all observers thus every other observer in the universe must be doing something at every proper time moment of any other observer's clock. Whatever is being done will be done in the exact same common universal present moment of all existence in the universe because the present moment is the only time that anything can occur because it is the only moment that exists and the only locus of reality.

The clock times of different observers can flow at different rates through this common present moment, but there is always a one-to-one correspondence between the proper times of any two observers throughout the entire universe. This is consistent with Universal Reality's proposal that the common universal present moment of existence is all that exists, and is the current moment of happening of all the computations of the universe. It is the current universal P-time tick of the entire computational universe.

Therefore the proper times of observers can be used to notate the passage of P-time, and their correspondences, when they can be determined, can be used to establish identical P-times among observers even though P-time has no intrinsic metric of its own. And if all else fails any two observers can simply communicate their current relativistic conditions and proper times to enable their P-time simultaneity to be calculated.

CLOCK TIME

Clock time is the rate at which events occur in the present moment in any particular relativistic conditions. But we must be careful to understand this correctly in terms of how the elemental computations of dimensionality take place.

We must also carefully distinguish between what we can call *actual* versus *observational* clock time dilation. For example clocks actually do run slower in gravitational wells (areas of strong gravitation) and during space travel along extended world lines. These are examples of actual clock time dilation because they produce permanent effects that all observers agree upon.

In contrast the apparent clock time dilation two observers moving rapidly with respect to each other see on each other's clocks is an observational rather than an actual effect. This effect is reciprocal as both observers see each other's clock running slower, and this effect vanishes when the relative motion ceases.

In Universal Reality happening is the processor that executes the computations of the universe in the universal present moment of existence. Happening computes all the processes in the universe and generates the passage of P-time. Clock time is the local observable rate of those processes as they are computed. This is how clock time rates emerge locally from the elemental computations of the universe.

Since the computations of the universe are pre-dimensional and dimensionality including clock time dimensionality results from these computations it's reasonable to assume there is a common processor rate across the whole universe. However at this point there is no clock time yet so it's somewhat meaningless to talk about processor *rates* without some measure that could be applied and the only available measures are clock times, which vary. Thus the rate of happening is the absolute reference clock that computes the rates of all local clock times. P-time has no metric of its own because it's the source of all metrics.

Clock time is the effective or observable rate of the computational changes produced by each processor tick. To understand how clock time rates are generated by happening in P-time we first need to examine an extremely important but little recognized principle of clock time hidden within relativity.

THE STc PRINCIPLE

One of the most important implications of relativity is that everything in the universe continuously travels through *spacetime* at the speed of light. That is the combined *vector* velocity of everything in the universe through both space and time is always equal to the speed of light with no exceptions. This is a fundamental principle that is always true for all observers in all cases. Universal Reality calls this the *STc Principle* and it's an essential key to understanding how relativity works especially with respect to how clock time behaves.

Note that the combined velocity through space and time is a vector sum rather than the simple addition of their speeds. By the Pythagorean principle of vector addition it's the square root of the sums of the squares of time and space velocities that's always equal to c. See the diagram in the notes of my book 'Reality' for details (Owen, 2013).

Even though this is one of the most important principles of physics it's almost totally ignored by physicists, who rarely even mention it, and then generally consider it only as a curiosity they don't understand the profound importance of (Greene, 1999, 2005).

However the STc Principle not only underlies most of special relativity but also provides a firm physical basis for the apparent mystery of the arrow of time and confirmation of a privileged present moment by relativity itself, a fact that neither Einstein nor most other physicists seem to have recognized.

To understand the STc Principle we need to understand what is meant by velocity through time. Relativity expresses velocities through time as relative clock rates times the speed of light. Multiplying by the speed of light puts velocity through time in the same units as velocities through space and enables them to be correctly compared as velocities through different dimensions of a single 4-dimensional geometry.

By definition an observer has no linear spatial velocity relative to himself so observers always experience all their spacetime c velocity through time at the speed of light. This is an observer's local frame view but to get the true picture he always has to add in the intrinsic spatial velocity of any fields present as explained in the next section.

The STc Principle is easier to understand in the case of observers in relative motion in empty space because there is no intrinsic velocity of a gravitational field in empty space. Observers in relative motion always see each other's time run slower as a function of their relative velocity. Whether this effect is actual and persistent or just observable and transient depends on whether the velocity is with respect to the absolute computational background or just to the other observer.

In relative motion the velocity through time of any observer on his own proper clock is 1c. In other words every observer is always moving through time at the speed of light on his own clock. And an observer always measures the time velocities of all other clocks relative to his own clock. Since the fastest possible velocity is the speed of light, all other time velocities will be in the range of 0 to 1 relative to the observer, which multiplied by c gives the velocity through time of any other clock relative to the clock of the observer.

The rate of every observer's own clock is always 1c, so he sees all other clocks running at 1c or slower disregarding gravitation. If another clock is running at less than 1c compared to his own clock that slower rate is called *time dilation* in relativity.

For example if I observe the coordinate time of a clock moving rapidly relative to me running at half the speed of my own clock then its velocity through time will be c/2 measured by my own clock. But if I compare my own clock rate to itself the relative rate is always 1, and 1 times the speed of light is the speed of light, so I myself am always traveling at the speed of light through time according to my own clock as is everything in the universe according to its own comoving clock.

If I observe a clock speeding through space relative to me I will see that clock running slower so the total velocity of that clock through both space and time always adds up to the speed of light. This will always be true for any observer observing any other clock in the universe whether it's moving or at rest when gravitation is disregarded. Thus the STc Principle is a fundamental principle of physics.

By definition nothing can move relative to itself in space, thus all the combined spacetime velocity of anything in the universe will always be only through time according to its own comoving clock, and everything is always moving at the speed of light through time according to its own clock.

This movement through time at the speed of light is what we experience as the passage of time through the present moment. Thus proper time always runs at the speed of light without exception. This is in contrast to coordinate time, which is the slower time I observe on a clock moving relative to me in empty space.

The STc Principle applies only to clock time; it doesn't apply to P-time, which has no observable rate since it's an intrinsic aspect of the virtual quantum vacuum, namely the processor cycle rate that is the same throughout the universe.

It is always some form of relative motion, of spatial velocity, that slows clock time rates when dimensionality is computed. Whenever any form of relative motion is being computed it gets computed more slowly and this manifests as a slower associated clock time rate.

The reason for this is there is a *fixed* processor cycle rate computing everything in the universe and some of this cycle rate is used to compute spatial motion, leaving less to compute rates of temporal change. The rate of internal change of state of any process manifests as its associated clock time rate, so the clock time rate slows as a function of the computation of relative spatial motion. This automatically manifests the STc Principle, which describes how clock time works for all processes. The processor cycles of the quantum vacuum are distributed between computations of relative spatial motion and the clock time rates at which local processes occur. This is how the STc Principle emerges in the computation of all processes.

Like other aspects of the virtual data of the quantum vacuum, the processor rate is not directly observable but is observable in the STc Principle and the resulting computational structure of the universe at the aggregate level.

TIME & GRAVITATION

To understand the STc Principle in the context of gravitation we simply recall that a gravitational field is a field of intrinsic velocity in space. Thus an object seemingly at rest in a gravitational field actually has an intrinsic spatial velocity because it's in a velocity density field.

A gravitational field is a field of intrinsic velocity with strength inversely proportional to the square of the distance from a gravitational mass. Thus the field has a intrinsic velocity gradient extending out from the mass and every point in the field has a resulting velocity vector towards the mass because the intrinsic velocity of the field is greater closer to the mass than away from it and the difference at any point produces a velocity vector pointing towards the mass.

Thus an object at rest in the field experiences a velocity vector towards the mass that its inertial motion tends to follow. This is the source of what we call gravitational attraction. There is no actual attraction; it's just a matter of an object in inertial motion traveling through space along its velocity vectors.

We confirm a gravitation field as a velocity field in our direct experience as the velocity we experience towards a gravitating mass along the velocity vectors it produces. And the force of gravity we experience standing on the surface of the earth is actually our acceleration against this inertial motion due to the surface of the earth blocking our motion. This is in agreement with Einstein's Equivalence Principle that states that what we feel as a gravitational force is actually an acceleration (Wikipedia, Equivalence principle).

So to understand how the STc Principle works in a gravitational field we simply add the intrinsic velocity density of the field to any linear velocity an object has. The velocity through time of the object then slows due to the combined spatial velocity of linear velocity and the intrinsic gravitational velocity of the field. This is why both linear velocity and gravitational fields produce time dilation. They are both forms of spatial velocity that slow the velocity of time by the STc Principle.

Thus a clock at rest in a gravitational field will observe a clock at rest in empty space running faster because its own clock rate is slowed by the intrinsic velocity of the field. The clock in empty space will likewise see the clock in the gravitational field running slower for the same reason. In contrast to clocks in relative linear motion, both observers agree on this effect; thus it's an actual relativistic effect and the difference in elapsed time produced is permanent.

The velocity density model explains the relativistic slowing of time by gravitational fields in the same way time is slowed by linear velocity. Gravitational time dilation and the time dilation of linear motion are both due to increased spatial velocity slowing temporal velocity in accordance with the STc and MEv Principles.

Thus the total spatial velocity of an object is its linear velocity plus the intrinsic velocity of any fields it's in. This gives its total spatial velocity, which combined with its temporal velocity always equals c.

The STc Principle holds in all circumstances when properly interpreted. Take two observers, A at rest in a strong gravitational field and B at rest in empty flat space. Because A is in a gravitational field it has the intrinsic velocity of the field even though it's at rest with no linear spatial velocity. Thus its velocity in time is slowed by both forms of spatial velocity so its total spacetime velocity remains equal to c. This is how gravitation dilates (slows) time as relativity predicts. On the other hand observer B in empty space isn't in a gravitational field and has no intrinsic velocity. Thus all its spacetime velocity is through time at c.

Both A and B agree on all this. A sees B's time running faster than his own time, and B sees A's time running slower than his. This is also the proper view because it's that of the absolute background computational space in which everything is computed and thus it reflects actual rather than observational relativistic effects.

This velocity density model of gravitation is equivalent to the curved spacetime model of general relativity and predicts exactly the same effects but it goes much further because it provides an actual mechanism for those effects that is lacking in relativity. That mechanism is the STc Principle and its source in the processor cycles that compute dimensionality.

Now relativity tells us that everything including light is affected by gravitation so light passing by observer A in the field must also take on the intrinsic velocity of the field. Thus the total spatial velocity of light which is always c is now its combined linear velocity plus the intrinsic velocity of the field. And since light has no velocity in time this means that the *linear* velocity of light in a gravitational field is actually less than c because some of its spatial velocity is the intrinsic velocity of the field.

Now this is the proper perspective from the absolute view of the computational space in which all this is being computed. However A's time is slowed by the same amount as the linear velocity of his light is slowed. Thus A continues to measure the local *linear* velocity of light as c as all observers always do. In gravitational fields the local linear velocity of light is always slowed by the same amount as the time of a local observer so all observers always measure the local linear velocity of light as c in all situations.

Not so for B however. B's time, in empty space, is not slowed so that he actually does see light moving slower than c at A's location in the gravitational field. Thus while the speed of light is always c when locally measured by any observer, observers can see light travel either faster or slower than c at other locations depending on the relative strengths of any gravitational fields.

For example A with his time slowed in a gravitational field sees light traveling faster than the speed of light at B's non-gravitational location. But this is an observational relativistic effect rather than an actual one because A's view is not that of the absolute background space in which everything is being computed.

An extreme example of light not actually traveling at the speed of light is a black hole. From the perspective of an observer in empty space light at the event horizon of a black hole appears to have zero velocity because it's never able to reach the observer. But conversely an observer right outside the event horizon would observe the velocity of light in empty space approaching infinity because his own time is slowed to next to zero. Thus observers routinely observe light not traveling at the speed of light at other relativistic locations.

The STc Principle is the uniform principle that resolves all these seemingly contradictory relativistic views because it's the view of the computational space in which they are all computed. If we just add the *intrinsic velocity* of any gravitational field to the linear velocity to get the total spatial velocity of any process and subtract that from c then we get the true resultant time velocity of that process relative to the computational space in which the dimensionality of the universe is computed. And all observers who do this will agree on this view of time.

Thus we get a very simple way of understanding general relativity that works for all observers in all situations. Observer A in the gravitational field just needs to add the intrinsic velocity of the field to any linear velocity he or anything else including light may have to correctly understand how both he and observer B do in fact obey the STc Principle, as do all observers in the universe.

THE VELOCITY OF SPACETIME

So the STc Principle is a universal fundamental principle but it must be understood that the c in the principle is not the actual observed speed of light itself, which can vary, but the fundamental velocity of spacetime itself, and since space itself is a field of intrinsic velocity the value of c in the STc Principle is by the METc Principle the fundamental combined fixed value of mass-energy (linear plus intrinsic spatial velocities) plus the velocity of time of any point.

Thus the speed of light should actually be understood as the intrinsic velocity of spacetime, or more fundamentally as the intrinsic velocity of mass-energy plus time at any point of computational space. It's mistakenly called the speed of light because light just happens to always travel at c when measured locally though it can appear to travel at different velocities when observed from a distance.

A more complete picture of velocity density is to recall that

Universal Reality models mass and gravitational fields as intense fine vibrations. Thus a gravitational field becomes a field of vibrations in the fabric of space. These vibrations are the actual intrinsic velocity density of the field. The intrinsic velocity density of a gravitational field consists of intense fine vibrations in the space surrounding masses and masses are themselves fields of these vibrations in the fabric of space. This is explained in detail at the beginning of the next chapter.

Because each volume of space contains vibrations the distance across it is further because the ups and downs of the vibrational wave must be traversed. This is completely equivalent to the curved spacetime model of general relativity because its curves could be compressed into vibrational waveforms in a flat Euclidean space and if they were stretched out again the resulting space would be curved.

In this model a mass is a field of minute vibrations in a flat Cartesian space center on the massive particle(s) producing it. This is much easier to visualize than the curved spacetime of general relativity. The beauty of the velocity density vibrations model is its flat Cartesian space is very easy to understand and work with and also much easier for the elemental program to compute.

In the vibrational space model the speed of light is actually the same for all observers everywhere in the universe. Light appears to move slower across vibrational space because actual distances across it including the ups and downs of the vibrations are much greater than they appear. Thus the actual distance traversed by light up and down the vibrations is such that light always does travel at c everywhere in the universe even in gravitational fields.

In the standard interpretation of general relativity this is due to the curvature of space in gravitational fields. Thus light beams are actually always traveling at c but when they have to traverse the curvature of space the distance traveled is greater than it appears so light appears to move slower than c. We don't directly see the curvature of space produced by gravitational fields because light travels along it but the actual distance is greater than the nominal distance between two points in a gravitational field. Vibrational spaces model gravitational fields as flat Cartesian spaces as we actually see them rather than the curved spaces general relativity proposes and this is another significant advantage of Universal Reality.

Thus we actually see the sun very slightly larger than we would if it didn't curve space near it and it's actually slightly farther away than its nominal distance measured along the curved space of its gravitational field. This effect is proportional to gravitational strength, which is why light appears to slow to zero speed as it nears a black hole and images of things pile up at the event horizon and fade out as they cross it.

The STc Principle and the equivalence of mass-energy (gravitation) with spatial velocity are the two keys to understanding general relativity. However our understanding is greatly improved by replacing the spacetime curvature model with an equivalent intrinsic velocity density vibrational space model.

Even though time travels at different rates depending on different relativistic conditions, there is a hypothetical standard clock time rate for the universe, which is the maximum rate clock time can flow. This is the time rate of stationary clocks in deep space far from any gravitational field though there is nowhere this is strictly true.

This is the clock time rate of empty space, which is a function of the intrinsic velocity of the zero-point energy of the quantum vacuum. Since all forms of mass-energy produce gravitational effects the zero-point energy also produces a gravitational effect. And since all gravitation fields are equivalent to fields of intrinsic velocity, and the amount of intrinsic velocity sets the balance of space and time velocities there should be a relationships between the values of c and the zero-point energy.

Thus the clock time rate of time in empty zero-point energy space is the standard baseline clock time rate of the universe. It would be the maximum possible clock time rate and all other clock time rates would be slower proportional to their relativistic circumstances. Nevertheless all observers measure their own proper time rates as the speed of light even if they are in a gravitational potential or moving through space as explained above.

THE ARROW OF TIME

One of the perennial mysteries of science has been the source of the arrow of time; the fact that time continuously flows in the forward (by convention) direction. Many scientists have sought for the source of the arrow in vain, often mistakenly attributing it to entropy (see the section on 'Entropy' for why this isn't correct), but the explanation is quite straightforward and a simple consequence of the STc Principle.

The STc Principle states that everything continuously travels at the speed of light through time on its own clock. Therefore it's quite clear that time must be experienced as flowing in a single direction by every observer and that every observer's own clock will always appear to have its hands moving at the same rate in the same direction as all others. Thus the STc Principle itself is the actual source of the arrow of time, and puts the arrow of time on a firm scientific basis. The arrow of time is a direct implication of the theory of relativity; another completely unrecognized fact of modern physics, and even actively denied by scientists who don't understand it.

CONFIRMING THE PRESENT MOMENT

Amazingly the necessity of a privileged present moment of time distinct from all other moments of time is another direct consequence of the STc Principle, again completely unrecognized by most physicists. In fact many physicists continue to deny the existence of a present moment mistakenly believing it's inconsistent with relativity. It is not. It is actually *required* by relativity and a direct consequence of the STc Principle that underlies special relativity.

By the STc Principle everything continuously travels through spacetime at the speed of light. This means everything must always be at one and only one point in time, and that point must be the current present moment of its actual existence. Thus the STc Principle requires the existence of a privileged present moment for all observers that is the current time of their existence, and the *only* time they are actually at, the time that defines their now.

Thus relativity itself absolutely requires a present moment that progresses in clock time and conclusively falsifies the nonsensical 'block universe' hypothesis in which all times exist at 'once' as a single static structure (Price, 1996, 12-13, 14, 15-16), (Wikipedia, Eternalism (philosophy of time)).

This present moment required by relativity is the same present moment we have already identified as the presence of existence whose presence naturally manifests as a single universal present moment in which everything exists. Thus Universal Reality is clearly consistent with and confirms a proper understanding of relativity in this respect.

Because existence exists it must have a presence and its presence manifests as a universal present moment in which everything exists. This present moment of existence is identical to existence and encompasses all that exists and there is no before or after that actually exists. The presence of existence is the source of the present moment that is now confirmed by the STc Principle to be consistent with relativity.

Physicists who deny the existence of the present moment should remember it's the most fundamental and persistent of all observations, and that the role of science is to explain observations, not to deny them.

SPACE TRAVEL

Due to the STc Principle a clock moving through space will run slower than a clock at rest, the slowing depending on the spatial velocity along its world line relative to computational space. Traveling along a world line in space will always take less time on the traveler's clock than the clock of a stay at home observer.

This slowing of a clock traveling through space can be enormous as its velocity approaches the speed of light, a fact that makes interplanetary travel theoretically feasible, at least with respect to the time required. Calculations show that a trip from earth to the center of the galaxy at a constant 1g (the equivalent of earth's gravity) acceleration for half of the trip and a 1g deceleration for the other half would take only 42 years on the clocks of the travelers, though well over 42,000 years, a little over the distance in light years to the center of the galaxy, would pass on clocks back on earth or at the galactic center (Misner, Thorne & Wheeler, 1973).

Of course a propulsion system that could produce a constant 1g acceleration for 42 years is not currently available and the difficulty of detecting and avoiding any intervening objects at close to light speed is near impossible. Nevertheless time dilation does make interstellar travel a theoretical possibility. So alien civilizations, if they exist, could just as easily travel to earth as well. The time it would take on their clocks would be quite acceptable even though it would take a very long time by our earth clocks and by clocks back on the aliens' home planet.

TIME TRAVEL

There are many misconceptions about time travel, especially when the significance of a present moment of existence is not understood. However when we understand that the present moment is the only actual locus of existence everything becomes clear.

First we, and everything else in the universe are already continuously traveling through time at the speed of light on our own comoving clocks (a clock on our wall or a watch in our pocket). So we are all constantly traveling in time and we are all time travelers in this respect. We can't not travel in time because the passing of time is precisely us traveling through time at the speed of light.

However we each travel at different clock time rates depending on our respective relativistic conditions, either because we have different relative motions or are in different gravitational potentials.

Though we are all traveling through clock time at different rates we all stay in the common universal current present moment. There is no possibility of traveling out of the universal present moment because it's all that exists. The present moment is the only locus of reality and of the entire actual universe.

These are the actual limits on time travel. No going back to the past or forward to the future. The future doesn't exist so there is nowhere there to go. Likewise the past doesn't exist so there is also nowhere to go in the past either. We all stay in the common universal present moment, but our clock times, and all associated physical processes including our aging, can progress at different rates within the common present moment.

So we certainly can travel in time at different rates in the forward direction of time's arrow. There is extensive observational proof of this and relativity describes it precisely. Twins can separate and meet up again with different ages, but this is not the same as actually traveling into the future or the past by either twin. Both stay in the same universal present moment at all times and can never leave it. One just ages faster than the other in that present moment. The notion of traveling to an actual past or future out of the present is simply impossible.

So unfortunately there is no going back in time to view dinosaurs, and also no going back in time to change things there that alter the present. Thus there are no possible time travel paradoxes. And as interesting as it would be, no arrival in the present of time travelers from the future. It's simply impossible because the future has never existed because it hasn't been computed and therefore doesn't exist. It doesn't exist until it's actually computed in the present moment and then it becomes the present.

We have all arrived in the present moment from the past, but we have never left the continually evolving universal present moment to do so. But it is true that the past a space traveler arrived from could be very far back in time by our clocks if his clock was running very much slower than ours. With the right interstellar flight plan he could have begun his trip when Nero was emperor of Rome and arrived back here on earth just yesterday not much older than when he left Rome. In the colloquial sense that would certainly be a person arriving in the present from the past, but he and no one could ever travel back to his or anyone's past since only the present moment exists and the Roman empire would be long gone when he arrived.

So we can certainly arrive at the same location in the present moment from different original past times, and that could certainly be very interesting, but everyone is continually in the common current universal present moment during the entire duration of his or her lives and travels. Some lives could be very much longer than others according to other clocks but only if they lived at much slower rates.

Our ancient Roman space traveler could arrive back on earth today to meet his 60th generation grandson and catch up on 2000 years of missed history. Again extremely interesting but at every second during those 2000 years he and the earth would have both existed in the same current present moment. Events on earth would just have been progressing at a much faster clock rate than aboard the Roman space ship.

It is also theoretically possible for you or I to travel to an arbitrary date in the future in the same colloquial sense by taking a space flight with the right speed and trajectory. But this is just a matter of slowing our clock in the universal common present moment relative to the rate of clocks at our destination. No one every leaves the common present moment but we could arrive there with much less elapsed time on our own clock. However this is impossible without a very high velocity space flight or intense gravitational field.

Because our spacetime is nearly flat on earth and we have a very low velocity relative to the background there is no way anyone else's time could be running appreciably *faster* than our own and there are really no effects to consider here.

Everything that exists always exists in the same universal common present moment at all times as it evolves but time travelers could certainly arrive in the present from deep in the past with first hand information and even photos and videos given the proper technology. We can only hope!

OBSERVER SINGULARITIES

Our location in spacetime is a singularity in the sense that clock time continuously flows in through the point of our location and out in all directions into the past. Thus only our own current location exists in the present moment on our own clock. Everything else in the universe is at some distance from us and thus exists at least slightly in the past relative to us from the perspective of our present moment. Thus every observer exists in his own clock time singularity.

Of course everything and all observers actually exist in the same universal present moment but that common existence is not directly observable due to the finite speed of light. Our actual experience of all other things and observers in our present moment is always a no longer extant past representation down the radial time dimension of the universe.

Clock time continuously flows in from non-existence through our singularity in the present moment. The future continuously becomes the present as the state of the universe is continuously recomputed at the point of our existence. But there is no actual future that we reach that then becomes the present. The present state of the universe is just continuously recomputed in the present moment and clock time is simply the local observational rate at which the results of those computations happen.

Though only the present moment has reality an observer can think of clock time as continuously emerging from nothingness into his singularity and then flowing out into the past into the distance along the radial time dimension in every direction. Everywhere we look in the universe we see the universe receding from us from the back of the moving train of time into the distance along the time dimension.

Everywhere we look we look into the past receding from our eyes, and nowhere we look do we see the future approaching except in our imaginations of it. Thus our location in space and time is the point of continual creation into existence, and once created the universe flows out into the past in all directions from us.

SEEING ALL 4 DIMENSIONS

There has been much discussion about how to visualize the 4-dimensions of the spacetime universe, of how to see the time dimension just as we see the 3 spatial dimensions. However the fact is we already see all 4 dimensions of the universe all the time laid out clearly before our eyes.

We can confirm the 4-dimensional geometry of the universe visually because we actually see it. We see down the time dimension into the past as distance in every direction from every point in our 3-dimensional space. This is called our light cone and it's our personal view of the cosmic geometry of the universe from the singularity of our location in our present moment of spacetime.

We see all 4-dimensions but there's a catch because the light cone we see is only a slice through 4-dimensions rather than the entire 4-dimensional universe. We see the past only as it existed at certain distances, and we see all of space only as it existed at particular times in the past. Thus we neither see all of space nor all of time, but only a slice through both centered on our singularity.

Our experience of the passage of time through the present moment is our direct experience of the fundamental process of the universe, the continual recomputation of the information state of the universe including the passage of clock time through the present moment. At any moment this manifests in our mental simulations of reality as the 4-dimensional spacetime we directly observe around us.

ENTROPY

Entropy is the tendency for the energy states in any isolated volume of space to reach equilibrium over time. For example in a completely insulated box objects at initially different temperatures will eventually all reach the same temperature. Thus presumably the entire universe will eventually reach an energy equilibrium in which no additional transfer of energy can occur and all processes will come to a halt (Wikipedia, Heat death of the universe).

However this energy equilibrium is not perfect nor is it necessarily eternal due to random zero-point energy fluctuations in the quantum vacuum, which are not subject to entropy and continually affect the state of the universe of actual particles. But these effects are statistically extremely unlikely to produce any large-scale energy imbalances that affect the process towards maximum entropy.

Because entropy appears to be a fundamental unidirectional process in time that seems irreversible some physicists have proposed that it's somehow the source of the arrow of time but this is not correct. We have already correctly identified the STc Principle as the source of the arrow of time, and more fundamentally the fact of the happening of existence, which continually computes the evolution of the universe, is the ultimate source of clock time and its arrow.

And entropy can't be the source of the arrow of time because it varies wildly from region to region. There are many areas of the universe in which entropy is decreasing due to incoming energy and there is certainly no reversal of the arrow of time in those areas. If entropy were responsible for the arrow of time it would have to be a universal rather than a local effect.

However there is no physical mechanism that could account for such a universal effect. For one thing entropy is entirely a *result* of physical processes rather than the *cause* of anything. And more importantly entropy depends entirely on the current mix of fundamental forces at any location.

Entropy states are not fundamental, as usually assumed, because they depend on the spatial mix of prevailing forces. For example cosmic scale entropy states reverse if gravitation reverses, and at smaller scales entropy depends on the distribution of the other three fundamental forces.

In an initially stable universe with only attractive gravitation the ultimate maximum entropy state will be a single black hole because all matter will eventually clump together. But in a universe with only repulsive gravitation the ultimate maximum entropy state will be a continually expanding universe in which all matter continues to fly apart forever. Thus entropy reverses if gravitation reverses.

In our universe where there is an apparent mix of attractive and repulsive (dark energy) gravitation and that mix seems to be changing it's unclear what the final maximum entropy state will be.

Thus cosmological discussions of entropy are almost always flawed because they fail to recognize that entropy itself is not fundamental. What is fundamental is the force mix that defines the measure of entropy. Entropy is meaningless without reference to the force mix it's relative to. Maximum entropy has to be redefined as a state of equilibrium *under the mix of prevailing forces*.

Thus entropy is not a fundamental principle as usually thought. It's entirely a *result* of the evolution of the actual fundamental computational principles. Like all emergent laws it describes reality but doesn't actually compute anything.

When the dependence of entropy states on force mix and distribution is understood it's also clear entropy has no causal connection to time and is certainly not the source of the arrow of time.

TIME REVERSAL

There are two cases to time reversal, first time itself reversing and second an individual object traveling backwards in forward flowing time.

First time itself can't run backwards. The direction of time is due to the sequential nature of the processor cycles that compute the universe. Without a reversal in the sequence time can't flow backwards so it always flows in the same direction. But it's really a moot problem since whatever direction the sequence proceeds that determines the direction of time. Since the sequence of processor cycles is what determines the direction of time it's meaningless to imagine it reversing because there is no background reference with which to measure its direction since it itself is the ultimate reference. By definition time always runs 'forward'.

So we can imagine processes running in different directions within time. But the very notion of time itself running backwards is nonsensical because whatever direction it runs is by definition the forward direction.

Some physicists have seen the apparent non-reversibility of certain processes as an unsolved problem. For example broken eggs never spontaneously reassemble into unbroken eggs. Water waves radiating from a dropped stone never reverse direction and converge to pop the stone back up out of the water, and people never reverse their aging process and start growing young again.

These are all irreversible temporal processes yet the equations of science seem to describe them perfectly well in both temporal directions, so traditional interpretations of science offer no explanation for their irreversibility.

In Universal Reality this is a pseudo problem. First the equations that describe these processes are not actually computing them but only describing them. All such processes are the emergent manifestations of elemental computational processes that are *not time reversible* since they all involve non-reversible choices among random possibilities at the quantum level. Once a single specific choice is made from a probability distribution of possible choices there is no way to reverse that choice into the probability distribution it was selected from. So I don't see any real problem here. It's just the way our computational universe naturally evolves as a consequence of the complete fine-tuning.

Another deeper problem of time reversibility is the nature of the time parity particle component. Elementary particles all have a number of particle components, among them space and time parity or chirality, whose nature and function are not well understood. They are clearly related to the difference between standard and anti-particles because their values are opposite in antiparticles. In fact to change particles into anti-particles and vice versa you just reverse their charges, and their spatial and temporal parities.

Parity is the handedness of a particle. An antiparticle is like a regular particle reflected in a mirror in all three spatial dimensions, and in a time mirror as well. But what this means is not completely clear.

However 4-dimensional parity seems to be one of the elemental components of reality necessary to make something real in our universe.

When particles and antiparticles meet and annihilate into energy their opposite parities cancel each other out into nothing. When particles and antiparticles emerge together out of the quantum vacuum the non-actuality of the quantum vacuum can be thought of as separating into opposite actualized particle components including opposite parities. It is as if all the something in the universe is just opposite amounts of nothing.

Thus it's reasonable to assume that the dimensionality of the entanglement network incorporates a largely normal parity as it's constructed and that the interactions of antiparticles with it involves the incorporation of opposite parities with respect to it. Thus spacetime itself seems to have an intrinsic spatial and temporal parity computed into its fabric with respect to which the parities of individual particle events is computed just as it embodies an absolute dimensional framework that linear motion and rotation are relative to.

In quantum physics, for example in their Feynman diagram representations, antiparticles must be thought of as moving backward in time to make sense of their interactions with normal particles (Wikipedia, Feynman diagram). Obviously the particle itself can't actually be moving backwards in time or it would disappear out of the present into the past, but it's as if it is *pointing* backwards in time while going forward in time, while normal particles are facing forward in time in the same direction time is flowing.

This backward facing direction in time is likely related to the problem of why there are comparatively very few antiparticles remaining in our universe when equal numbers of particles and antiparticles were presumably created together in pairs out of the quantum vacuum in the big bang. There doesn't seem to be any evidence that all those missing antiparticles are hiding anywhere.

Why this is true is uncertain but it seems like particles facing in the wrong direction in time have a much tougher time than particles facing in the direction they are going.

Note also that P-time has no meaningful time direction other than the one it creates by its own existence that defines the single possible direction of clock time's arrow. Computations just occur as happening computes them and that then creates a direction to the clock time it produces. Clock time is the effective relative rate at which processes occur in relativistic situations. It is not the rate at which P-time computes them but the rate at which their temporal aspects occur after their relative spatial motion has been computed so that the total rates of both always equals the speed of light.

So there is no reason to believe a reverse flowing clock time is even computable by the sequential processor cycles of happening. The only way a reverse clock time could be observable would be as some physical process running backwards, but it would just be running backwards in a forward flowing clock time like a movie played backwards. Time wouldn't reverse, only the movie.

So it seems there is no possibility of clock time actually running backwards and even the concept appears meaningless. If the clock time of the entire universe ran backwards everything would appear the same because all observers' clocks and minds would be running backwards as well. And if some particular process was observably running backwards it would have to be running backwards relative to the forward flowing time of observers so clock time would still be running forwards. If there was an actual reverse direction of some local clock time, it's not clear that would be observable even if it existed.

The reverse time parity of antiparticles is an indivisible unit and single units don't actually progress in time, they can only be pointed in time. Only multiple successive units can be understood as having a direction in time. A single frame from a movie is equivalent to a still photo and has no direction in time. Thus an antiparticle with backward facing time parity doesn't actually move backwards in time. It's analogous to a still from a movie made to play backwards but viewed in a time moving forward. In itself one can't tell, but only when it interacts with another particle does its temporal posture become evident.

There are other problems with individual processes moving backwards in forward flowing time. Suppose you are moving forward normally in time but at some point reverse your direction in time and begin to move backwards. First this is impossible as the current present moment in which everything exists continues to move forward and leaves you behind in a non-existent past.

But assume for a moment you could travel backwards in time. What would you see? Assuming you stepped slightly to the side to avoid a collision you would see yourself standing beside you pointing forward in time as you moved backwards. For every second you retraced you'd see yourself as you were at that time getting younger reliving your life backwards.

So this raises a number of problems. First it's unclear that you would 'see' anything, as the photons from other things would now be leaving your eyes back towards them. Second it seems to imply the instantaneous creation of another you, the appearance of formed matter out of nothing, and it's not clear which of the two you's you actually would be, or are you 'now' both of you? Third would you be getting younger along with your other, or would you be getting older as he got younger? And it implies the reversibility of random quantum processes, which seems impossible.

But now consider being your original self in the same situation not yet having turned around in time. Now what do you see? Now at every second you see your turned around self moving forward in time beside you, but facing backwards in time getting younger! And you should see him quite well because light is bouncing off him into your eyes normally. It's still an impossible scenario that can't exist because the current present moment in which everything does exist is now far advanced from you and your anti-you.

This impossible scenario is mentioned only because it may shed light on the nature of antiparticles because it could explain how antiparticles moving backwards in time seem to remain in the present moment facing backwards in time rather than zipping by us into the past. They may in some sense actually be traveling backwards in time from a creation point in the future. If so we would view them traveling along with us forward in time as our forward moving time retraced their past into the future, and they would appear to be facing backwards in our forward flowing time.

It's not clear how this could be consistent with a universal present moment but it appears there may be something here. It is clear there is something deeper hidden in the nature of spacetime parity but for now it remains a mystery.

If for example there were some mechanism by which forward flowing time simply moved past the future creation dates of antiparticles they would vanish leaving the great preponderance of normal particles observed in the universe today.

The other possibility we can imagine is just our consciousness moving backwards in time. In this scenario there is no second body created, we simply observe ourselves living backwards getting younger. But this is clearly impossible as the flow of data through consciousness is a product of normally time flowing neural processes. It's impossible in the same sense as experiencing some other being's consciousness when their consciousness is a product of their being.

COMPUTING SPACETIME

MASS VIBRATIONS & GRAVITATION

We begin this chapter with an outline of the complete theory of vibrational mass and how it explains the relativistic effects of gravitation in an equivalent but much easier to understand manner than general relativity. This model is superior to the curved space model of general relativity in that it models spacetime as a flat Cartesian space, which is how we actually see it. It also provides an explanatory mechanism for how the presence of mass affects space that is missing from general relativity, which never explains why or how mass actually curves spacetime. And it's also part of a unified theory that includes quantum theory as explained in the chapter on Quantum Reality.

1. By the METc Principle mass-energy and space are two aspects of the same thing. They are both forms of spatial velocity in the form of in-place vibrations in the fabric of computational space at the minimum dimensional scale.
2. Empty space is a field of minimal amplitude vibrations corresponding to the zero-point energy of the quantum vacuum, which is likely related to the c value of the speed of light.
3. Gravitational fields are fields of increased amplitude vibrations centered on massive particles.
4. Massive particles are concentrated detached units of space packaged in particle component sets.
5. Packaging in particle component sets enables particles to move relative to the empty space background.
6. The mass of a massive particle is a field of space vibrations centered on the particle component package. The vibrational field is an inseparable part of the mass of particles.
7. Gravitation is homogeneous (same effect in all directions at any point) thus the vibrations must have the same form in all three directions. In a reduced 2-dimension representation they can be visualized as standing symmetrical wave peaks oscillating up and down.
8. The gravitation produced by multiple masses reinforces but never cancels therefore the peaks produced by multiple masses are evenly distributed in the fabric of space in the same locations and

multiple masses additively amplify (increase the height of the peaks) the vibrations which corresponds to the strength of the field.
9. Thus the fabric of space consists of a uniform field of velocity vibrations of various amplitudes corresponding to the presence or absence of masses and their vibrational gravitational fields.
10. The locations of individual vibrational peaks can be taken to define unit cells in the fabric of space. These cells can be considered the individual quanta of space. They represent the minimal computational granularity of space and are far below the scale of quantum interactions.
11. Vibrational space is a flat Cartesian 3-dimensional computational space in which individual cells can have different velocity densities.
12. Vibrational space is composed of cells of unit volume. These cells are the basic computational entities of space and are encoded by the cells of the computational array in which dimensional space is represented and computed.
13. Every cell has an individual vibrational *amplitude*, which is its gravitational strength.
14. A cell's vibrational amplitude gives it a proportional intrinsic spatial velocity, which is the strength of its gravitational field.
15. Every cell has a nominal Cartesian width, which is the same for all cells, and an actual traversal width that is greater in a gravitational field when the ups and downs of its vibrational peaks and valleys are considered. The nominal cell width is the same for all cells but the traversal width varies with the amplitude of its vibrations, the strength of its gravitational field.
16. By the STc Principle each cell has a slower time velocity due to the intrinsic spatial velocity of its vibrations. This is the source of gravitational time dilation.
17. These three effects characterize every cell in computational space and are greater in cells in gravitational fields since they have greater vibrational amplitudes. They account for all general relativistic effects just as the curved space model of general relativity does.
18. The curved space model of general relativity is completely equivalent to our vibrational space model. If the vibrations of a volume of vibrational space were frozen and stretched out that volume would be dilated and the space there curved as a result. Conversely if the space curves of general relativity are compressed into a landscape of peaks and valleys and set into vibration we get our vibrational model. So essentially our vibrational space model replaces the curved space of general

relativity with vibrations in the individual cells of a flat Cartesian space.

19. Both models give exactly the same relativistic effects but our model is superior on three counts. It models space as the flat Cartesian space we actually observe; it provides a unifying explanatory mechanism for gravitation missing from general relativity (why the presence of mass curves space is unexplained in general relativity); and it also models quantum reality as explained further on.
20. Gravitational fields are fields of intrinsic spatial velocity due to their vibrational amplitudes. Thus by the STc Principle an object in a gravitational field experiences the intrinsic velocity of the field and its velocity in time is slowed. This is the source of gravitational time dilation.
21. The total spatial velocity of an object is now its linear velocity plus the intrinsic velocity of its location in a gravitational field. By the STc Principle this total spatial velocity vector subtracted from c reduces its time velocity. Thus the time dilation of linear motion and that of gravitational time dilation are now revealed as two aspects of the same thing, an insight missing from relativity.
22. Objects travel farther through vibrational space than its nominal Cartesian dimensions because the ups and downs of all its vibrations must be traversed.
23. This includes light itself. Light always travels at the speed of light but since it must travel farther through denser vibrational space it appears to be traveling slower than c from the point of view of an observer in empty space. In the extreme of a black hole the apparent speed of light drops to zero for this reason.
24. However the velocity of time of a *local* observer is slowed proportionally to the greater distance traversed thus all observers always measure the local speed of light as c in all cases even though it may appear different to remote observers.
25. Thus all observers experience all their combined spacetime velocity as through time at the speed of light. This is their local frame view. However to get the true picture they must recognize any intrinsic spatial velocity they have and subtract that from their apparent time velocity to get their actual time velocity.
26. This gives the true universal view of all relativistic processes in the observable universe because it's the view of computational space in which they are all actually computed.
27. A gravitational field is a field of vibrational density that falls off by the square of the nominal distance from a gravitating mass. Thus every point in the field has a velocity density gradient with the velocity density greater towards the gravitating mass than

away from it. This gradient produces a velocity vector pointing towards the mass that inertial motion tends to follow. This is the actual source of gravitational attraction, which is missing from relativity, which doesn't properly explain why an object at rest in curved space begins to move.

28. By the MEv Principle the mass of a massive particle is modeled as a field of fine spatial vibrations centered on the nominal particle. It's the intrinsic spatial velocity of these vibrational fields in aggregate that slows the time velocity of objects in the field by the STc Principle and so accounts for the effects of general relativity.

29. The vibrations of a particle's mass have amplitude and frequency. The amplitude is the intrinsic velocity in space of the field, which is the source of its gravitational effect, and its frequency is its clock time velocity, it's internal clock time rate.

30. By the STc Principle the vector sum of a particle mass's amplitude and frequency, its intrinsic velocity in space and its velocity in time, must always equal the speed of light c. Thus particles with different rest masses will have vibrations with slightly different amplitudes. Thus we must assume they will also have slightly different frequencies in accordance with the STc Principle.

31. The masses of particles are so miniscule any difference in their velocity through time, their frequencies, will likely be undetectable. However it just might show up as anomalies in particle half-lives or other aggregate effects.

32. The rest mass of a particle at rest is fixed. However if it gains linear velocity this adds kinetic energy to the vibrational energy of the particle so its total spatial velocity (linear kinetic energy plus intrinsic spatial velocity of vibrational amplitude) is increased and its time velocity is slowed. The slowing of its time velocity (frequency) increases the particle's intrinsic spatial velocity (vibrational amplitude) so the total vector sum of its vibrational velocity in space and time is conserved. Since the frequency is slowed the amplitude is increased and since amplitude is observational mass the rest mass of a moving particle increases as relativity predicts. Thus the mass of a moving particle increases because its clock time rate is slowed and this increases its own intrinsic velocity in space, which is its observational mass.

33. The same is true in gravitational fields. A particle experiences the intrinsic spatial velocity of the field, which reduces its internal vibrational clock rate. This in turn increases its own intrinsic spatial velocity in the amplitude of its mass vibrations. This increases the observational mass of the particle, which in the case

of gravitational fields is the particle's weight. This is why objects in gravitational fields have increased weight.
34. Thus the increase in observational mass with linear velocity and the increase in observational mass in a gravitational field are both caused by time dilation and both revealed as the same effect, an increase in weight.
35. The frequency of the mass vibrations of particles are their internal clock rates, and their amplitudes are their observational masses. The vector sum of both is computed by the fixed number of processor cycles allocated to computing velocity in space and velocity in time. Additional linear velocity reduces the frequency of the vibrations, which increases their amplitude, which is the observational mass of the particle.
36. Photons are massless and have no internal vibrational structure to compute and thus no internal clock time rates. Therefore all the processor cycles that compute photons go to computing their linear wave motion through space and photons automatically always travel at the speed of light because they have no internal velocity through time. In contrast massive particles have internal vibrational structures that must be computed and so can never travel at the speed of light because they always have some velocity through time.

This outline explains the essentials of how fields of mass vibrations in space produce the effects of general relativity. We can now explain in greater detail how relativistic spacetime is actually computed.

RETHINKING SPACETIME

It will be useful to first take a moment to examine the concept of spacetime before proceeding. The notion of a single fixed universal spacetime within which all things exist and all events occur is absolutely fundamental to science and our common sense view of the world, yet there are very good reasons for thinking it simply doesn't exist, and no evidence that it actually does exist.

For one thing we certainly never observe or measure any such fixed empty space. All we actually observe and measure are the dimensional relationships between events, specifically the events of measurements and observations of things by us. Try as we may, we simply cannot observe or measure empty space itself. However we try we always end up observing or measuring the dimensional relationships of objects and events 'in' space. Thus our concept of an empty space in which things exist is actually a logico-mathematical construct inferred from the dimensional relationships among the events we measure rather than an actually observable physical structure.

For example we never see the actual empty space between objects and us. What we see is an object's apparent size on our retinas, which is then processed by our brains to compute a distance relationship between us and the object based on a mental model of the presumed and apparent sizes of the object. It's this dimensional relationship that's observed rather than any actual empty space between us and the object. And this is true in all cases without exception.

So the apparently physical space around us is actually our brain's projection of the dimensional relationships generated in our simulation of reality back out into an apparently external world of its own creation. And we know this is a fiction that exists only as data structures within our neurons. The only thing true about it is that these dimensional relationships in aggregate do form a consistent 3-dimensional logico-mathematical structure in which objects can be meaningfully placed by our simulation.

For example if one object is 10 feet away from us and another 30 feet in a straight line then our brain can correctly compute the objects are 20 feet apart. So spacetime is the consistent logico-mathematical structure that emerges at the aggregate level of dimensional relationships rather than an observable physical structure.

If spacetime is actually a logico-mathematical construct then it needn't exist as the physical entity science assumes, and can just as easily be an information structure that emerges from aggregates of dimensional relationships. So the apparently physical spacetime of science could just be the overall mathematical structure of how dimensional relationships emerge at aggregate scales. There is no way to demonstrate this isn't true. And a logico-mathematical structure rather than a physical one is all we need for science to keep working as it always did.

General relativity has already gone part way to this understanding. General relativity imagines no single universal space that is valid for all observers. It conceives of space in terms of *manifolds*, which are views of curved space from the perspectives of individual observers (Wikipedia, Manifold). And relativity tells us that there is no single curved background space that all individual manifolds exactly map to. The spacetime manifold of every relativistic observer can theoretically be different.

Though relativity conceives of manifolds as *views* of spacetime rather than actual different spacetimes, relativistic manifolds are inherently inconsistent with each other, which casts considerable doubt on the concept of a single universal spacetime. Certainly a single universal spacetime that is the same for all observers doesn't exist in general relativity.

Of course general relativity can model single cosmological spaces quite effectively, but only when the views of individual observers are artificially ignored and other generalizations are made (Wikipedia, Friedmann–Lemaître–Robertson–Walker metric). So we can very reasonably conclude that what general relativity is really telling us is that there isn't a single universal spacetime valid for all observers. In any case it's clear that a computational spacetime consisting of multiple independent spacetime fragments can be consistent with general relativity.

In view of this evidence it is clear that the single universal spacetime background that science assumes is one more questionable *interpretation* of science, rather than an observable fact of science.

Universal Reality has no difficulty in dealing with the lack of a single universal background spacetime. In Universal Reality spacetime is not a single separate pre-existing container *for* physical events but is the dimensional information we already know is computed *by* events.

Taking spacetime as a logico-mathematical structure that emerges from events rather than a pre-existing container for events solves a number of important conceptual problems and doesn't seem to diminish the applicability of science to explain and predict natural phenomena.

HAPPENING & THE P-TIME PROCESSOR

Universal Reality proposes a computational universe in which a relatively simple elemental program is executed in a manner that automatically generates a universe that obeys the rules of both general relativity and quantum theory. The key to this unity is that both material structures and spacetime are computed together as a single integrated data structure. This model is based on trial implementations of model universes using the XOJO programming system on a Power Mac.

The universe consists solely of the data of particles and their particle components, and the program that computes them. It executes in a computational space identified with the quantum vacuum to update all the data of the universe at every P-time tick, each of which recomputes the complete data state of the observable universe in a current universal present moment.

Happening animates the universe and drives its temporal evolution. Happening operates as a universal processor, which is an innate aspect of the quantum vacuum, the medium or substrate of existence in which all the data that constitutes the universe exists.

The data states of all processes in the universe are computed simultaneously at every P-time tick because all the data of the universe exists simultaneously in the quantum vacuum whose happening is its processor. Dimensional spacetime including the computations of all local clock time rates are computed simultaneously in the current P-time present moment by the processor of happening for all processes in the observable universe.

In the XOJO model this is done in a clock event executed for loop on all particles, but in the actual universe the data of all particles is computed simultaneously because all data exists simultaneously in the processor in computational space. The reasons for identifying computational space with the quantum vacuum and its details have been explained in previous chapters.

In every P-time tick there is an application of the processor to recompute every separate process in the observable universe. All the applications of the processor occur simultaneously in the current P-time tick so that all processes in the universe are computed simultaneously. Each processor application contains sufficient processor cycles to fully recompute each process in the universe. There are a fixed number of processor cycles allocated between computing velocities in space and velocities in time so that the combined spacetime velocity of all processes is always the speed of light, c.

THE UNIVERSAL REFERENCE BACKGROUND

The computational space of reality is non-dimensional in the same sense that computer programs define non-dimensional computational spaces. The computations themselves are observable only through their effects on the data that makes up the observable universe. All data, dimensional and structural, is computed within the computational space of the quantum vacuum. This data is then interpreted as a physical universe of material structures in a physical spacetime in observers' simulations of reality.

Computational space and P-time together are a preferred unobservable background of purely numeric (dimensional data stored are numbers) structure and dimensionality within which all observable dimensional frames are computed consistently. This preferred computational frame neatly solves the problem of what absolute rotation (for example Newton's bucket) is with respect to and what actual world lines are with respect to. They are both with respect to the preferred background frame in which they are actually computed. This is in distinction to the purely observational effects of mutual relative motion, which vanish without lasting effect when the motion stops.

Observers observe the universe in terms of measurements relative to their own frames. All observable measurements ultimately reduce to particle interactions and only particle interactions produce observable data. The actually computed values are with respect to the background frame in which they are computed and are non-observable. Observable values are generated by particle interactions and are always observed with respect to observer frames.

The relativistic effects of motion through space are experimentally confirmed and widely used to correctly calculate the trajectories of bodies in space, but there is a deeper mystery at the heart of relativity that hasn't been solved until now. Namely what is actual spatial motion relative to? There is nothing in relativity theory itself that explains this because relativity claims that all frames are equivalent and none is preferred over any other. This is why it's called relativity and is considered a basic principle. But this is incorrect.

In the twin example why is motion with respect to the earth actual, and produces actual agreed effects, but the motion of earth with respect to the traveling spacecraft isn't and doesn't? The equations of relativity provide no answer, and this has been an unsolved dilemma for over a century since relativity first appeared.

Thus there must be a fundamental assumption in relativity itself that goes largely unrecognized and is even actively denied. There has to be a single absolute fixed background with respect to which actual as opposed to purely observational relativistic effects occur. But the whole original idea underlying relativity was that all motion is relative. So if all motion is relative how can its effects only be actual with respect to some absolute notion of spacetime that relativity can't even properly define? This is an important problem that clearly requires a solution.

This problem arises both in determining the actual spatial lengths of world lines and the actual time dilation of clocks traversing them, and in determining what actual rotation is relative to as explained in the following section.

There is clearly some absolute reference with respect to which actual spatial motion and actual rotational motion are relative to but what is it? Relativity tells us that all coordinate systems are equally valid so why couldn't we pick a coordinate system moving along with the traveling twin's ship and have the clocks back on earth actually slow down rather than the clocks on the spaceship?

There has to be an absolute frame with respect to which actual motion occurs and purely relative motion doesn't. If there wasn't logical contradictions would occur when space travelers meet because they would both see each other's clocks still ticking at different rates standing right next to each other.

This would lead to all sorts of problems with the laws of physics. Which clock rate would actually describe which laws of physics at that location? The twins would both age at different rates in front of each other's eyes and physical processes could run at wildly different rates in the same location with disastrous results. Thus some sort of absolute spacetime is required to maintain the logical consistency of the computations of the laws of nature and keep the universe from tearing itself apart.

As Ernst Mach pointed out in the case of rotation it appears this absolute frame is more or less aligned with respect to the total mass of the universe and this is roughly in accord with observational results. But why? This is a fundamental question few physicists have attempted to understand so it's usually just ignored if even recognized.

More recently the cosmic microwave background (CMB)

radiation is being used a reference frame for cosmological motion but the fundamental problem is still the same. Preferred frames have crept back into modern physics over the protests of strict relativists, even though the very concept is antithetical to the originating concept of relativity.

It's also difficult to understand how a uniform effect throughout the universe could be the result of the obviously non-uniform distribution of mass in the universe at galactic scales. How could the total average mass of the entire universe cause the effect and not the much nearer distribution of mass in our galaxy or even our solar system? It doesn't seem to make sense.

However the reason for Newton's bucket and the privileged alignment of spacetime with the aggregate mass of the universe is a natural consequence of how spacetime emerges computationally in Universal Reality.

The reason there is a privileged background for both effects is because spacetime is created by quantum events as explained in the chapter on Quantum Reality. At the largest scale the interactions of all particles in the universe form a universal network of entangled dimensional relationships with respect to which the dimensional relationships of all subsequent events naturally align as they are computed with respect to it.

This network of dimensional relationships is a consistent logico-mathematical framework with respect to which the dimensional relationships of subsequent particle events automatically align. This is the solution to Newton's bucket and the reason why world lines with respect to the aggregate mass of the universe is approximately the correct choice that produces actual as opposed to observational relativistic effects.

Relativistic events take place with respect to a more or less absolute spacetime background because they are computed with respect to the dimensional entanglement network. So instead of being an empty *physical* structure, spacetime is instead a *logico-mathematical* structure in which subsequent computations automatically align with the ones from which they are computed. The mass distribution of the universe is an aspect of the entanglement network so subsequent dimensional computations occur with respect to it.

Thus the dimensionality of the entanglement network becomes an absolute reference background with respect to which actual as opposed to merely observational relativistic effects are relative. Without this absolute reference background there could be no notion of actual as opposed to observational relative linear or rotational motion and relativity itself would not be consistent. Effects such as the twin paradox and Newton's Bucket would lead to contradictions and the observable universe would likely cease to exist.

Since the absolute reference background is continually recomputed at the local level there could be small inconsistencies from location to location and from era to era across the universe, which might produce slight anomalies in the rotations of gyroscopes or the expected time dilation of long space flights or other relativistic effects. See the upcoming section on Dimensional Drift for more on this.

So the correct definition of absolute rotation or linear motion is with respect to the dimensional consistency of the *proximate* absolute reference background, the dimensional alignment of the entanglement network. In general this will be very closely through not necessarily exactly aligned with the total mass distribution of the universe.

The absolute reference background may have other implications. Given their intimate connection with spacetime could the values of c and G, the speed of light and the gravitational constant, be somehow a function of the total mass of the universe including that beyond the particle horizon and could the size of the observable universe then be determined from those values? Could c and G both be emergent effects of the dimensional entanglement network rather than intrinsic constants of the complete fine-tuning?

NEWTON'S BUCKET & MACH'S PRINCIPLE

This universal reference background is the key to understanding two fundamental problems of relativity. It provides the solution to the problem of Newton's Bucket and the question of what relativistic world lines are relative to.

When a bucket of water is rotated the water begins to rotate and climb the walls due to centrifugal force. This is easy to understand but there is a hidden mystery involved. What is the water rotating with respect to? A bucket of water that is not rotating has a flat surface, but what is the water still with respect to in that case (Wikipedia, Bucket argument)?

At first we might suspect the rotation is with respect to the surface of the earth but that isn't true because we have the same effect in the rotation of gyroscopes in deep space. They are either rotating or they aren't, and they always have a rotation that must be relative to something but there is no known physical mechanism in science to explain what.

It's clear there is some absolute reference spacetime that more or less aligns with the total mass of the universe, and the rotation of the bucket is with respect to that, but there is no law of physics that specifies why actual rotation is always with respect to the CMB or total mass of the universe rather than any other coordinate system.

Ernst Mach proposed that the rotation was with respect to the distribution of inertial mass of the universe and was some as yet unknown effect of gravitation but he could offer no scientific reason for it, and there isn't any law of science that provides any reason why this should be true (Wikipedia, Mach's principle). And just making up such an important fundamental law that has no connection with any other law and has no other apparent effects is clearly unjustified.

Once again the explanation comes naturally to Universal Reality from its concept of an absolute reference background in which the universe is actually computed. It seems quite obvious that actual rotation must be relative to the established consistency of the frame in which it's computed, as opposed to any observer frame. If for example an observer is riding a rotating carousel in empty space dark the carousel appears to be motionless but the presence of centrifugal force proves it isn't.

So the solution to Newton's bucket is simple. Actual as opposed to observational relative motion is always with respect to the dimensional consistency of the universal computational space in which all motion is computed.

The quantum of rotation is spin. Spins are the elemental units of rotation and angular momentum. Thus as dimensionality is computed the resulting network of entanglement relationships gains an absolute dimensional *orientation* due to the progressive alignment of the spin orientations of all non-zero spin particles with respect to the computational background. As a result the dimensional network of the observable universe becomes a common absolute orientation reference as it's computed. All rotation is relative to it because all successive computations are computed from previous ones in terms of them.

This doesn't mean the axes of spin orientations are all pointing in the same direction, but that they are all pointing in *some* direction relative to a common reference standard of orientation across the entanglement network. If there was no absolute reference standard for orientation rotations couldn't even be compared.

So the existence of spin is another little necessary ingredient of reality because in aggregate it generates a universal absolute reference orientation standard with respect to which all rotation is relative. So it may be only the existence of the spin particle component that in aggregate builds a universe in which rotation makes consistent sense.

Thus Universal Reality's computational approach to spacetime provides a solution to the problems of Newton's bucket, what world lines are relative to, and why there is an absolute underlying frame in spacetime. And it does this in a manner that leads to a unification of relativity and quantum as explained below.

This same approach also resolves the apparently paradoxical nature of quantum processes, as we will see in the chapter on Quantum Reality. The universal reference background that emerges computationally from quantum events provides a dimensional model that explains general relativity but is fuzzy enough to accommodate quantum effects as well.

DIMENSIONAL DRIFT

Dimensional drift is the hypothesis that the absolute background relative to which actual relativistic effects occur may not be completely consistent from location to location because it's computed locally by quantum processes that are inherently dimensionally fuzzy. It also clearly changes over time with the redistribution of particles in the universe due especially to the expansion of space and this could also be a source of anomalies in its consistency from one time to another.

Thus at very large scales the exact dimensional parameters of the universal background frame may not be exactly cross-consistent or correctly known from earth. Attributes such as scale, reference motionlessness, and orientation of the absolute background could theoretically vary from one location in space to another. So it's possible we might observe unexpected anomalies in relativistic measurements from one location to another that could provide a conclusive and falsifiable test of Universal Reality versus general relativity.

However such anomalies might be difficult to detect because our dimensional measurements could be subject to the same anomalies depending on their nature. However if one location of the absolute background was either stretched or moving slightly with respect to another and a space probe went from one location to the other its signals might indicate unexplained relativistic effects.

Such small anomalies have in fact been actually detected. For example the two Pioneer spacecraft seem to be slowing slightly more than relativity predicts as they leave the solar system. While the currently accepted explanation of the slowing is the thermal recoil force from onboard generators this doesn't seem to explain small variations in the effect and there have been a number of other explanations proposed (Wikipedia, Pioneer anomaly).

Since the dimensionality of the observable universe is computed locally at the particle scale it's certainly reasonable to assume the alignment of the absolute background with the average mass distribution of the universe is only approximate and may vary slightly from location to location. The total average mass of the universe should provide the same absolute reference frame everywhere in the universe, but the absolute background could vary slightly from area to area. What should be definitive is the logico-mathematical consistency of the entanglement network whose dimensionality is the absolute background at the local scale. This is the reference with respect to which spatial motion actually occurs, and that may or may not be in exact alignment with the total mass of the universe or the CMB.

The absolute reference background is not a fixed pre-existing spacetime container, nor even a fixed data structure, but is the aggregate logico-mathematical consistency of all individual dimensional events as they are continually recomputed over time. Thus the consistency persists but the dimensionality that supports it clearly evolves over time as the universe is recomputed and expands. The current Hubble expansion is very slow, however in the inflationary period the expansion seems to have been enormous and near instantaneous and its dimensional effects might persist long after.

The first particle events began occurring with the big bang and inflation and since these earliest events began consistently the logico-mathematical background of their aggregate consistency immediately emerged as an implicit information structure relative to which the dimensionality of all subsequent events could be said to occur.

Thus the standard reference background of computational spacetime with respect to which absolute linear motion and rotation are relative in the sense of producing actual relativistic effects is constructed piecewise by individual quantum events whose dimensionality is inherently fuzzy.

So relativistic dimensionality should have manifested from the beginning or at least almost from the beginning without any problems, however the overall reference consistency it's relative to has clearly changed over time with the dimensional evolution of the universe.

The absolute reference background reflects the aggregate dimensional consistency among all events in the observable universe, but the distribution of events continually changes over time and varies across the universe. On average the dimensional background remains the same but it's possible that some local differences in the background might sometimes develop. The overall homogeneity of the dimensional background could be subject to local distortions and dimensional drift.

As space expands and the distribution of galaxies changes there could be some measurable drift of the background dimensionality with respect to which absolute motion along world lines occurs. This might be detectable as anomalous astronomical effects in particular with respect to absolute linear and rotational motion.

This is something Universal Reality predicts could occur, and if it does would provide good evidence for Universal Reality's computational theory of spacetime and an absolute background reference dimensionality with respect to which actual motion occurs.

If in fact there are areas with sparse enough or different enough computational connections that have slightly different dimensional background references there is surely a computational process that reconciles them when they intersect. Only in this manner could a consistent computational universe be maintained. But of course the quantum vacuum already has a mechanism to create and choose among probability distributions via the process of random choice so it's to be expected that it must have some general computational self-correcting mechanisms to enforce its overall logico-mathematical consistency.

Most likely any such anomalies on the cosmic scale are currently misinterpreted as either slight inconsistencies of measurement or incorrect dimensional measurements. But if interconnected effects were to be compared and inconsistencies discovered this type of error might be detectable.

Another possible source of dimensional drift may be due to the granularity of dimensionality in a digital universe. The universe should be granular and digital at its minimum scale because only exact digital data can be consistently computed. But this precludes infinite precision in dimensional values and their calculations. Any inconsistencies here are presumably reconciled by taking numeric averages at the finest scales. However over time rounding errors could conceivably accumulate to dimensional discrepancies at observable scales even though the scale of granularity is likely many orders of magnitude below that of measurability.

It is not altogether clear the ways dimensional drift might show up. A uniform expansion or contraction of the background wouldn't be observable but we see the past as well as the present so if it changed significantly over time observable discrepancies in relativistic measurements might be detectable between different eras in time.

It is clear from relativity itself there is a preferred background frame with respect to which actual relativistic effects occur. This frame is established by the logico-mathematical consistency of all computations of the entanglement network, and thus closely aligned with the distribution of matter in the universe as Planck surmised. However all individual events are local so absolute motion should be with respect to local areas of the overall logico-mathematical consistency. Ultimately any discrepancies of consistency should depend on variations in the flux density of particle interactions over time across the universe.

The big question is to what extent this background reference and its actual relativistic effects is local to the mass-energy distribution of our galaxy or perhaps even to some extent our solar system rather than to the average distribution of all mass-energy in the universe. This likely depends on the density and connectedness of events across the universe relative to what degree mass-energy distributions have changed over time. This should lead to testable predictions.

Thus if an observer is stationary with respect to at least the local consistency of the logico-mathematical background as human observers are for most practical purposes the relativistic effects they observe are actual with slight corrections for their own relative motion with respect to it.

An absolute background and dimensional drift are clearly testable proposals. If we find the absolute background reference dimensionality against which actual world lines are measured is slightly inconsistent across different regions this would be strong evidence for Universal Reality's model of reality.

UNIFYING RELATIVITY & QUANTUM THEORY

A major goal of Universal Reality is to outline a conceptual model that unifies quantum theory and general relativity. The fundamental inconsistency between quantum theory and general relativity derives from their different concepts of spacetime. Quantum theory considers spacetime a fixed pre-existing container for events that remains unmodified by events. However the spacetime of general relativity is dynamic and is curved by the presence and movements of mass-energy even though it's also a pre-existing container for events.

Thus it's clear that to unify the two theories their divergent views of spacetime must be reconciled. The key to achieving this is to recognize how quantum events actually create dimensionality so that the dynamic spacetime of general relativity naturally emerges from them. In this model spacetime is the dimensional relationships among events, and is computed simultaneously with the events.

This approach is suggested by several lines of reasoning, two of particular importance. First when we carefully consider the nature of spacetime we find it reduces to the dimensional consistency among measurements. In other words we never actually observe empty space but only the dimensional relationships among mass-energy structures. All we actually observe is particle-based events with dimensional values. This includes the measurements our senses make as we experience the world. Thus what we think of as spacetime is actually the logico-mathematical consistency among dimensional measurements and there is no observational evidence that empty space as an encompassing contain for events even exists.

Second quantum events conserve mass-energy and due to the non-proportionality of particle masses energy excess energies must be expressed in terms of velocities of emitted particles. Thus the non-proportionality of particle masses requires the creation of dimensionality for events to occur. Particle events must create dimensional spacetime to be able to occur so total mass-energy can be conserved. The conservation of mass-energy in quantum events must be expressed in terms of velocities and their resulting positions, which are defining elements of spacetime. Thus by creating interrelated positions and velocities vast chains of particle events create the network of dimensionality observers interpret as spacetime.

Taking these points together in the context of a computational universe in which everything is ultimately numeric data, suggests that quantum events are actually creating spacetime on the fly as a mechanism to conserve mass-energy.

And only if spacetime is dynamically created by events rather than being a pre-existing container for events, can the dynamic spacetime of general relativity be generated by quantum events. So this is a very promising approach.

This enables the wavefunctions, uncertainty, and vacuum energy fluctuations of quantum theory to be reinterpreted as descriptions of spacetime in the process of creation rather than descriptions of fuzzy particles with respect to a fixed pre-existing infinitely exact spacetime. The quantum equations, in so far as they are accurate, remain the same but their interpretation is turned on its head.

In this model quantum fuzziness or indeterminacy isn't the innate nature of particles but the innate nature of spacetime as its being computed. This in turn implies that particles and particle events could be fairly simple arithmetic data, and the complexities of quantum theory emerge only as dimensionality is computed. Note that this view is consistent with the underlying trajectories and time evolution of wavefunctions, which are exact. This implies a more fundamental exact data description on which dimensional fuzziness is overlaid.

So if reality might be numerically exact at the fundamental level, why is dimensionality fuzzy at the quantum level? What is the mechanism that produces quantum indeterminacy when dimensionality is computed?

It turns out the manner in which the processor computes the data of the universe provides a reasonable mechanism for both the STc Principle that underlies relativity and quantum indeterminacy. The allocation of a fixed number of processor cycles between calculations of spatial velocity and time velocity is the source of the STc Principle, and fine scale processor cycle oscillations between space and time can account for quantum fuzziness.

PROCESSOR CYCLES & THE STc PRINCIPLE

In every P-time tick the processor of happening runs the fundamental program against each coherent data state in the universe to compute a process. Each process creates an event in each P-time tick. Particle interaction events produce potentially observable values but most events are virtual and merely compute the unobserved evolution of particle data. The elemental program analyzes each data state to determine which of its subroutines is required to compute the current process and branches to that subroutine.

A separate *application* of the elemental program computes each separate process in every P-time tick. Each separate process is computed by many processor cycles in each P-time tick to produce its next current present moment data state. A fixed number of processor cycles (perhaps more accurately sets of cycles but for convenience referred to as cycles) in every P-time tick are allocated to compute the total spacetime velocity of every process. This fixed number of cycles is allocated between computing the velocity in space and velocity in time of each individual process. The fixed number of velocity allocated processor cycles in each P-time tick sets the value of c, the speed of light.

These velocity allocated processor cycles are the computational source of the STc Principle; that everything in the universe continually moves though combined space and time at the speed of light. More precisely the vector sum of the space and time velocities of everything in the universe is always c as required by relativity. Thus the fixed cycle rate of the universal processor is the source of the value of c and its allocation between space and time velocities the source of the STc Principle that underlies the relativistic nature of reality.

The total spatial velocity of a process, its linear velocity plus the intrinsic velocity of any fields, is computed first, and the cycles left over then compute the internal evolution of the process. The rate of internal evolution of a process manifests as how fast the process is happening, which is observationally its proper clock time rate. Fewer processor cycles left over to compute the evolution of internal processes produces a slower clock time rate for the process. This is the computational source of clock time and its relativistic time dilation.

The processor of happening computes all local clock time rates process by process simultaneously in each universal P-time tick. Thus all the different local clock time rates in the universe are computed simultaneously in the current universal present moment common to the entire universe.

The processor cycle rate can have no explicit rate itself because it's the source of all relativistic clock time rates and can only be measured in terms of the clock time rates it produces. The processor computes all processes in the universe to have a fixed speed of light velocity. This total velocity is distributed between their spatial velocity (linear velocity plus intrinsic velocity of any fields) and the internal rate at which they evolve which is its clock time rate, its velocity in time.

PROCESSOR CYCLES & QUANTUM RANDOMNESS

This computational model in which the universal processor allocates a fixed number of cycles to computing velocity in space and velocity in time also explains wavefunctions and other aspects of quantum indeterminacy in a straightforward manner.

Assume the allocation of processor cycles between space and time computations at the quantum level is not precise but subject to very fine random oscillations. Effectively the dimensional background in terms of which dimensional values are computed is not fixed but continually vibrating. The background rulers and clocks used to compute dimensional values are continually oscillating and exchanging identities during each P-time tick.

As a result dimensional values involving space and time related variables (for example energy and time, or position and momentum) are computed in terms of a fluctuating standard and are always mutually uncertain by a minimum amount described by the Uncertainty Principle. So the intrinsic indeterminacy of space versus time at the quantum scale as dimensional values are computed is the source of the Uncertainty Principle.

Modern physics derives the Uncertainty Principle from the zero-point energy fluctuations of the quantum vacuum and this is consistent with our processor cycle model (Wikipedia, Uncertainty principle). The quantum vacuum is a direct manifestation of the virtual reference background for spatial dimensionality. The zero-point energy fluctuations are the direct manifestation of the processor cycle oscillations between space and time.

As a result each point of the quantum vacuum appears to continually oscillate between energy (velocity in space) and time. Energy excitations (velocities) in space continually appear as virtual particles but on time scales (velocities) so short they vanish in most cases before they can fully actualize. As a result the quantum vacuum is a fluctuating sea of transient virtual particles with an average energy determined by the spatial aspect (velocities) of the processor cycle oscillations.

These same processor oscillations can also explain why quantum particles appear to behave as wavefunctions. Whenever the exactly conserved energies of particles is translated into dimensional values of positions and velocities the processor cycle oscillations give them indeterminate values in a manner that evolves according to the Schrödinger equation.

The processor cycle oscillations for each separate process are independently computed so that the processor imposes oscillations of different phases on each process scaled by the intrinsic velocities of each process, such as the internal vibrational velocities of mass, and the nominal trajectories and velocities of the underlying numeric particle modeled by the wavefunction. As a result each process develops a scaled indeterminacy around its nominal trajectory that evolves along its nominal trajectory and this indeterminacy takes the mathematical form of a wavefunction. The overall velocity and trajectory of the wavefunction is calculated by the apportionment of processor cycles, while the indeterminate waveform of the wavefunction is calculated by processor cycle space versus time oscillations.

Thus the universe at the quantum scale is accurately described by wavefunctions that are mistakenly thought to encode the dimensional indeterminacy of particles with respect to a fixed exact spacetime background. Modern physics interprets wavefunctions as particles fuzzy with respect to a fixed spacetime background. However Universal Reality interprets wavefunctions as fragments of dimensionality inherently fuzzy with respect to each other due to the manner in which processor cycles compute dimensional values.

This new interpretation of wavefunctions as fluctuations of spacetime itself as it's computed rather than particles smeared out in a fixed preexisting spacetime frees us from the model of spacetime as a fixed container for events. This in turn enables spacetime to be seen as emerging from the dimensionality of mass-energy structures and computed with them as a single unified structure. What we interpret as a dimensional spacetime now consists of dimensional relationships among particles that are computed by quantum events. This enables a single unified process to compute a universe that obeys the laws of both quantum theory and general relativity, and mass-energy structures and their dimensional relationships as a unity as well.

Like relativity all types of quantum indeterminacy are a natural result of the allocation of processor cycles in our computational model. As we will soon see both relativity and quantum theory now emerge automatically and consistently from this single computational model, general relativity at large scales, and quantum phenomena at fine scales. This model also explains the underlying phenomena of quantum theory in a completely non-paradoxical manner.

FOUR COMPUTATIONAL LEVELS

Our proposed model of the universe can be better understood in terms of four computational levels. Each level is a perspective on the unified mass-energy spacetime universe as it's computed. This four level view is conceptually useful though everything actually exists as a single unified process.

0. At Level 0 the universe consists entirely of the exact numeric data of all particle components and particles being computed by applications of an elemental program that consists of fairly simple logical and arithmetic operations. These computations take place in the universal processor of the quantum vacuum in accordance with the virtual data of the complete fine-tuning. This level is not directly observable.

Every particle interaction, including particle field interactions, is computed as an individual process. There is no physical universe and no actual physical motion or dimensionality or mass-energy structures at this level. Everything exists as numeric and relational data states whether dimensional or non-dimensional. The actual fundamental universe is computational and completely non-physical and non-material.

All the data that constitutes the universe exists simultaneously in the quantum vacuum, which acts as a universal processor that continually computes its current state from its previous state. Each universal computation manifests a current universal present moment P-time common to all processes. This universal computational space and P-time establishes an absolute reference background for all the dimensionality and mass-energy structures of the subsequent levels.

At every P-time tick a separate application of the single fundamental program uses multiple processor cycles to recompute each coherent process in the universe. Each P-time tick manifests as the current universal present moment of the universe in which all clock times, mass-energy states, and their dimensional relationships are computed individually on a process-by-process basis.

An important consequence of this model is that nature doesn't have to store data as equations as it would have to do if wavefunctions and dimensional fragments were actual data components of the universe. Everything is now stored and computed as its exact Level 0 data values. Thus the observable universe can consist only of exact data values without any equations necessary. This greatly simplifies a computational universe and it's reasonable to assume nature strives for simplicity.

1. Level 1 is the quantum universe that physics describes with wavefunctions, the Schrödinger equation, and other quantum equations. The quantum wavefunction picture of reality is due to the processor allocating the cycles it uses to compute velocity in space and velocity in time in a manner that oscillates between space and time at the quantum scale. This results in a spacetime indeterminacy of quantum processes that manifests as wave-like probability distributions of space versus time of complementary dimensional variables such as position and velocity (momentum).

As a result whenever any dimensional value is computed it's inherently uncertain with respect to its space and time dimensionality at quantum scales. In effect the space versus time rulers with respect to which events are measured are continually oscillating back and forth between space and time.

These processor cycle oscillations have a number of important effects with respect to space and time dimensionality. They are responsible for the apparent wave nature of particles, and for the Uncertainty Principle and the zero-point energy fluctuations of the quantum vacuum from which it derives. The processor cycles themselves are also the source of the vibrational velocity that characterizes all forms of mass and energy. In other words they are an analogue of the Higgs field as explained in the next section on the equivalence of mass-energy and space.

2. Level 2 is the observable universe. It consists of the totality of all measurements of observables. Measurement of observables includes the decoherence values of all particle interactions which are effectively mutual measurements by particles of each other's dimensionality. All scientific measurements and observations ultimately reduce to individual particle decoherences. Level 2 also includes all types of sensory observations, which also ultimately reduce to particle decoherences.

Note that while wavefunctions and the wave nature of particles are useful mathematical structures they are not directly observable and thus not part of the observable universe. For example the observables in the double slit experiment are individual point particle impacts on the back screen (Wikipedia, Double-slit experiment). The waves that produce them are never directly *observed* but only *inferred* from the pattern of the point particle impacts. Thus the wave nature of particles is inferred from patterns of particles behaving as particles and is not an actually observable phenomenon.

The observable universe consists of an entanglement network of observable values produced by the total interconnected network of particle events. The entanglement network is produced by the conservation of particle components through all particle events. This produces consistent relationships among the particle component values of particles emitted by events, and by extension of all particles in the universe since all observable particle values are backwards connected by events through the entanglement network into the single consistent structure of the observable universe.

The entanglement network is the combined spacetime mass-energy data structure of the observable universe. It incorporates the combined data of all mass-energy structures and their dimensionality in a single consistent logico-mathematical structure. By a consistent logico-mathematical structure we mean that the relationships among all observables obey the consistent logico-mathematical laws of physics insofar as they are known.

General relativity is incorporated into the quantum entanglement network with the addition of two elements. First the dimensionality that emerges from quantum measurements is scaled by the velocity equivalents of mass-energy. The equivalence of spatial velocity and mass-energy is explained above.

Second the allocation of a fixed number of processor cycles per P-time tick between computing velocity in space and velocity in time results in time dilation (slower clock time velocity) with increased spatial velocity in accordance with general relativity even for quantum processes. When all forms of mass-energy are taken as various types of spatial velocity proper clock time rates are computed locally for each process simultaneously and the curved spacetime (velocity density in our model) of general relativity is automatically produced.

So the incorporation of general relativity into the entanglement network of quantum measurements derives from two principles. First treating all forms of mass-energy as various forms of spatial velocity, and second the allocation of a fixed number of processor cycles to computing first velocity in space and then allocating the left over cycles to computing velocity in time. When gravitational mass fields are modeled as fields of vibrational velocity centered on particles this correctly produces the dynamic spacetime of general relativity.

Thus in this model dimensional spacetime is computed together with the dimensionality of mass-energy structures as an integrated system. By abandoning the notion of spacetime as a preexisting empty container for events and letting quantum events compute it we arrive at a spacetime compatible with both quantum theory and general relativity.

So a simple computational process combined with a processor that allocates a fixed number of cycles between computing space and time is sufficient to compute the entire observable universe as all complexity emerges automatically from the complete fine-tuning as the fundamental program computes the interactions of elementary particles and their particle components.

Level 2 is still not a physical universe. It's the observable universe of scientific measurement that consists only of the consistent universal network of all values of individual observables. It does however consistently include all quantum observables in an entanglement network that implicitly incorporates the curved spacetime of general relativity.

3. The 'physical' universe. The apparently physical universe in which we seem to experience our existence is not actually physical or material at all. Its apparent physicality is our mind's interpretation and generalization of the logico-mathematical consistency of the observable universe of Level 2. Spacetime is a basic aspect of the simulation of reality produced by the human brain and projected outward around us. It's a reification of the data structure of the observable universe of Level 2. Basically the dimensional values of all observables are laid out in the simulation and sorted by their values and an imaginary empty spacetime is interpolated around them to connect them and mistaken for an actual physical entity.

We experience our simulation of the universe as a physical spacetime filled with material objects but this is a convenient illusion. It's a highly successful evolutionary adaptation that greatly simplifies our mental computations of reality that is shared among humans and in various forms by other species. This is explained in detail in the chapter on The Simulated Universe.

THE EQUIVALENCE OF MASS-ENERGY & SPACE

Universal Reality models all forms of mass-energy as different types of relative motion (vibrational, waveform, or linear) whose values are their spatial velocities. Only if all types of mass-energy are forms of the same thing can they be converted into each other and conserved. Energy is conserved in all cases as equivalent amounts of velocity are converted from one form to another.

Further space and all types of mass-energy are forms of the same underlying entity, velocity (near null velocity in the case of empty flat space). Kinetic energy is linear velocity, mass is vibrational velocity, and photon energies are wave frequencies. Empty space consists of the fluctuation velocity of the quantum vacuum, which effectively opens space up and gives it expanse. The various charges of particles including mass are equivalent to localized vibrational fields in space, and space is the underlying substrate or raw stuff of particle masses. Quantum field theory (QFT) also treats particles as local field excitations (Wikipedia, Quantum field theory).

Particle mass is modeled as a fine vibrational velocity that adds velocity to its surrounding dimensionality in the form of a velocity field centered on the particle. Thus gravitational fields are areas of increased spatial velocity surrounding masses that fall off by the square of the radius due to the 3-dimensional geometry of space. Particle charges including mass are not points but diffuse spatial fields of vibrational velocity centered on the nominal location of their point particles. Space itself is the underlying stuff of which charges including mass are made. Space itself is a single universal field of relative motion, of spatial velocity. Energetic charges are just fields of the same stuff as space with increased localized spatial velocity.

Thus space and spatial velocity are aspects of the same thing. Velocities are energy excitations in space and empty space is the zero-point energy of the quantum vacuum. Particles are localized excitations of the quantum vacuum, of space. This is demonstrated by the Unruh effect in which the acceleration of an observer in empty space manifests as the appearance of particles in surrounding space (Wikipedia, Unruh effect).

While the Unruh effect is an observational effect due to the relative velocity of an observer, it does confirm the equivalence of space and mass-energy velocity that can manifest as particles when instantiated in valid particle component sets. Actual particles are combinations of relative velocity of various forms in space corresponding to the type of charges they carry. Charge fields are fields of various forms of vibrational velocity in surrounding space that add intrinsic spatial velocity to points in the field which reduces the velocity of time in accordance with the STc Principle. Thus other particles passing through these velocity fields experience the time dilation effects of general relativity.

The concentration of sufficient energy (velocity) at points in the quantum vacuum (space) causes actual particles to appear. A particle – antiparticle pair is created if the velocity energy is equal or greater than the mass of the particle pair created so there is sufficient energy velocity to be converted into mass velocity. The concentrated velocity energy is converted into the mass of the new particles with any left over converted into the kinetic energy of linear velocity of the particles. When particle-antiparticle pairs appear all other particle components are conserved since antiparticles have opposite particle component values to regular particles so they all sum to zero. Thus the quantum vacuum is a reservoir of the particle components necessary to make particles when sufficient spatial velocity (energy) is added.

This is an example of the equivalence of mass-energy and space that demonstrates they are aspects of the same thing. All mass-energy is just space with relative velocity of one type or another, and empty space is the diffuse medium of virtual particles from which actual individual particles can be formed.

Fields are aspects of particle charges (including mass as the gravitational charge), rather than separate entities. The spatial velocities of charges are not points but fields centered on the location of the nominal point charge. Due to the fact that particles are energetic excitations in space (velocities) the fields of the four forces can equivalently be modeled as particle exchanges. For example modern particle physics models electromagnetic fields as exchanges of virtual photons (Wikipedia, Quantum electrodynamics).

The equivalence of mass-energy and space when all forms of mass-energy are treated as different forms of spatial velocity takes us a considerable distance towards a new interpretation and better understanding of general relativity.

If particles are not actually points but velocity fields in space centered on points this eliminates the problem of how mass curves space. There is no more mysterious 'action at a distance' in relativity in which mass somehow curves space distant from it. Every massive particle extends out through space to the limit of its effects. Particles are velocity effects in space centered on localized points of particle component sets.

A NEW MODEL OF GRAVITATION

In Universal Reality all relativistic effects derive from the equivalence of mass-energy and spatial velocity, and the fact that the vector sum of space and time velocities is always c. This enables us to model combined relativistic mass-energy and spacetime in an entirely new and easy to understand manner. This model is much superior to the usual curved space model of general relativity because it reflects spacetime as Euclidean as we actually observe it and seamlessly incorporates spacetime and mass-energy in a unified structure wholly equivalent to the curved space model of general relativity.

In Universal Reality masses, like all charges, are very fine localized vibrations or excitations of space itself. Thus masses are vibrational spatial velocities and that's all they are. The gravitational fields of masses are very fine vibrations in space surrounding to location of their particle component sets. Charges and fields are not separate entities but inseparable aspects of charges. Thus a mass is a field of very fine vibrations in space centered on the point of spatial vibration where the nominal particle mass is located.

Note that representing charges as ultra fine spatial vibrations is reasonably consistent with String Theory's representation of elementary particles as vibrating strings though in Universal Reality it's charges rather than particles that are vibrations, and they are different vibrational forms in 3-dimensional space rather than extra compacted dimensions (Wikipedia, String theory).

Because the field itself consists of vibrations in the fabric of space the field increases the intrinsic velocity density of space within it. It adds intrinsic spatial velocity to all points within the field and by the STc Principle this automatically slows time so that total spacetime velocity remains equal to the speed of light, c.

This is a simple and easy to understand model of how gravitational fields produce time dilation. They slow time because they actually are fields of intrinsic spatial velocity. Thus the total time dilation and other relativistic effects of any process is now just a matter of adding its linear velocity and gravitational vibrational velocity. Linear velocities and gravitational fields produce the exact same relativistic effects because they are both forms of spatial velocity subject to the STc Principle.

This is an entirely new and revolutionary understanding of general relativity derived from Universal Reality's METc Principle. By recognizing mass as vibrational velocities in space this principle reveals that all relativistic effects derive from the fundamental fact that the total velocity of all processes is always c, and that space itself, in particular its mass excitations are forms of intrinsic velocity that simply add to the total spatial velocity of any process to reduce its time velocity.

This suggests a much simpler conceptual model of spacetime than the curved spacetime of general relativity. We can now model computational spacetime as Euclidean (flat) with each point being characterized by a spatial velocity density. In empty space this is the velocity due to the zero-point energy and the presence of any gravitational or other energetic fields just adds to this velocity density.

Every point in this flat Euclidean space has an intrinsic spatial velocity. This indicates the proportion of processor cycles allocated to compute the time versus space velocity of a particle at that point. So adding this intrinsic velocity to the linear velocity of a particle at this point gives total spatial velocity, which determines the resulting time velocity, the time dilation, of the particle. All relativistic effects become simply a function of the sum of linear and intrinsic spatial velocity of an object at any point.

So we arrive at a very simple and easily understood but completely accurate model of general relativity's spacetime. It looks exactly like ordinary flat space but a gravitational velocity density characterizes each point. The velocity density can be visualized by a velocity meter which indicates the tilt angle of total spacetime velocity from all time velocity at 90° straight up to all in velocity in space at 0° horizontal. The relativistic gravitational effects on particles traveling through points in space depend on the tilt of the meter.

Thus the value of zero-point energy sets the reference point for the tilt in empty space to 90° which corresponds to c, the maximum possible velocity though time. The zero-point energy can be thought of as the resistance to velocity of space or more accurately the maximum possible velocity through spacetime. It's closely related to the speed of light c value of the universe because it's related to how fast space and time can be traversed.

This flat Euclidean model is topologically equivalent to the curved spacetime of general relativity because either can be distorted into the other without tearing but it's much easier to visualize and understand. The curves of curved space can be compressed into vibrational peaks, and the vibrational peaks of velocity density space can be smoothed and stretched out into the curves of general relativity's space.

Velocity dense space looks exactly the same as flat space but particles traveling through it have to move farther because they have to ride the ups and downs of the vibrational ripples just as they have to travel further around the curves of spacetime in the general relativity model. As a result the apparent speed of light through both velocity density and curved space appears slower from the outside though the speed of light always remains the same when measured locally.

One great advantage of the velocity density model is that it reflects the way we actually see gravitational space in the real world as Euclidean. Even if general relativity takes space as curved it still appears Euclidean because light beam trajectories ride the curvature of space. And velocity density space is also amenable to being represented and computed simply in terms of a standard data array in Computational Space.

All in all our velocity density model is much easier to visualize and understand than the equivalent curved spacetime model of general relativity. In fact our model is the most accurate model of the seemingly flat spacetime we observe around us which also appears Euclidean. It's the curved spacetime model of general relativity that while useful and accurate, is misleading and essentially impossible to visualize.

GRAVITATIONAL ATTRACTION

It's clear how the intrinsic velocity of a velocity density field surrounding a massive object causes clocks to slow in accordance with the STc Principle, but how does this explain gravitational attraction? Why do other masses tend to move towards areas of greater velocity density?

The same question arises in general relativity's curved space model of gravitation. Relativity states that inertial motion follows the lines of curved space inward around gravitational masses and this is correct so far as it goes. But why would a stationary apple released above the earth begin to move even if space there is curved? What is it about curved space that causes motion in the first place?

Gravitational attraction becomes easy to understand in the velocity density model. A gravitational field is a velocity density field that falls off by the square of the distance due to the geometry of 3-dimensional space. This means that the velocity density field is a velocity gradient field in which the intrinsic velocities at any point are greater in the direction of the gravitating mass. Since the intrinsic velocity of surrounding points is greater in the direction of the source this produces a velocity vector at every point that points toward the gravitating mass and objects in the field tend to move in the direction of the velocity vector.

Thus by understanding mass as spatial velocity we have a natural explanation for the fundamental nature of the gravitational force. Such an explanation was entirely lacking in Newtonian gravitation's attraction at a distance, and is still completely lacking in the curved space of general relativity. General relativity tells us that the presence of mass curves the surrounding spacetime but offers *no explanation at all* for why it curves it! In contrast, Universal Reality provides a clear and convincing answer for both the force of gravity *and* its relativistic effects. Masses actually are fields of velocity density gradients in space that inertial motion tends to follow, and because masses actually are fields of intrinsic spatial velocity they automatically slow time and produce other relativistic effects according to the STc Principle.

And conversely space itself consists of a field of velocity density. In flat space the density is the same everywhere so there are no resulting velocity vectors and inertial motion is just the continuation of unaccelerated linear velocity. In the absence of gravitation in flat space there is no net velocity vector in any direction, but around masses there is a field of resulting velocity vectors pointing inwards towards masses that defines inertial motion in that area. This is what a gravitational field actually is.

Apples fall towards the earth because of the velocity vectors produced by earth's velocity density field. And we stand on the earth because our inertial motion towards earth's center along its velocity vectors is blocked by its surface.

What we feel as the force of gravity is the intense fine vibration of the earth's mass increasing the velocity density of the spacetime around us, or equivalently the tug of the resulting velocity vectors on our bodies.

During the inertial motion of a free fall along the curve or velocity vectors of spacetime an observer experiences no force. It's only when that motion is interrupted and an observer stands on the surface of earth that he experiences a continual acceleration against his natural inertial motion. It is this feeling of resisting an inertial motion along a velocity density gradient in spacetime that is commonly but mistakenly called the force of gravity.

Though the inertial trajectory of a falling object seems substantial from the point of view of an observer standing next to it on the surface of earth, the actual trajectory is not just through space, but through time as well. The distance traveled in *space* by a falling object is miniscule compared to the length of its world line through *time*.

For example the distance in space traveled in a 1.3 second fall to earth is a little over 24 feet, but the distance traveled in time is 1.3 light seconds, equal to the 240,000 mile distance to the moon. So the falling object's actual world line extends a distance through spacetime equal to that from the earth to the moon and is almost perfectly straight since it deviates only 24 feet in 240,000 miles, or one foot per ten thousand miles. Actually using a light beam to define a straight line the world line hasn't really curved, rather the spacetime it travels through is curved by that amount. So almost all of anything's spacetime velocity will usually be through time rather than space unless it begins to approach the speed of light in space.

In this example the slowing of the falling object's clock is too small to be measured. But in a stronger gravitational field it would become apparent. An observer falling with a clock experiences its proper time still ticking at c since his clock is falling along with him and is slowed by the same amount he is. Relative to himself and his own clock he's not moving so all his own motion appears to be through time.

Note there is a very slight difference in the time rate of a stationary clock in a gravitational field and a clock falling past it. Both experience the same slowing due to the gravitational field since they are both at the same position within the field. But the falling clock is also slowed slightly more due to its motion relative to the background of computational space. This second effect depends on the velocity of the falling clock as it passes the stationary clock. On earth the effect is negligible except for very fast moving mesons produced by cosmic rays whose half-lives increase due to their internal clock rates slowing.

To summarize, the mass of elementary particles consists of vibrational velocity, and the elemental computations these particles are involved in produces an entanglement network containing intrinsic vibrational velocities that are equivalent to the dilated spacetime of general relativity. It's this velocity density field that produces the slowing of clocks and the velocity density vectors that produce the gravitational attraction of objects towards the source of the field.

The velocity density model should also be consistent with frame dragging, tidal forces, and other relativistic gravitational effects. For the model to be accepted these effects must all be understood in terms of velocity density effects propagating across the entanglement network at the local speed of light, which is the rate at which all computational changes propagate across the entanglement network. Ultimately this model must correctly incorporate all the spacetime effects of the components of the stress-energy tensor in the Einstein field equations (Wikipedia, Einstein field equations).

Universal Reality's vibrational fields can be considered gravitational waves. Thus gravitation itself is actually gravitational waves produced as vibrations in spacetime by the presence of mass vibrations. However the recently detected gravitational waves predicted by general relativity are *changes* in these waves produced by rapid changes of position of extremely large masses (Wikipedia, Gravitational wave).

The velocity density model of gravitation holds for the other force charges as well. Each type of force can be modeled as a particular mode of vibrational velocity that forms a velocity density field centered on its charge(s). In particular positive and negative electromagnetic charges and poles take the form of fields of opposite helical rotations that reinforce or cancel each other depending on the direction of the twist as explained in the chapter on The Other Forces. Because electromagnetism is also a form of energy it also produces gravitational effects in the same manner than mass does.

Because mass vibrations are such fine scale in place relative motions they appear the same in all (non-relativistic) frames and particles are said to have fixed rest masses. This is also true of the wave velocities of photons, which appear largely the same to all observers. This is in contrast to relative linear velocity, which clearly depends on even small observer velocities.

Though the velocity density around a single mass-energy particle is miniscule each additional particle adds its amplitude to the velocity density already present. So the cumulative effect around large collections of massive particles generates the extensive velocity densities in the surrounding space and clock time around stars and planets.

The vibrational velocity model of mass is equivalent to the conservation of mass and energy as transformations of equivalent amounts of relative motion from one form to another. Otherwise there is no explanation of how or why mass-energy should be conserved. The conservation of mass and energy makes sense only if they are different forms of the same thing. Thus the conservation of mass-energy always involves the transformation of equivalent amounts of various forms of relative velocity from one form to another.

The beauty of our model of mass-energy as relative motion is that it explains both the conservation of mass-energy and gravitational time dilation and other relativistic effects. By the METc Principle mass-energy is relative motion and by the STc Principle relative motion causes time dilation and other relativistic effects. Thus the vibrational velocity of mass increases the velocity density of space around it and results in time dilation and other gravitational effects.

Thus a concentration of massive particles is a source of intense fine vibration that propagates outward through the surrounding entanglement network, and observers experience an intrinsic relative motion that slows their clocks in accordance with the STc Principle.

Thus if gravitation is understood in terms of the mass-energy that produces it, and mass-energy is recognized as a form of relative motion then the STc Principle becomes a truly unifying principle that describes both the time dilation of relative motion and gravitational time dilation as aspects of a single process.

So relative velocity produces gravitational time dilation just as it produces the time dilation associated with linear motion. In one case it is the kinetic energy of linear motion, in the other it's the vibrational mass-energy of elementary particles in aggregate. Both are just different forms of relative motion, which is why they are interchangeable and energy can be conserved through all its forms.

The vibrational relative velocity of mass and the relative velocity of linear motion are both forms of energy and each can be converted to the other, or to the relative wave motion of electromagnetic radiation. Particle interactions are little computational factories that convert one form of relative motion to another.

The fixed c value of total spacetime velocity is produced by a fixed number of processor cycles some of which are used up in computing spatial motion, leaving fewer to compute rates of temporal change. This automatically manifests as the STc Principle.

All forms of mass and energy are velocity in or of space; mass is space in relative vibrational motion to the velocity background of empty space. Space itself is the vibrational motion of the quantum vacuum as expressed in its zero-point energy value. On a large scale empty space is uniform relative motion, which can't be detected because there must be something to move relative to something else for motion to manifest. Thus particles must be created out of the quantum vacuum as particle component sets for there to be something to move relative to the background.

So space is just the presence of motion, including the uniform lack of observable motion, and velocity is its measure. But motion takes time as well as space so spacetime is the presence of motion, of happening, and ultimately of life.

THE CLOCK POSTULATE

Universal Reality explains gravitational relativistic effects exclusively in terms of the total *velocity* at any point, the magnitude of the velocity being the combined linear velocity of an object plus the intrinsic velocity of the gravitational field being traversed. However some discussions of relativity assume *acceleration* is necessary to actually dilate time on the basis of Einstein's Equivalence Principle, which notes the equivalence of gravitation and an accelerating elevator in empty space (Wikipedia, Equivalence principle).

Though time dilation can be correctly analyzed in terms of acceleration to produce an equivalent result, the simplest approach is based on the rather ambiguously named 'clock postulate'. This states that the rate of a clock doesn't depend on its acceleration but only on its instantaneous velocity over all points on its world line. This was discussed in Einstein's original 1905 paper on special relativity as well as in subsequent kinematical derivations of the Lorentz transformations (Wikipedia, Twin paradox).

This means that we can ignore acceleration completely in the analysis of time dilation produced by spatial motion. Acceleration has no effect other than how velocity changes as a result. Thus we can modify the usual twin example to exclude acceleration altogether and still get the same result that the traveling twin's clock slows relative to the earth bound twin.

We can demonstrate this with a simple thought experiment by completely excluding acceleration from the twins example by adding a third traveler going in the opposite direction from the traveling twin who synchronizes his clock to the clock of the twin traveling away from earth as they pass. We find that when this third traveler returns to earth his clock will be slowed exactly the same as if the twin had instantaneously turned around and returned. Thus the acceleration of the twin turning around has been eliminated with no effect.

To complete the picture we can eliminate acceleration altogether if the outgoing clock doesn't take off from earth but simply synchronizes its clock with earth clocks as it passes at constant velocity to begin its journey, and finally by simply comparing the incoming third clock to earth clocks as it passes the earth without decelerating to land. In this example there is no acceleration whatsoever, but the combined passage of time on the moving clocks is still less than on a stationary earth clock. The traveling twin (and his surrogate) has still aged less than the earth bound twin due entirely to non-accelerated velocity over a longer world line.

So acceleration per se has no direct effect on time dilation. Its effect is due only to the fact that it changes spatial velocity. Thus even for an accelerating clock, time dilation can be correctly calculated just by integrating the changing instantaneous velocities all along the path of travel to get the total velocity along the world line.

By the STc Principle two clocks in relative motion will each observe the other clock running slower. By the STc Principle everything always has a constant combined velocity through space and time equal to the speed of light. Because part of the constant spacetime velocity of each clock is now seen as their relative motion through space, each sees the other's clock slow down to compensate for its increased relative velocity through space. The vector sum of velocity through space and through time is still c, some of that velocity is now through space so less is through time.

This is an *observational* effect each observer sees in the other's clock since each is moving at the same speed relative to the other. The two observers each see the other's time slow by the same amount proportional to their relative speed. But once the relative velocity ceases and they are stationary with respect to each other this effect vanishes and their clocks are seen by both to be running at the same rate again.

However when the relative motion ceases there can still be a real and agreed difference in the amount of clock time that elapsed during the motion. This *actual*, as opposed to *observational*, effect depends on the extent of actual travel through space, on the actual spatial length of the respective world lines and the actual spatial velocities along those world lines.

So the unifying principle of relativistic time dilation is the STc Principle. The coordinate time of any clock is always *observed* as slowing from c proportional to its *relative* motion through space. However this effect is only *actual,* in the sense it's lasting and agreed among observers, if the motion was an actual motion through the computational background space as opposed to just an observed relative motion.

For example if clock A remains on the earth and clock B travels through space and returns they both see each other in relative motion during the trip with each other's clocks running slower, but when B returns and they compare clocks they both agree that only B's clock has actually slowed with respect to A's. This is because only B actually traveled in space and A didn't. (The motion of earth in its orbit can be ignored since it's miniscule compared to the world line of B.)

Thus the key to understanding the relativistic slowing of time due to motion is that coordinate clocks always slow when in spatial motion, but it's only an actual permanent effect to the extent the spatial motion was actual motion through space as opposed to just motion relative to an observer who was moving himself. And acceleration has no intrinsic effect at all, the only thing that counts is the velocity of the spatial motion not whether it's inertial motion or varies due to acceleration.

This difference between actual and observational relative effects is what demonstrates there must be an absolute spacetime background relative to which actual spatial motion occurs. This can only be the actual computational space in which all motion is computed.

MASS VIBRATIONS & THE HIGGS FIELD

In our theory it's the processor cycles that compute the observable universe that convert numeric mass values into the fine vibrational velocities of mass. This means that the processor cycles are an analogue of the Higgs Field that physics suggests gives particles their masses (Wikipedia, Higgs boson). The processor cycles translate the numeric mass values of Level 0 into vibrational velocities as they are dimensionalized. Thus the processor cycles act as the Higgs field that gives particles their observational masses.

The amplitude of these vibrations is the gravitational strength of the particle mass. Thus more massive particles would have greater vibrational amplitudes than less massive particles. In the case of massless photons there are no intrinsic vibrations but instead extrinsic electromagnetic waves whose frequencies are their energies.

Gravitation is only additive and never cancels so the amplitudes of multiple particles can reinforce but never cancel. Thus they must be modeled as vibrational peaks in the elemental cells of space itself so they are all in the same position. And since the vibrational frequencies are the internal clocks of massive particles the frequencies of all additive vibrations of any given cell must all have the same frequency because any point cell in space can have only a single intrinsic space and time velocity ratio to as to obey the STc Principle. This explains why they the intrinsic vibrations of mass reinforce but never cancel.

So the presence of additional masses simply increases the total velocity density of the field, which in turn produces the correct relativistic gravitational results. This effectively increases the density of spatial vibrations that must be traversed in any volume of space, which is equivalent to a greater spacetime curvature.

Thus it's the processor cycles that compute all processes that acts as the Higgs field that converts numeric mass values into vibrational velocity densities and gives particles their observational masses. Since the fixed number of processor cycles determines the c value of the speed of light the value of c must be intimately related to the strength of the Higgs.

THE INCREASE OF MASS WITH VELOCITY

The processor cycles remaining after the computation of spatial velocities involved in an event go to computing the event's temporal velocity, its time dilation. The event's temporal velocity is the proportion of cycles used to compute the evolution of its internal details. This includes both the intrinsic clock rate of the particles themselves based on the presence of any internal vibrations of charges of the particles and by extension the rate at which particles interact.

Mass and other charges are modeled as spherical fields of spatial velocity densities. The respective charges are specific forms of vibrational velocity that produce the fields. Thus charged particles have internal detail in the form of their vibrations that is computed by the fixed allocation of processor cycles. Massive particles have the internal detail of their mass vibrations and as a result can never travel at the speed of light because there must always be some time allocated processor cycles to compute their internal vibrational detail.

In contrast massless photons and bosons have no mass and thus no internal detail to be computed by temporal processor cycles, and as a result time doesn't pass on their internal clocks. Photons have no velocity in time and as a result all their velocity is through space and they always move at the speed of light through space. All the processor cycles that compute the evolution of photons are devoted to computing their spatial velocity since they have no internal details to be computed by any temporal cycles. In contrast their electromagnetic wave frequencies are computed externally rather than internally in the same sense linear velocity is.

Of course massless photons take time to move through space on the clocks of external observers but their commoving proper time clocks never advance. Time never passes on their own clocks, and as a result all their c velocity is through space. In contrast the vibrational relative motion of the masses of massive particles prohibits them from ever moving at the speed of light.

Changes in fields including the gravitational field also propagate through the computational background at the speed of light since their *propagation* is a massless process that has no internal structure even though the fields themselves are fields of mass-energy that do have internal structure.

Thus our computational model consistently explains why massless photons always travel at the speed of light due to the allocation of processor cycles between computing velocity in space and velocity in time, while massive particles must always travel slower than the speed of light because they have internal temporal details that must be computed.

If we assume that mass at Level 0 where it's conserved is simply a numeric value, then it's the processor cycles that convert this numeric value into dimensional vibrations. By the MEv Principle, which states that mass-energy is spatial velocity, the amplitude of these vibrations, their velocity in space, is a particle's observational mass. Their amplitude is their intrinsic spatial velocity, their gravitational strength, and their frequency is their internal clock time rate.

By the STc Principle a particle's total spacetime c velocity is the vector sum of its time and space velocities, the vector sum of the frequency and amplitude of its mass vibrations. Thus any slowing of a particle's internal clock rate (vibrational frequency) will increase its observational mass (vibrational amplitude).

We can define a particle's *nominal rest mass* as the amplitude of its mass vibrations when it's at rest relative to an observer. The nominal rest mass is the rest mass noted in tables of particle data. In contrast a particle's *observational mass* is its rest mass when measured by an observer whether or not the particle is in motion relative to the observer. By special relativity the observational mass of a particle increases with its relative spatial velocity.

An isolated single particle at rest in empty flat space will have only a miniscule intrinsic spatial velocity in the amplitude of its vibrations and nearly all its vibrational energy will be expressed as a speed of light velocity through time in the frequency of its vibrations. In this case its observational mass will be its nominal rest mass.

However if the particle starts moving with spatial velocity this reduces its velocity in time. By the STc Principle linear velocity will significantly reduce the number of processor cycles available to compute the frequency of its vibrations and their frequency per P-time tick will decrease. Because the vector sum of frequency and amplitude must always equal c the observational rest mass (the amplitude of the particle's own intrinsic mass vibrations) increases and this is the source of the increase of mass with linear velocity predicted by special relativity.

When the particle is in a gravitational field the amplitude of its own intrinsic spatial velocity is added to the intrinsic spatial velocity of the field. The particle also 'feels' the total intrinsic velocity of the field it's in. By the STc Principle this further slows its own velocity in time, the frequency of its vibrations, in accordance with general relativity. This in turn increases its own intrinsic spatial velocity, the amplitude of its vibrations. So a massive particle in a gravitational field experiences a slowing of time due to gravitational time dilation and its observational mass also increases.

The increase in observational mass manifests as the increased *weight* of the particle in a strong gravitational field. This explains why masses in stronger gravitational fields are heavier because their weight, their observational mass, is increased by the slowing of their internal time velocity by the intrinsic spatial velocity of the field. Thus the STc Principle explains both gravitational time dilation and the weights of masses in gravitational fields as aspects of the same computational process.

The particle in the gravitational field experiences the intrinsic spatial velocity of the field. This reduces the frequency of its vibrations (its proper time velocity) which in turn increases their amplitude so that the vector sum of the particle's space and time velocities remains equal to c by the STc Principle. This is of course also true of all classical level objects, which are all composed of individual massive particles.

The frequency and amplitude of a particle's vibrations depends on the allocation of processor cycles used to compute them. In flat empty space with nearly all processor cycles used to compute velocity in time, the mass of a particle will be minuscule. However if the particle is in a gravitational field or gains linear spatial velocity this reduces the processor cycles used to compute its time velocity. This in turn makes more processor cycles available to compute the particle's own intrinsic spatial velocity increasing the amplitude of its vibrations which manifests as increased observational mass.

Thus the apparent or observational mass of a particle or object composed of particles increases when it experiences increased linear velocity or the increase intrinsic spatial velocity of a gravitational field. The increase in mass with linear velocity is an observed consequence of relativity (Wikipedia, Special relativity). So our model correctly explains the observed increase of mass with relativistic velocity as well as the nature of weight itself as aspects of the same computational process. The apparent increase in mass of an object with relativistic velocity is simply increasing its weight in the same way the weight of an object in a strong gravitational field is increased.

All forms of mass-energy are relative motion (spatial velocity) of one type or another corresponding to the type of force charge involved. Because of zero-point energy resistance to relative motion with respect to the absolute background could be said to produce a resistance friction that manifests as mass. Relative linear velocity is converted to the relative intrinsic velocity of mass as the speed of light is approached.

All forms of mass and energy are forms of spatial velocity. Space itself is the minimal spatial velocity of the zero-point energy. Particle masses are elemental units of space (spatial velocity) crystalized around valid particle component sets. The mass particle component consists of spatial velocity in the form of very fine scale vibrations in the fabric of space. The amplitude of these vibrations is the particle's intrinsic spatial velocity, which is its observational mass. The frequency of these vibrations is the particle's internal clock time rate, its velocity in time. By the STc Principle the vector sum of the amplitudes and frequencies of these mass vibrations always equals the speed of light c.

Any additional spatial velocity a particle experiences reduces its velocity in time, its vibrational frequency, and increase the amplitude of its vibrations and thus increases its observational mass. This manifests as an increase in mass with linear velocity, and as the increase in observational mass that we call weight in a gravitational field.

DIMENSIONAL SPACETIME

The preceding discussion is more evidence that clock time and dimensional space are not an external physical spacetime framework within which events occur, but are computed as the dimensional aspects of events themselves.

What we call spacetime is the logico-mathematical consistency among observer measurements. Human observers interpret this dimensional consistency as a preexisting empty physical spacetime within which things exist and events occur. However this is a projection of the logico-mathematical consistency of measurements into an apparently physical world constructed in our brain's internal simulation of reality. Physical spacetime is a reification of the logico-mathematical consistency of dimensional observations.

This logico-mathematical consistency is generated by the conservation of particle components in particle interaction events. Particle interactions seem to be relatively simple logico-arithmetic computations that conserve the total amounts of particle components including dimensional particle components such as total mass-energy and spin orientation. Particle component conservation entangles the particles emitted by events on their conserved particle components, meaning that the values of particle components of emitted particles bear fixed relationships to each other.

Sequences of events form entanglement chains and all events together form an entanglement network of particle components in which the values of all particle components in the universe are related across particles. It's the internal dimensional consistency of this entanglement network that human observers interpret as an enveloping spacetime container for events.

For events to exactly conserve particle component values of energy and momentum the dimensional fuzziness of particles must be decohered to exact (within limits of Uncertainty) dimensional values by events so they can be exactly apportioned among emitted particles.

These exact dimensional and other particle component values constitute measurements of particle observables. It's the observational results of these measurements that are the observable data of the universe. The actual computational universe and its values are unobservable. The observable universe exists only as observer compilations of observed values ultimately deriving from decoherences and particle component values.

OBSERVER FRAMES

All observable values are measurements relative to some observer frame. The relativistic equations describing these frame views have the same covariant forms as they do with respect to the preferred universal background frame in which they are actually computed. The difference is that the frame they are actually computed in becomes the preferred universal frame with respect to which actual rotation and actual world lines are relative to.

To the extent that observer frames are not aligned with the actual background frame, their relativistic effects are observational rather than actual and vanish without lasting effect as soon as any relative motion ceases. The equations have the same general relativistic form, but the effects are observational rather than actual. And of course observers don't use the same number system, scales or units in which the universe is actually computed but which remain unobservable.

'Physical' spacetime is the logico-mathematical consistency among observations projected onto the simulated universe produced by an observer mind, and more generally the shared simulation of a group of like-minded observers such as humans or physicists.

Logico-mathematical consistency means that particle observations obey the predictions of general relativity at large scales and quantum theory at small scales. However the various interpretations of each theory, all based on a pre-existing spacetime within which events occur, are outmoded. The computational interpretation of Universal Reality in which both theories can be united is much more fruitful, reproducing general relativity at large scales and quantum theory at small scales.

Another major difference between the actual computational frame in which the universe is computed and observer views is that observers invariably incorporate past dimensional observations into their dimensional simulations whereas the actual computational frame of the observable universe includes only current values as they are computed.

Because observers are part of the universe they and their dimensional views are all ultimately computed by the elemental computations so all observer views are consistent aspects of the actual computational universe. Observers are part of the entanglement network and thus part of its numeric consistency. The elemental program that computes the consistency of the whole network also computes the observers that exist within it and the relativistic effects they observe.

Though actual observers are all emergent structures that view the entanglement network through their simulations of it, relativistic observer views can be analyzed in terms of imaginary single point observers and frames if we are careful. This is fine because all observations ultimately reduce to single particle events though they are typically mediated by chains of events through laboratory instruments that scale results up to the classical level, or through the even more complex perceptual systems of observers.

It must be remembered that the numeric consistency of the entanglement network is only established through the computations that generate the numeric values and their relationships. The numeric consistency is a matter of the equations of the quantum vacuum consistently computing exact numeric values from previous ones.

The quantum vacuum also computes observers as part of the entanglement network so observers are part of its numeric consistency. So the view of the entanglement network that any observer sees is also computed by the elemental program of the quantum vacuum.

The same equations that actually compute reality in its own background frame are also used to compute views of the entanglement network from moving frames within it. The relative motions and coordinate origin points of their frames define individual observers.

Because observers view the entanglement network from the perspective of their own frames they ascribe their own relative motion to the entanglement network. Thus the quantum vacuum computes the relativistic views of observables as if they had the inverse of the observer's own relative motion.

Because this is all calculated by reality it's the real actual view of the observer, but it is not a view shared by other observers with different relative motions. Every observer with a different relative motion will have a different view of reality computed by the quantum vacuum that seems correct to him but isn't shared by other observers.

All frames in relative motion will have different relativistic views and these are the real actual views of observers from those frames. However only relative motion with respect to the background frame in which the entanglement network is computed has actual lasting relativistic effects. All other relative motion produces relativistic effects that are observational though they are quite real and genuine in the frames of observers. This means that when observers meet and compare relativistic effects they aren't lasting or agreed unless they were produced by relative motion with respect to the actual background frame in which everything is computed.

For example observers in relative motion will each see the other's clock run slower. But when they meet only the clock(s) that moved with respect to the entanglement network background will show less actual elapsed time. And both observers will then agree on this.

Similarly rotational motion will only produce actual centrifugal effects to the extent it's with respect to the background frame of the entanglement network. Take an observer at the center of a merry-go-round in deep space. Visually he has no way to know if the merry-go-round is rotating or not because there is no visual background. However if the merry-go-round is rotating with respect to the entanglement network loose objects tend to fly off it whether or not the observer is rotating with it or not. So the view of the observer is observational but the effects with respect to the actual computational background frame are actual. The effect depends entirely on the rotation with respect to the computational background and not with respect to the observer.

The relative motion of observers can compute identical *observational* effects but these are local and change with the observer's relative motion. The absolute computational background is necessary to maintain the logico-mathematical consistency of the universe and thus its existence.

Educated observers are able to discover the equations reality uses to compute a relativistic entanglement network and invent the laws of general relativity to convert among the views of observers but this occurs at the emergent level of their simulations of reality.

Ultimately the quantum vacuum even computes the simulations of observers as emergent aspects of the entanglement network, but at this point it makes sense to consider the simulations themselves as emergent programs that perform the computations.

Observer views are all part of the super consistency of the universe, the fact that the universe is logico-mathematically consistent across all observer views at all levels of emergence, and this is all due to the super consistency implicit in the design of the complete fine-tuning.

The simulation programs of observers are imperfect models of the computational structure of the universe. They selectively filter information and fragments of its logico-mathematical structure and apply them to modeling situations of importance to the individual observer. Because they are based on small amounts of highly filtered input data their computational results are inexact but good enough in general for the observer to function and survive as part of the actual entanglement network that computes it.

The simulation programs do have significant adaptive advantages in that they are able to store and compare past data states to infer causality, and they are able to compute in terms of individual things and events and relationships extracted from vast floods of raw data. Even though not exact and often inconsistent from thought to thought this program is highly adaptive and easily switches among small-scale models to compute reasonably effective actions. It enables on the fly comprehension by redefining things as it goes, something the quantum vacuum is unable to do at the level at which it actually computes the entanglement network. Nevertheless it's the entanglement network that ultimately enables all this through the super consistency of its complete fine-tuning.

THE OTHER FORCES

ELECTROMAGNETISM

The electromagnetic force is quite interesting because the relationship between its electric and magnetic components behaves much as energy, space and time do. Just as energy is space in relative motion, so magnetism is electricity in relative motion. Electricity and magnetism are two orthogonal (90°) components of a single underlying entity just as space and time are, and in both cases each is transformed into the other via relative motion.

In physics, a magnetic field is the relativistic part of an electric field, as Einstein explained in his 1905 paper on special relativity. When an electric charge is moving from the perspective of an observer, the electric field of this charge due to space contraction is no longer seen by the observer as spherically symmetric due to relativistic shortening along the axis of motion, and must be computed using the Lorentz transformations. One of the products of these transformations is the part of the electric field that only acts on moving charges which is called the magnetic field (Wikipedia, Electromagnetism portal).

This similarity between electricity and magnetism and space and time underlies the Kaluza-Klein Theory in which electromagnetism is modeled as a 5th compacted dimension and an electric charge is a standing velocity in that 5th dimension (Halpern, 2006). The beauty of this theory is that when this 5th dimension is added to the 4 dimensions of general relativity, Maxwell's equations of electromagnetism automatically emerge (Wikipedia, Kaluza-Klein theory).

Thus one can consider electricity as a fundamental force and magnetism as electric charge(s) in motion. This motion can take several forms. At the elemental level of particle components all electrically charged particles have an intrinsic half integer spin, which effectively rotates the charge about an axis.

Because spin gives its associated electric charge rotational motion spin manifests as magnetism and particle spin is the intrinsic underlying unit of magnetism. Since spin about an axis produces an orientation of the axis spin is equivalently an intrinsic underlying unit of dimensional *orientation* and angular momentum relative to the computational background. This is critical to understanding Newton's Bucket as we have seen.

The quantum mechanical velocity of electrons in atoms produces the magnetism of permanent ferromagnets. Ferromagnetism is due primarily to the alignment of the spins of ionic electrons in atoms. In most materials the spins of particles are randomly aligned and tend to maintain their random alignments just as spinning gyroscopes do.

Materials made of atoms with filled electron shells have a total dipole moment of zero, because every electron's magnetic moment is cancelled by the opposite moment of the second electron in the pair (see the section on Bound Entanglement in the next chapter for how this works). Only atoms with partially filled shells (i.e., unpaired spins) can have a net magnetic moment, so ferromagnetism only occurs in materials with partially filled shells.

These unpaired dipoles (often called "spins" even though they also generally include angular momentum) tend to align in parallel to an external magnetic field, an effect called paramagnetism. Ferromagnetism involves an additional phenomenon; the dipoles tend to align spontaneously, giving rise to a spontaneous magnetization, even in the absence of an applied field (Wikipedia, Ferromagnetism).

A fundamental characteristic of magnetism is because it's due to rotational spin about an axis and the poles of the axis are spinning in opposite directions from the point of view of the exterior, magnets always appear to have equal and opposite magnetic poles. Because magnetism is fundamentally a product of axial rotation there can be no isolated magnetic monopoles. Magnetism is always dipole.

However magnetic poles are actually an illusion because magnetic field lines continue through the interior of a magnet and just emerge at the other pole in the opposite direction. Thus magnetic field lines always form closed loops as opposed to electric field lines, which radiate outward from electric charges. And magnetic poles are simply a name given to where a denser concentration of field lines enters or exits a magnet.

So the magnetic force is actually along the field lines proportional to their density, which is greater at the 'poles' of a magnet. Thus it appears the poles are doing the attracting or repulsing but it's actually the density of field lines themselves. The opposite poles are due to the field lines pointing inward as they enter at one pole and pointing outward as they exit at the other.

Thus magnetism doesn't really have positive and negative poles. It's just a matter of which direction the lines of force are pointing and how dense they are. So for example the magnetic field around a straight current carrying wire has *no poles* because the field lines are all circularly concentric around the wire. So it's the density gradient of the lines that is greater towards the wire that exerts a magnetic force either towards or away from the wire.

Magnetism is different from electricity in this respect, which does always come in positive or negative charges. And electric charges are always isolated to individual particles.

Again with electricity other charges are attracted or repelled not so much by the charges themselves but because the electric field is a velocity density gradient in spacetime with a velocity vector at every point. As with mass the field is an inseparable part of the actual charge and other charges tend to move along velocity vectors in the field gradient.

The second form of magnetism due to the movement of electric charges is due to the orbital motion of electrons in atoms. This orbital motion produces quite a strong magnetic force but like the spins of most particles it's randomly oriented and mostly cancels out.

The third form of magnetism is due to the motion of electric charges in currents. When electric charges move through a wire they generate a magnetic field encircling the wire according to the Right Hand grasp rule. When many wires are wrapped tightly in a coil (a solenoid) the magnetic field generated within the coil is multiplied and when the current is properly modulated will rotate an iron rotor. This is of course the principle of the electric motor.

The magnetism generated by particle spin can be easily understood by analogy to that generated by a moving current in a wire. If we slice the spinning particle open on one side and lay it out flat we see that the magnetic field encircles the particle just as it does a wire.
There is also an opposite effect in which a changing magnetic field produces an electrical current as in a generator. The principle, Faraday's law, is that an electromotive force is generated in an electrical conductor that encircles a varying magnetic flux. Motors and generators

are similar in form and many motors can be mechanically driven to generate electricity and frequently make acceptable generators.

So all three of these magnetic effects are manifestations of electric charges in motion. Magnetism is electric charge in relative motion and this relative motion can be either that of the electric charges themselves or of an observer relative to them. Thus magnetism is a clear everyday example of relativity in action. Whenever we experience magnetism we actually experience a relativistic effect.

An observer at rest with respect to a system of static free electric charges will see no magnetic field. However if either the charges *or* the observer begins to move the observer perceives it as a current and an associated magnetic field. A magnetic field is simply an electric field seen in a moving coordinate system. It doesn't matter whether the electric field or the observer is moving; all that counts is their relative motion. How this works as an effect of Lorentz contraction along the direction of motion is depicted graphically at (Schroeder, 1999).

However recall that actual relativistic motion is with respect to the computational space in which it's computed, as opposed to observational relativistic motion, which is simply relative motion between observer and what is observed. So there will be an absolute transformation of electric to magnetic force from actual motion with respect to computational space, but only observational effects in the frame of observers due to their own relative motion with respect to computational space.

So magnetism is actually a relativistic effect of electricity, and electricity is transformed into magnetism by relative motion just as velocity in time can be transformed into the mass-energy of velocity in space by relative motion. Both are examples of the Lorentz transform, which is simply the Pythagorean theorem describing the orthogonal projections of a single vector onto orthogonal coordinate axes. Thus the electric and magnetic fields are 90° orthogonal projections of a single underlying entity just as space and time velocity are.

Thus when we play with a magnet and observe its effects we should realize it works because of the enormous in place velocities of its electric charges at the particle and atomic levels. The energy within matter is enormous and thus the in place velocities are enormous. It's only the near exact balance of forces that holds the energy of particles together into the seemingly ordinary and trivial objects around us.

THE ELECTROMAGNETIC FIELD

Though electric and magnetic fields are usually considered separate but related entities that can produce each other, there is actually only a single integrated electromagnetic field that is best understood in terms of the electromagnetic tensor (Wikipedia, Electromagnetic tensor). This tensor describes the relationships between the spatial vectors of the electric and magnetic fields.

In the electromagnetic tensor the individual electric and magnetic fields change with the choice of the reference frame, while the tensor itself doesn't. The electromagnetic tensor takes the form of a 4 x 4 matrix in any particular coordinate basis where the components of all but the diagonal cells are the values of the magnetic field along each coordinate axis and the values of the electric field divided by c along each coordinate axis. Thus there are 6 independent components of the tensor; E_x, E_y, E_z (the electric field) and B_x, B_y, and B_z (the magnetic field).

Thus every point in the field has a vector for E and for B along all three coordinate axes in whatever coordinate basis is being used. The values can be either plus, minus or zero for each vector component.

When the electromagnetic tensor is multiplied by a metric tensor expressing a change in coordinate basis, such as relative motion to an observer, the individual B and E values change. The form of the tensor gives the correct transformations of electricity into magnetic fields and vice versa for any observer frame, though there's still only a single electromagnetic field manifesting as a combined electric and magnetic field depending on the frame of reference of the observer.

This is expressed by the fact that the tensor has an invariant that doesn't change with transformations of coordinate basis. This invariant is $\mathbf{B}^2-\mathbf{E}^2/c^2$, which means that the total electromagnetic force is conserved in all coordinate transforms no matter how the individual electric and magnetic forces transform into each other (Wikipedia, Electromagnetic tensor). This is analogous to the conserved transformation of space and clock time velocities into each other expressed by the STc Principle.

The solution to the tensor gives the force vectors of the two fields at every point and the total of all force vectors for all points traces the combined lines of force for both forces.

The electromagnetic tensor has another invariant 4/c (**B**·**E**), the dot product of the magnetic and electric force vectors, which roughly means that electricity and magnetism are orthogonal manifestations of a single electromagnetic force.

The invariance of the space-time four-vector is associated with the fact that the speed of light is a constant. The invariance of the energy-momentum four-vector is associated with the fact that the rest mass of a particle is invariant under coordinate transformations and the invariance of the electric-magnetic four-vector is associated with the fact that the total electromagnetic field is invariant under coordinate transformations. In these invariances we see fundamental principles of the universe at work.

Thus time turns into space with increasing velocity, and electricity turns into magnetism with increasing velocity. And increasing spatial velocity is increasing mass-energy. Thus increasing mass-energy turns time into space and electricity into magnetism.

Now if we just take the lines of forces as consisting of helices in accordance with our representation of field energy as various forms of velocity vectors in space we get a proper attraction and repulsion model of plus and minus poles and charges of these two separate but interrelated forces.

THE HELICAL FIELD MODEL

Electromagnetism is another form of energy and thus according to Universal Reality a form of relative motion, with a strength equal to its velocity density. Like mass electric charges are fields of spherical velocity densities in spacetime radiating from the center of the charge. These fields of spacetime distortion alter the proportion of space and time distances and velocities at points in the field and the field gradient produces velocity vectors that induce relative inertial motion in other particles.

Since charged particles also have mass they are associations of two kinds of velocity density, one produced by their mass, and another by their charge. The fields are easily visualized as spherical areas within a *flat Euclidean* space in which the relative distances and velocities of time and space are shifted at every point in the field as described in the previous chapter.

Both gravitational and electromagnetic fields fall off as the square of the distance due to the simple fact that in 3-dimensional space the area of the surface of a sphere increases by the square of the radius. Thus the strength of fields falls off inversely with distance. Thus the constant strength of the field is simply diluted by the increasing volume of 3-dimensional space as the distance from the center increases.

This model of mass and gravitation suggests a similar model for electromagnetism. This is a very neat theory original to Universal Reality with a lot of explanatory power. It also provides an excellent explanation for how the standard theory of electromagnetic fields as virtual photons works.

The difference in the vibrations of mass and electric charge is in the form of the vibrations. Electric charges are spherical fields of *helical* spacetime distortions in the surrounding dimensional fabric. In other words charges produce a field of miniscule corkscrew twists in the surrounding spacetime that form the field lines of both electric and magnetic fields and increase the velocity density of points in space. Of course the actual fields are continuous and fill all space, the field lines are just a graphical sampling of the entire actual field.

These spacetime distortions produce velocity vectors felt mainly by other charged particles whose own fields couple to them since their helical distortions tend to reinforce or cancel each other out depending on their direction of twist. Electric and magnetic field lines are modeled as separate orthogonal projections of these helical distortions in space.

The transformation of electric force into magnetic force with the spatial velocity of charges occurs as the helical vortices of the electric field begin to tilt orthogonally into the helical vortices of the magnetic field according to the right hand grasp rule. With greater and greater velocity (current flow) the helices tilt more and more and become a magnetic field that appears as field lines of magnetism perpendicular to those of the electric field lines.

So current velocity transforms electricity into magnetism by flipping the helical vortices of the electric field into the orthogonal direction where they become helical vortices of the magnetic field. If the actual velocity of the charges approached the speed of light all the electric field vortices would flip over into magnetic field vortices and the electric field would become entirely a magnetic field due to the conservation of the total electromagnetic field.

These helical spacetime distortions are generated by individual charges and rotate in two possible directions, either in the clockwise or counterclockwise direction. These correspond to positive and negative electromagnetic charges or positive and negative magnetic poles. The fact there are only two possible rotational directions for helices neatly explains why there are only two electromagnetic charges and two magnetic poles.

The effective diameter and density of twists of these helices is fixed since the charges or spins generating them are the fixed plus or minus electrical charges and spins of elementary particles. The densities are additively scaled to produce the measured values of electric and magnetic forces produced by multiple particles.

The spins of the elementary electromagnetic charged particles clearly produce the helices of their electromagnetic fields, which extend outward from the spinning charges.

While the vibrations of mass come in different amplitudes and frequencies due to the non-proportionality of particle masses and the additive nature of the gravitational force, the helical vortices of electromagnetism have identical forms because the strength of their elemental charges are identical. Presumably multiple charges just add to the number of identical helices to produce a denser field. This enables the helices of electromagnetic fields to cancel or reinforce depending on whether they are turning in the same or opposite directions.

The relativistic effect of linear motion on mass vibrations is a tilting of time velocity into space velocity while the relativistic effect of linear velocity on the electromagnetic helices tilts them from the parallel to electric field lines towards the perpendicular magnetic field line orientation.

So the electric force is analogous to velocity in time and the magnetic force analogous to velocity in space. Both velocity in time and the electric force tilt into their alter egos with linear velocity and the vector sum of each with its alter ego is conserved. This similarity in form is why the electromagnetic force can be modeled as a compacted 5th dimension in Kaluza-Klein theory though our helical model is preferable.

If the charge producing the field of helices is set in motion relative to an observer the individual helices begin to tilt from parallel to the lines of electric force towards perpendicular to them. The perpendicular projection of the tilted helices becomes the magnetic field and the parallel projection to the lines of force is the electric field. This is the relativistic source of the magnetic force and its field lines.

Helices cancel each other out when they are rotating in opposite directions and reinforce when they are rotating in the same direction. Thus helices rotating in opposite directions cancel where they are pointing in the same direction, and reinforce where they are pointing in opposite directions. And helices rotating in the same direction cancel when they are pointing in opposite directions, and reinforce when they are pointing in the same direction. This is the key to understanding magnetic attraction and repulsion, and the repulsion and attraction of electric charges.

Thus in areas between poles or charges of the *same* sign the helices cancel each other out and they reinforce in areas outside the charges or magnetic poles. Thus the velocity density of spacetime is increased in the areas outside and unaffected between. Thus the velocity vectors at the points of the poles or charges are directed away from each other and this is the source of the *repulsion* of identical magnetic poles and identical electric charges.

And in areas between *opposite* poles or charges the helices reinforce and cancel beyond producing an area of strong velocity density between the charges so the velocity vectors where the charges are located point towards each other and this is the source of the *attraction* between opposite charges and opposite poles.

In areas external to two opposite charges or poles their field helices will be rotating in opposite directions and will almost completely cancel each other out. This is why there is no net magnetism in most materials because their collective helices cancel each other out in external areas. Only in materials where slight spatial imbalances of charge polarity exist and are aligned will external helices not be completely canceled. In such cases magnetic effects will be present and the materials will be magnetic.

Thus Universal Reality suggests a simple and elegant new model of electromagnetic attraction and repulsion in terms of the spacetime dilation generated by a helical velocity density just as it did for the similar gravitational effects of mass in the form of simpler vibrations.

And the specific forms of the spacetime distortions produced by mass and electric charges neatly explain why there are two opposite electromagnetic charges and only a single gravitational charge.

Both theories are explained in terms of the same spacetime velocity density model, and both are complementary distortions that can exist together in the same spacetime volumes, which are observer views of the computational data structure of the entanglement network.

Thus both mass and charge can produce their different distortions in spacetime simultaneously. They work together naturally in standard 4-dimensional spacetime. Electromagnetic effects propagate across the curved spacetime of general relativity, and gravitational effects are produced by a simpler form of velocity density. Gravitation and electromagnetism are simultaneous distortions or velocity densities in spacetime of different forms. But why the intrinsic velocity of a gravitational field doesn't also tilt the electric force into magnetic force is an open question.

As forms of energy both mass and electromagnetic charges increase the velocity density of spacetime in their particular ways and this affects the relativistic behavior of objects within spacetime but most of the velocity density of electromagnetism consists of helical distortions that largely cancel each other out but also strongly couple to the helical fields of other charges.

Thus Universal Reality produces a simple and elegant theory of both mass and gravitation, and of electric charges and magnetism in terms of different forms of relative motion that both work according to the same underlying MEv Principle of all types of mass-energy as forms of spatial velocity.

And this model is quite easy to visualize and understand in terms of a flat Euclidean spacetime containing fields of the two forms of velocity density that produce velocity vectors in the direction of slower time. This Neo-Euclidean model is equivalent to the curved spacetime of general relativity, but much easier to comprehend because it directly reflects our actual flat Cartesian view of spacetime.

Think of each point in this Euclidean spacetime having a fixed c value (speed of light) of combined time and space velocities. The relative motion of either mass or charge just distorts the proportion of time and space velocities so that time slows and distance lengthens. We can still think of the overall spacetime as Euclidean but the relative velocity density of space and time at every point is distorted by mass or electromagnetic charge.

Both the vibrational distortions of mass and the helical distortions of electromagnetism produce spacetime distortions proportional to the relative velocity of their energy. Thus both produce velocity vectors that determine the inertial motion of test particles in their fields that are different depending on whether or not the test particle is charged.

Gravitation is a much weaker force than magnetism, the strength of the relative motion produced is 10^{-36} times weaker than that produced by electric charges, and thus the field densities and vector velocities are much less for a particle of mass than a unit of charge.

However because mass has only one charge (there are no negative masses) masses are all attracted to each other and tend to clump without limit into planets, stars and galaxies and produce very large gravitational velocity density fields.

On the other hand equal electric charges repel each other and cannot clump (except in very small units under the influence of the strong force in nuclei). Opposite charges do clump at a distance in the form of atoms and molecules, but their opposite helical spacetime distortions almost entirely cancel each other out beyond the clumps when they do, so most of the large scale structure of the universe is due to gravitational velocity density fields.

Thus though electromagnetism is intrinsically far stronger than gravitation, gravitation rules on cosmic scales and electromagnetism mainly just holds atomic matter together with external effects largely cancelling out. There are large magnetic fields on cosmological scales but their strength is generally much less than gravitation.

As a form of energy the helical spacetime distortions of electric charges do have some gravitational effect on uncharged particles as predicted by the stress-energy tensor of the Einstein field equations, but these are generally negligible compared to those of mass. Since charged particles are normally paired the helices largely cancel in areas external to the particles and the velocity density between closely adjacent points will be minimal thus the velocity vector will be relatively small compared to that on a coupled charged particle which is effectively canceled in one direction and doubled in the other.

This is how our velocity density model addresses the initially vexing question of why the gravitational force is so much weaker than the electromagnetic force but curves space so much more.

The gravitational force is due to the vanishingly *small difference* in velocity density on either side of each particle along an axis towards its source mass. Thus the resulting velocity vector towards the gravitating mass is also extremely small. However the helical velocity densities of the electromagnetic force couple to those of other charges and so completely cancel on one side and completely reinforce on the other along their common axis. This effectively doubles the *entire value* of the velocity density at the point location of the particles. The difference in gravitation's small *gradient* in velocity density at the particle scale and electromagnetism's doubling of the *entire* velocity density is enormous. Thus the electromagnetic force is much stronger than the gravitational force.

The velocity density produced by electromagnetism is considerable, but since charges are normally paired ambient velocity densities effectively cancel and the difference on proximate and opposite sides of *uncharged* particles is effectively nonexistent so gravitational effects will be vanishingly small. Note also that mass produces a much greater gravitational effect than energy since by $E=mc^2$ the equivalent amount of mass m in a unit of energy is E/c^2 which is an extremely small number.

Another difference between electromagnetic and gravitational fields is that electromagnetic fields can be shielded but gravitational fields can't be. Again this is due to the fact that the helical velocity densities of electromagnetism come in two opposing rotations. Thus it's generally possible to construct a shield that either damps or diverts an electromagnetic field but adding any shield made of mass or energy just adds to a gravitational field rather than blocking it.

Thus our model of electromagnetism seems reasonably consistent with standard scientific theory and explains its basic concepts well in terms of velocity densities as it also does mass and gravitation and the conservation of mass-energy.

PHOTONS

Photons of electromagnetic radiation are now easily understood as units or quanta of the electromagnetic field that break away from the charges that generate the field and fly off on their own taking some of an electron's orbital energy with them. They can also be absorbed by orbital electrons and kick them into higher energy orbitals by increasing their orbital energies. Photons are essentially free quanta of electromagnetic fields that either break away or are absorbed back into fields in the process of emission or absorption of orbital energy.

Thus orbital photons are units of orbital velocity converted into the helical waveform of the electromagnetic field. This is an excellent example of how the conservation of mass-energy always involves the conversion of one form of spatial velocity into another.

The emission of photons of electromagnetic radiation is the result of the acceleration of an electric charge(s). This can occur either due to a free electrical charge or system of charges changing direction as in an antenna, or when an electron falls to a lower orbital and emits its lost orbital energy as a photon of electromagnetic energy, or absorbs a photon which accelerates it into a higher orbital.

Because they are quanta of helical electromagnetic fields photons can be described as helical vortices in space just as electromagnetic fields are, but in the form of localized packets. The helical packets of photons are no longer attached to the particle generating the field and thus fly off at the speed of light. As explained previously photons always travel at the speed of light by the STc Principle since they have no internal structures to be computed and thus no internal proper clock time velocity.

So long as a mass or charge is present its velocity density effects propagate across the entanglement network (spacetime) at the speed of light. This is because all computational effects propagate through the entanglement network at the speed of light. Remove the mass or the charge and the surrounding velocity density field dies off at the speed of light.

Thus if an individual helix breaks away from a charge it will naturally propagate through spacetime at the speed of light. Photons of light, and other forms of electromagnetic radiation, are just individual electromagnetic helices freely propagating through space at the speed of light because they aren't tied to a source charge. This typically occurs when electrons transition to lower orbital energies and emit the excess energy in the form of a photon.

Photons are actual individual helical distortion packets moving linearly through spacetime as opposed to fields of helical velocity densities in all directions around charges. They carry energy proportional to the frequency of their helical rotations. By contrast the helical spacetime distortions around charges can be considered to be composed of virtual photons, as they are not transmissions of energy but do produce energetic effects when interacted with. They both consist of helical waves of the same basic form though the electromagnetic waves of photons are orthogonal combinations of electric and magnetic waves. Together these oppositely oscillating waves result in a helical spacetime distortion traveling at the speed of light.

Thus Universal Reality naturally explains light as a form of electromagnetic energy and naturally explains why it moves at the speed of light through the spacetime entanglement network. Again though, the speed of light is actually the speed of clock time, the speed at which computational effects propagate through the dimensional entanglement network that humans interpret as spacetime. Photons of electromagnetic radiation are fundamentally computational effects.

The helical distortions in spacetime directed outward from charges have fixed amplitudes and frequencies and wavelengths since the basic units of electrical charge and spin that produce them have fixed strengths. They differ only in the direction of their helical twists corresponding to the sign of the charge or pole that produces them.

However the electromagnetic radiation of light is not anchored to a fixed charge so its helices can be produced in a more or less continuous spectrum of frequencies proportional to the amount of relative motion converted to produce them. Thus the energies of photons range from gamma rays through visible light to radio waves.

Like the helices extending from charges, those of electromagnetic radiation can also rotate in either a clockwise or counterclockwise direction. This accounts for the possible clockwise and counter clockwise circular polarizations of light. In most cases, such as the light from the sun, light beams are a mixture of the two polarizations.

The helical waves of photons generally don't cancel each other out or cause attraction or repulsion because they tend to be mixtures of many different frequencies and due to their great velocities effects tend to be more or less instantaneous and immediately over. However coherent beams of photons of the same frequency such as those emitted by lasers do interfere if correctly tuned (Wikipedia, Laser).

Like all relative motion these helical waves are dimensional aspects of the entanglement network and thus extend perfectly well across the vibrational fields produced by mass. Thus they are naturally subject to relativistic effects. They follow and add to the lines of spacetime curvature produced by the presence of mass. So both models together appear to correctly model electromagnetic fields in the curved spacetime of general relativity.

ELECTRICITY

It's common knowledge that opposite electrical charges attract and identical charges repel each other. This is called electrostatic force and is experienced at the classical level in static attraction and repulsion of some common everyday materials. This is caused by the buildup of free electrons on surfaces and may lead to electric sparks as the electrons jump from surface to surface to balance the charges (Wikipedia, Electrostatics). In ordinary materials the charges of electrons and protons are almost perfectly balanced and bound in atoms and molecules except in ions or where electric currents are present.

So the electrical force is simply the repulsion or attraction between the equal or opposite electrical charges of elementary particles. This is the force that binds particles together and creates all the matter in our universe. It's also the force that holds identical charges apart so matter doesn't all collapse in on itself. These binding energies give matter the precise structural balance it needs to create the specific atoms and molecules possible in our universe and all the chemistry and life that emerges from it.

So the electric force along with the strong force that overcomes it to bind positively charged protons together in nuclei are the two main forces that make atoms and molecules and everything made of them possible in our universe. The atoms that make up all the matter in the universe are electrical balances of identical numbers of negatively charged electrons in orbitals surrounding an equal number of positively charged protons held together by the strong force in the nucleus.

Electricity, in the everyday sense, is not fundamental but emergent. Electricity is simply the *movement* of electrical charges through space. This can be in the form of loose electrons moving from atom to atom as a current through a wire, or as plasma where free electrically charged particles (ions) are moving through space as in lightning. The movement of loose electrons through a wire to form a current can be a continuous flow in one direction as in direct current or a back and forth flow of electrons in alternating current.

The electric energy that powers all the devices of our modern world is not the conversion of the intrinsic (internal) relative motion of electron charges but the energy of linear velocity of the electrons themselves moving through space through some electrical device. Electrons themselves are not used up from an electrical circuit as they power our appliances; that would involve the *mass* of the electrons being converted into a form of atomic energy and would produce an enormous unsustainable charge imbalance that would tear the atoms in the wire apart. The energy drawn from an electric current is some of the energy of their *flow* through the circuit. It's some of the relative motion of their flow that is converted into other forms of relative motion such as light, heat or mechanical motion.

And of course there must be an original external source of energy from a generator that is converted into driving the flow of the electrons through the power lines in the first place. Generators can be driven by the conversion of various forms of relative motion such as the energy of moving wind or water, the relative motion of heat from burning coal or nuclear fission, or the absorption of solar photons.

There is not nearly enough space here to describe all the common electromagnetic effects in terms of our theory. However note that when charged particles move their standing helical velocity density fields move with them as the change in motion propagates through the field at the speed of light. Changes in velocity (accelerations) of charges produce electromagnetic radiation because they manifest both changing electric and magnetic fields, which is what electromagnetic radiation consists of.

An electric current in a wire creates a corresponding circular magnetic field around the wire. Its direction (clockwise or counter-clockwise) depends on the direction of the current in the wire. This is the principle of the electric motor in which charges moving through a solenoid (current carrying wire coil) around a rotor magnetically rotates the rotor converting the relative motion of the electrons into the relative motion of the rotor.

The effect is reciprocal. A current is induced in a loop of wire when it's moved with respect to a magnetic field; the direction of current depends on that of the movement. Thus when a magnet is mechanically rotated with respect to a coil of electrical wire, or vice versa, its moving field of helical distortions acts on the wire in the coil to move its electrons along and generate a current. This is the principle behind electric generators.

There must be a closed (continuous loop) circuit for loose electrons to flow along a wire. Otherwise they would tend to pile up at one end and repulse each other there, and leave net positive charges at the other end that would tend to attract them back to that end. Thus electrons can't flow along a disconnected wire without being added at one end and discharged at the other.

The movement of electric charges either as the flow of loose electrons through a wire or as the movement of a magnet consisting of aligned charges, is basically two views of the same phenomenon. The movement of electric charges at one location induces the movement of electric charges in the other through their intermediary magnetic fields. If the charges are fixed in the material the material itself will move, while if the charges are loose they will move within the material. Both motors and generators operate as a result of the single principle that electric charges induce velocity vectors in adjacent electrical charges through intermediary magnetic fields causing them to move.

THE STRONG & WEAK FORCES

The strong force, or strong interaction, is the force that holds neutrons and positively charged protons together to form atomic nuclei. It's approximately 100 times stronger than the electromagnetic force, which it must be to overcome the mutual repulsion of the positive charges of the protons. This force is carried by gluons, which bind the quarks that comprise protons and neutrons. These quarks are bound by exchanges of gluons, analogous to how electrons are attracted to protons via exchanges of virtual photons, however there are 6 different 'color' charges for the strong force as opposed to the two plus and minus charges for electromagnetic force, and the single positive mass charge of the gravitational force.

The weak force or weak interaction is the last of the 4 fundamental forces of nature. The weak interaction is responsible for the radioactive decay of subatomic particles and it plays an essential role in nuclear fission. It is carried by the ±W, and Z bosons. It's of particular interest that the weak force is the only force that doesn't produce bound states.

The strong and weak forces and the electromagnetic force, are well understood in terms of the Standard Model, which predicts most but not all, of what occurs at the level of elementary particles. Notably it says nothing about dark matter or dark energy, and is unable to incorporate gravitation into the theory. Thus clearly a deeper more comprehensive theory is needed (Wikipedia, Standard Model).

Universal Reality suggests that progress will come from reinterpreting all these phenomena as computational processes which in concert generate the entanglement network science currently interprets as material structures in dimensional spacetime. We have suggested ways this might occur with respect to gravitation and electromagnetism and expect the strong and weak forces can be best explained in a similar manner.

Thus the strong and weak forces will likely be elemental computations on the data of elementary particles that conserve particle components so as to produce observational results in accordance with the standard model and whatever extensions are necessary to complete it.

Since all the forces of nature are energetic processes and Universal Reality proposes that all forms of energy are various forms of relative motion that dilate spacetime in a manner that transmits the relative motion to other particles in the field, it's reasonable to expect explanations of the strong and weak forces in terms of additional forms of velocity densities. In fact it's possible that the 4 forces somehow correspond to the different possible modes of vibration of space just as electron orbitals correspond to the possible standing harmonic waves in 3-dimensional space in atoms as explained in the section on Bound Particles in the next chapter.

It should be noted that the science of the standard model is already taking small steps in this direction. It agrees in one important respect with Universal Reality's proposal that mass is actually a form of vibrational relative motion. Both protons and neutrons are composed of quarks but the masses of the constituent quarks account for only about 1% of the rest mass of the proton and neutron. Thus 99% of the mass of protons and neutrons is actually composed of the relative motion of the massless gluons that hold the quarks together (Wikipedia, Proton). Universal Reality naturally interprets this relative motion as an additional form of velocity density computed at the level of the entanglement network.

Here we have a clear case in which modern science has now discovered that mass actually is at least mostly relative motion as Universal Reality predicts. And since protons and neutrons make up almost all the mass of the visible matter in the universe, that means close to 99% of the mass of the visible universe is relative motion even according to contemporary scientific theory.

QUANTUM REALITY

EVENTS & THE ENTANGLEMENT NETWORK

At its most elemental level the universe is computed in terms of particle events. At every P-time tick the data state of every particle in the universe is simultaneously recomputed. When particles interact including when particles interact with fields, including the field of empty space, each interaction is computed as an individual process by a separate *application* of the universal processor. This includes the computations of bound particles in atoms and molecules. In this way all data states of all particle interactions are recomputed at every P-time tick.

Every event conserves the total amounts of all properly defined particle components of all particles. This means that when events transform particles into other particles the conserved amounts of all particle components are exactly redistributed among the emitted particles.

Particle component conservation *entangles* all emitted particles on each of their particle components separately because each is separately conserved. Entanglement simply means that the values of all particle components of emitted particles bear fixed relationships to each other. This relationship is numerically exact when only two particles are emitted or relational when multiple particles are involved. It's important to understand that entanglements are not isolated occurrences in scientific laboratories as some might assume but relationships among particles generated by *every* particle interaction in the universe.

Thus the entire universe consists of an *entanglement network* of entangled particles because all particles are part of a universal network of chains of particle events. All particles in the universe have event connections with all other particles going back to the original event of the big bang. Thus the observable universe itself consists of the entanglement network of all its actual particles, or more precisely of the entanglement network of all its particle components. This entanglement network contains all the information of the observable universe including the data of all particle structures and their dimensional relationships. Observers interpret the entanglement network as a material universe in a pre-existing physical spacetime in their simulations of reality even though it actually consists entirely of the data of particle component relationships and is constructed on the fly by particle events.

For example if a total mass-energy of e is conserved between only two emitted particles then knowing the mass-energy of e_1 allows us to know the mass-energy of the other exactly since $e_2 = e - e_1$. However if there are 3 or more particles emitted the relationship between any two particles is no longer exact since $e_2 = e - (e_1 + e_3 \ldots)$ and the individual mass-energies of e_1, e_3, etc. could vary and still satisfy the equivalence. However the possible variation is highly constrained by the equivalence so there still is considerable entanglement. This is true of the entanglement of each type of particle component since each type is conserved and entangled separately.

Because each type of particle component is conserved separately the entanglement network consists of separate layers for each type of particle component. The separate layers are joined together at nodes representing events, and the layers connecting nodes represent particle components packaged as individual particles. All the layers of each line connecting event nodes taken together represent a particle since each particle is composed of its particle components.

Entanglement is easy to understand, as it also occurs at the classical level. If we take a cookie and break it in two, the two halves of the cookie are entangled on their masses because the sum of their masses equals the mass of the original unbroken cookie. If we measure the mass of one half we immediately know what the mass of the other half is. So the basic concept of entanglement is quite simple. *Every* conserved quantity is automatically entangled when divided into multiple parts or reapportioned among different entities. Entanglement due to conservation is a universal law.

All the components that make up particles are conserved and entangled by particle events. These include particle identity (lepton and baryon number), mass-energy, the other charges, spin, weak isospin, and space and time parity (handedness in space and time). Particle components are the little bits of what it takes to make something real and actual in our universe because particle components make up the elementary particles that compose all structures.

Particle components are numeric data entities rather than physical entities. They, rather than particles, are the actual fundamental elements of the observable universe because it is they rather than particles that are conserved through all particle events. Particles are composed of predefined valid sets of particle components stored as data templates in the complete fine-tuning.

Momentum is also conserved within the limits of the Uncertainty Principle but importantly the positions and velocities of particles that are used to dimensionally express the conservation of mass-energy and momentum are not conserved. Total mass-energy is conserved due to the equivalence of all forms of mass and energy as different forms of relative motion, of spatial velocity.

Successive particle events entangle more and more particles in entanglement chains. And since particle events typically involve multiple interacting and emitted particles each of which may interact in subsequent events all particle events taken together create a vast entanglement network that includes all particles in the universe and stretches via common entanglement links back to the original event of the big bang.

The *current* present moment surface of the entanglement network is the entire actual universe in the current present moment. The current data state of the entanglement network is the actual universe, which is a logico-mathematical data structure. It's not a physical universe but it contains the data we humans interpret as a physical universe.

The entanglement network is a data structure that is computed by the fundamental program as it simultaneously computes all individual particle events. It's the data of the relationships among all particles in the universe produced by elemental events and thus it includes both the mass-energy structure and the spacetime dimensionality of the universe in a single unified structure. It encodes the logico-mathematical structures of both mass-energy and spacetime in a single universal information structure. In this way spacetime and mass-energy structures are computed together by quantum events. Spacetime is not a preexisting container in which events occur or are even computed; it's the consistent set of all dimensional relationships among particles computed by events.

THE SOURCE OF QUANTUM INDETERMINACY

A major unanswered problem of physics is why the universe, in particular its dimensionality, is 'fuzzy' or indeterminate and random at the quantum level. Universal Reality provides a simple and convincing answer in terms of the same space versus time allocation of processor cycles that computes a relativistic universe at the classical level.

Universal Reality proposes that all forms of quantum indeterminacy are the result of random space versus time *oscillations* in the processor that computes all events in the universe. Specifically the zero-point energy fluctuations of the quantum vacuum, the related Uncertainty Principle, and the wavefunction representations of particles are all manifestations of very fine scale oscillations between processor cycles allocated to computing velocity in space and velocity in time.

As the processor computes the space versus time evolution of processes in accordance with relativity as explained previously there is a simultaneous fine scale oscillation of these cycle allocations. This manifests as an indeterminacy of space versus time at the quantum level. In effect the reference background against which observable dimensional spacetime values are being computed is itself continually oscillating between space and time and this results in the observed spacetime 'fuzziness' of quantum phenomenon. This makes sense because all such fuzziness reduces to indeterminacies of space versus time upon examination.

This is demonstrated most clearly in the Uncertainty Principle (Wikipedia, Uncertainty principle). The uncertainty of energy, which is spatial velocity, versus time, and that of momentum (again spatial velocity) versus position are clear examples of a source in a minimal but fundamental conflation of space and time. If observable values are always determined with respect to a dimensional background that is itself oscillating between space and time at the quantum scale the observed values will inevitably be uncertain in space and time around the value of the oscillations.

So the allocation of processor cycles between computing space versus time velocities provides a very simple and convincing mechanism to explain both quantum fuzziness and general relativity at the classical scale. Both quantum and relativistic processes become part of the same computational system.

WAVEFUNCTIONS & DECOHERENCE EVENTS

The entanglement network consists of event nodes connected by particle that link them. At Level 0 events are exact calculations of the conservation of individual particle components. However conserved energy values must be translated into dimensional positions and velocities to observably manifest. Whenever dimensional values of space and time are computed the oscillation of processor cycles results in uncertainties of space versus time variables and these space versus time uncertainties take forms describable by wavefunctions.

Due to the processor oscillations the computation of space versus time is inherently uncertain at the quantum level even though there is an underlying Level 0 numeric exactitude demonstrated by the exact trajectory of the wavefunction itself and its exact deterministic time evolution as described by the Schrödinger equation (Wikipedia, Schrödinger equation).

The exact trajectory of the wavefunction itself reveals the exact underlying Level 0 numeric data entity trajectory of the particle. However the form of the wavefunction as it travels represents the probabilistic mapping of that entity into observable dimensional values. So at every P-time tick the exactly conserved energy values computed by events are expressed as positions and velocities whose space versus time values are conflated at the quantum scale by the random oscillations of processor cycles between computing velocities in space and velocities in time.

The computational evolution of the particle's exact trajectory into observable dimensional space and time values derives from the tick-by-tick oscillations of the space versus time processor oscillations. These oscillations occur randomly at each P-time tick for each processor application computing a separate process but the form of the oscillations over time is that of the probability amplitudes that wavefunctions describe.

Thus wavefunctions and the Schrödinger equation correctly describe the time evolution of particles but what they are actually describing is the time evolution of the allocation of the processor cycles that compute that evolution. Wavefunctions don't describe the fuzzy form of particles with respect to an exact pre-existing spacetime but the actual computation of dimensional spacetime from particle events. It's the spacetime itself that is intrinsically fuzzy at the quantum scale as it's computed because the processor cycles that compute dimensionality continually oscillate back and forth between computing space versus time in a random manner at the quantum level.

Thus Universal Reality provides a huge paradigm shift in the understanding of quantum processes. They are not composed of particles that are intrinsically fuzzy with respect to an exact universal pre-existing spacetime but spacetime itself in the process of being created by the conservation of mass-energy by quantum events that is created as dimensional fragments that are intrinsically fuzzy with respect to each other.

Every entanglement event creates dimensional relationships among its emitted particles to conserve their total mass-energies. The dimensional relationships among all the particles emitted by an event form a *dimensional fragment* that continues to be computed as a single process by dedicated applications of the universal processor. Because the time evolution of all the particles emitted by an event are computed as parts of a single process the entire process remains *coherent* and all the entanglement relationships are preserved.

Thus there is an on going computational connection between all entangled particles and this is the secret to the apparent spin entanglement paradox. Entangled particles are computed as aspects of a single process and this is how a measurement of the spin of one of those particles immediately affects the others. This is explained in more detail shortly.

The exact numeric dimensional values of particles at Level 0 is not observable as previously explained. The totality of all such values constitutes the absolute computational space with respect to which all linear and rotational relativistic motion is relative to.

It's the allocation of processor cycles to the time evolution of exact particle trajectories that computes the observable dimensionality of particles. Observable dimensionality consists of particle measurements including measurements in the generic sense of particles measuring each other's dimensionality via mutual events. All our human scientific and sensory observations are ultimately scaled up dimensional values produced by individual particle-particle interactions in which particles measure each other's dimensionality by interacting.

What humans interpret as a continuous physical spacetime is actually a reification of the logico-mathematical consistency of all dimensional measurements in which they participate. There simply is no actual empty physical spacetime container in which events occur. It's a convenient illusion produced by our simulations based on the internal consistency of individual particle measurements. Our minds pin the dimensionality of individual observations together and extrapolate an imaginary spacetime container on that basis with us at the center.

Getting back to the entanglement network upon which this is all based, events conserve all particle components including total mass-energy. For particle components to be conserved they must have exact values. This is not a problem because the processor cycles are always computing exact dimensional values in every P-time tick even though those values vary probabilistically from tick to tick.

Assume that an event occurs when two particles from two separate processes both take on the same random dimensional values. This is a simplification as events can occur when colocations are not exact but the process is essentially the same. Because all processes are being computed simultaneously in the universal processor the processor immediately 'knows' when particles collocate.

The type of event is determined by the energy, degree of colocation, and types of particles involved. There are two general types of events, scattering events in which the particles themselves acquire new trajectories, and particle events in which the particles break apart into their particle components to form new particles with a new distribution of the same particle components.

Events cannot occur unless the total amounts of all particle components can be validly allocated among emitted particles. However when alternate sets of particles could satisfy this requirement all those alternate sets are a possible outcome. This is referred to as the 'Totalitarian Principle' (Wikipedia, Totalitarian principle). In this case nature decides randomly which set to produce based on their probabilities, which generally has to do with the most efficient (entropic) distribution of energy.

This is demonstrated in the very high-energy collisions of particle colliders such as the LHC where scores of particles may be produced by a collision of only two hadrons. Many different sets of particles could be produced so long as all particle component totals are conserved.

With high enough energy new particles are literally created out of the vacuum to conserve the incoming energy. For this to occur particle components must be actualized in particle-antiparticle pairs out of the vacuum. In this case the total particle component values of the newly created particles sum to zero so the original totals are conserved.

When an event occurs a new process begins that is now computed by a new application of the processor. In general the entanglements the particles had with the other particles of their previous particles are instantly resolved as the particles are separated from their previous processes to become parts of the new process that computes the new event.

Thus the new process begins with the dimensional values of the event and these values are then converted into the total mass-energy of the process so it can be conserved and redistributed among the dimensional variables of the particles emitted by the event.

This 'freezing' of the random dimensional values of two interacting particles by the interaction is called decoherence (Wikipedia, Quantum decoherence). It is often incorrectly interpreted as the instantaneous 'collapse' of wavefunctions by a measurement.

Once an event occurs the process begin all over again. All particle components of the event including the total mass-energy are conserved and redistributed among the emitted particles, which continue to be computed as aspects of a single new coherent process. The conservation entangles each of the particle components among all the emitted particles and the coherent oscillations of processor cycles in the process continue to exactly relate the dimensional and other particle component values of all particles in the process. The result is an evolving dimensional fragment that carries the coherent dimensionality of the entangled particles in the process.

Because the dimensionality of all the particles in a coherent process are computed by a single application of the processor the random values the processor produces at every P-time tick are all interrelated so as to perverse all the conserved values of the originating event. This is how the processor keeps track of the entanglement relationships among all the particles of each coherent process. It's because they are all being computed as parts of a single process no matter how they may be separated in observable spacetime. Nonlocality is no problem at all in a computational universe because everything always exists together in the non-dimensional computational space in which it's computed.

The processor that computes all processes automatically computes probabilistic dimensional values for all particles at every P-time tick. This is required so particles can be tested for colocation because otherwise the processor wouldn't know when or where to initiate events.

These random dimensional values computed at every tick are in effect *virtual decoherences*. They are exact dimensional values produced randomly at every P-time tick by the processor cycle oscillations for all particles. They are produced randomly on the basis of the probability distributions of how the oscillations evolve from tick to tick which again is accurately described but misinterpreted by the wavefunction model.

In the case of particles the magnitude of dimensional fuzziness is not only a function of the oscillations but also of the individual particle components. This is how different particles in the same dimensional fragment can have the varying dimensionalities their different wavefunctions describe. The dimensional fuzziness itself is always due to the processor oscillations but its magnitude is affected by the particle components for example the masses of the particles.

So there is an exact numeric dimensional background that consists only of exact particles with exact dimensional values. This is the dimensionality of the computational space in which everything is computed and it consists only of data and it is completely unobservable except by its effect on observable dimensionality.

Observable dimensionality is computed probabilistically with respect to this exact background due to the random processor oscillations between space and time cycles. This results in observable dimensionality being fuzzy between space and time at the quantum level and this manifests as the well-known dimensional fuzziness of quantum processes in wavefunctions, the Uncertainty Principle, and the related zero-point energy fluctuations.

Each separate application of the processor to an individual dimensional fragment is characterized by its own probability distribution of possible randomness over successive P-time ticks. This is accurately described by the quantum wavefunction model, however Universal Reality interprets wavefunctions as descriptions of the time evolution of processor cycle oscillations rather than particles smeared out in a fixed continuous pre-existing spacetime because such a pre-existing spacetime simply doesn't exist.

Thus the entanglement network that encodes the data of the entire observable universe consists of exact (within uncertainty limits) event nodes connected by particles whose observable dimensionality is probabilistic due to the fine space versus time oscillations of the processor cycles that compute it. The universe consists of a vast dynamic entanglement network of more or less exact particle events linked by dimensionally probabilistic particles whose probability distributions move along exact trajectories.

Due to the non-proportionality of particle masses, a portion of conserved event energy must be expressed in terms of velocities and positions, so an observable dimensionality is created in which velocities and positions can manifest the excess energy as the kinetic energy of motion. This is how and why quantum events create spacetime, as the only available way for events to exactly conserve mass-energy. It's the non-proportionality of particle masses that requires the creation of a dimensional spacetime in which mass-energy can be conserved in terms of velocities and positions.

The result of this process is the entanglement network whose current P-time surface is the entire actual universe in the current present moment. The observable universe of science and the senses consists entirely of observable values ultimately produced by particles measuring particles in decoherence events.

The observable universe is a consistent logico-mathematical data structure in that the relationships among all observables are internally consistent according to the laws of physics. Even though particle dimensionality evolves probabilistically between events it does so in a logico-mathematically consistent manner. And at classical scales the covariance of general relativity is part of this consistency.

QUANTUM GRAVITY

A major problem in modern physics is the apparent incompatibility between quantum theory and general relativity. This is due primarily to the inconsistent manner in which the two theories view spacetime. Quantum theory considers spacetime an exact static preexisting container within which quantum events occur, while the preexisting spacetime of general relativity is dynamic and shaped by the presence of matter.

Universal Reality proposes a completely new approach in which there is no preexisting spacetime and what we call spacetime is the consistency of dimensional relationships produced by quantum entanglement events. In this theory spacetime is computed by quantum events as the dimensional relationships among mass-energy structures as they evolve.

To incorporate relativity into this model quantum events simply scale the dimensionality they compute proportional to the mass-energies of the particles involved in the event and this automatically produces the dynamic curved spacetime of general relativity or in our model vibrational density. Every quantum event simply computes the masses of the particles involved as fields of vibrational density in the surrounding fabric of space. In this manner the myriads of ongoing quantum events of all the particles of say a planet are additively computed as a gravitational field consisting of mass vibrations in the space surrounding the planet and as we have seen this field of vibrational density produces the relativistic effects of a gravitational field.

The relativistic effect is extremely weak at the quantum level due to the relative weakness of the gravitational force so the Schrödinger equation accurately describes events at the quantum level without it, but for large aggregates of particles the simple scaling of dimensional relationships by the presence of mass-energy produces the spacetime of general relativity in a rather straightforward manner as fields of velocity density. Once the paradigm of quantum events creating spacetime in the form of an entanglement network is accepted this practically suggests itself. Observable spacetime is then the interpretation of the dimensional aspects of the entanglement network in observer simulations.

In relativistic terms this is how mass tells space how to curve. The reciprocal effect, how space tells mass how to move, is due to the allocation of fewer processor cycles to computing velocity in time after some of the fixed number of cycles are used to compute spatial velocity. And the total spatial velocity is simply the sum of linear velocity and the velocity density of gravitational fields as previously explained.

So the presence of mass (energy) in wavefunctions scales the dimensionality that is created by quantum events by adding velocity density as a surrounding gravitational field, and this intrinsic velocity of the field in turn scales the allocation of processor cycles to computing the velocity in time of events in the field. This gives us a neat approach to explain how quantum events create general relativistic spacetime by scaling the dimensionality of mass-energy structures as they create them.

A VISUAL MODEL OF THE UNIVERSE

Universal Reality proposes the computational mechanics of the processor that computes the universe is the source of both general relativity and quantum phenomena. There is a very useful visual model that makes this clear when correctly interpreted.

Imagine the universe with one spatial dimension suppressed for clarity as a 2-dimensional surface. This surface represents a section of the 3-dimensional space of the universe in the current universal present moment, the only moment that actually exists. At every P-time tick the current state of the surface is completely recomputed to produce a new state for the entire surface. The entire evolutionary history of the universe corresponds to the succession of all past surfaces of which only the top one is the real and actual universe in the current present moment.

If we wish we can visualize the entire recomputed surface rising with each P-time tick so the entire stack of surfaces represents the evolution of all the data of the universe through time. However only the topmost surface corresponds to the entire actual universe in the universal current present moment.

Now to be clear the universe is not actually the physical dimensional structure this visualization implies. It's a graphical visualization of what is actually entirely a data structure. It's a model of the observable dimensionality of the data of the universe in exactly the same sense that dimensional data in a computer program consists only of numbers in memory locations but can be graphically displayed as objects in space on a computer screen. The same is true of the actual universe which consists of logico-mathematical data in the form of the entanglement network in computational space but which observer minds visualize graphically as a physical spacetime universe.

Note also that even observable spacetime consists only of individual observable dimensional values rather than a continuous encompassing physical space. So the surface model is misleading as observable dimensional values are points that a surface would connect if it existed which it doesn't except in our mental simulations of the universe. It just so happens those points always fall where a surface would be if it existed. Our minds note the consistency of where points fall and extrapolates that into a continuous empty space, but this empty space is a fiction of our simulation of reality rather than part of reality itself.

The data of a computer model exists in the computational space defined by the program. Similarly the data of the universe exists in the computational space of the quantum vacuum, which is the medium or substrate of existence and thereby gives all data within it actual existence as a real actual universe.

A major difference between the two types of models is that the quantum vacuum in which all the data of the observable universe exists acts as a single universal processor that computes all the data of the universe simultaneously, whereas in computers the processor(s) computes data stored outside the processor sequentially fetching single units of code and data into the processor to compute them. In the quantum vacuum all the data of the universe always exists within the universal processor that computes it. Only in this manner can internal consistency among all ongoing processes be guaranteed.

So our visual model of the universe is just a graphical visualization of how spacetime could be interpolated from observable dimensional values as they are computed from elemental data. However the model is extremely useful in understanding how the universe is computed if we keep these caveats in mind.

Each successive surface layer of the universe is computed from the data state of the previous layer. This takes place by separate applications of the universal processor acting on the data states of all the individual processes in the universe. Thus each individual process in our model is computed by an individual *application* of the universal processor. In this manner every coherent process in the surface is in a continual state of recomputation.

Now imagine each individual computational process of the universe occurring within the surface as it progressively rises with each P-time tick. The surface is not a pre-existing physical spacetime but represents the unobservable background reference space in terms of which observable dimensional values are computed. Rather than being a dimensional space itself it's the standard reference with respect to which dimensionality is computed.

Now imagine each point in the surface is finely oscillating back and forth sideways (in space) and up and down (in time). Thus the dimensional values of the process being mapped to the background reference become intrinsically uncertain at the scale of the oscillations. The process itself (the particles involved) is exact but its dimensionality is inherently fuzzy because the reference space and time rules that measure it are continually oscillating.

Now visualize each individual computational process as a separate plane because each process is computed by a separate application of the processor, by a separate call on the elemental program. Thus each individual process will be subject to a different processor oscillation sequence from tick to tick as it's being computed. All aspects of each individual process are computed as part of a coherent whole so that each part has an exact relationship to all other parts in this respect.

However because each process is an individual application of the elemental program computed by its own application of processor cycles, every individual process and dimensional fragment can be visualized with respect to a separately oscillating reference background with respect to which its dimensionality is being computed. So there is an different oscillation pattern for each coherent dimensional process that is also a function of the internal details of each individual particle of the process.

Each dimensional process computes what can be considered a separate dimensional fragment because its dimensionality is inherently fuzzy with respect to all others.

We can retrieve the overall view of the model by collapsing all the individual process planes of each P-time surface back into the surface so long as we remember that each individual plane corresponding to an individual process and dimensional fragment is actually oscillating differently because it's being computed by a separate application of the processor with it own individual oscillatory pattern.

The result is that entangled particles are still well described by their individual wavefunction equations. But we also have a non-paradoxical explanation for entanglement and quantum indeterminacy in general in our model that is lacking in quantum theory. And because dimensionality is computed along with particle structures our model is also consistent with general relativity as well as quantum theory.

Though the dimensionality of all individual processes oscillates differently at the quantum scale the *classical scale* dimensionality of each process's background surface appears identical because the oscillations are not detectable at the classical scale and so the classical universe behaves in a relativistic rather than quantum manner.

Thus Universal Reality neatly unifies quantum theory and general relativity in a single model by demonstrating how they are both computed by the same space versus time velocity allocation of the processor cycles that compute the mass-energy and dimensional structures of the universe in the form of an entanglement network of the particle components that make up elementary particles.

RESOLVING THE SPIN ENTANGLEMENT PARADOX

The processor oscillation model explains the coherence of entangled particles in a dimensional fragment. They were created together as part of a single process and continue to be computed together as part of that single process so long as they remain entangled.

For example spin entangled particles initially created with equal and opposite spins in a single process of spin conservation, a single application of the elemental program, continue to move apart as part of a single process that computes the evolution of both particles with the same processor cycle oscillations at each P-time tick.

The spin entanglement paradox is of course considered a major unsolved problem of quantum theory, which assumes quantum events occur separated in a physical spacetime. But in a computational model everything is *automatically local* because everything is computed in a single universal computational space with no physical expanse so computing non-locality and spin orientation entanglement is no problem at all so long as a single coherent process computes it.

The apparent paradoxes of quantum reality all turn out to be with respect to a mistakenly assumed pre-existing spacetime contained that doesn't actually exist. By deprecating this pre-existing spacetime container and showing how dimensionality is actually computed by quantum events all quantum paradoxes are resolved. The same model simultaneously unifies quantum reality and general relativity.

In our computational universe all quantum paradox is resolved because there is no actual spatial distance between the data of separate particles. Everything (all the data of the universe) is always in the same place (the single processor) at the same P-time. Thus the entangled values (for example spin orientations) of two particles at different locations are automatically related because they are computed as a single coherent process with the same pattern of processor oscillations.

In a computational universe there is no problem of non-locality or faster than light transmission of information because everything is always in the single processor that computes the universe at every moment. This straightforward non-paradoxical explanation of spin entanglement is excellent evidence in support of our theory.

So there is an inherent computational connection between the spins of spin entangled particles, as there very obviously must be. This occurs automatically because they are computationally related by the conservation of the event that created them. Since they are computationally connected and conserved then it's easy to see why decoherence of either one to an exact orientation automatically decoheres that of the other to the exact opposite orientation because only this preserves the original conservation of spin.

Thus if the spin orientation of either particle is measured the process that computes the measured orientation is already simultaneously computing the spin orientation of the other particle and when the orientation of the other is fixed so is that of its partner since they were always being computed together at every step of the way since their creation. The spin entanglement paradox is no longer paradoxical but the straightforward consequence of a computational universe in which entangled particles are computed as aspects of a single process as separate dimensional fragments.

In effect the decoherence of the spin orientation of either particle via a measurement automatically decoheres the spin orientation of the other. This is a general principle in Universal Reality; that a decoherence of any of the entangled particles in a dimensional fragment automatically decoheres the others because they are all being computed as aspects of a single process by a single application of the elemental program and processor.

If there are only two particles the decoherence is complete but if there are multiple other entangled particles in the dimensional fragment their entanglements are adjusted by the decoherence of the measured particle which drops out of its dimensional fragment to become part of a new dimensional fragment computed by the process computing the new event.

To be more accurate all the particles in a single dimensional fragment *virtually decohere* to exact dimensional values at every P-time tick. This means their virtual decoherence values are all interrelated so as to preserve the conservation of the event that created them. Thus the positions and velocities of all particles in the dimensional fragment are exactly related at every P-time tick.

However if one particle's virtual decoherence is converted into an actual decoherence by a colocation with a new particle the dimensional fuzziness of the other particles in the dimensional fragment then continue on as before adjusted by the fact of the actually decohered particle dropping out to join the newly created dimensional fragment of the event that decohered it. However in the case of entangled spin orientations of only two particles the actual decoherence of the spin orientation of one particle produces a permanent decoherence of the spin orientation of the other particle in its dimensional fragment.

This solves the problem of why spin orientations are always opposite, but why do they both remain fixed once one is measured? This is simply because spin orientation is not intrinsically affected by processor oscillations because it's an orientation in space that has no time component unlike a velocity. Therefore the oscillations of temporal position and velocity don't affect it and neither do oscillations of spatial position and velocity.

But why aren't spin orientations determined when they are created? Why are they created equal and opposite but with no specified orientation to the background? This is because total spin would be conserved by any orientation of the paired particles so long as they were equal and opposite. Thus there is no computational rule to determine any specific paired orientation and whenever there is no computational rule either a choice is immediately made randomly if it can be or is deferred until specific rules apply.

In the creation of paired spins there are no determining rules and the choice is deferred until it is made in a subsequent event that has rules that force the decoherence of some actual spin orientation. The new event must decohere the spin to an exact value so as to be able to conserve the total spin of the new event. This is an important general principal that underlies all quantum randomness.

FIELDS & VIRTUAL DECOHERENCE

All particles must *virtually decohere* to exact values (within uncertainty) at every P-time tick. This enables the elemental program to test for any colocation that would trigger an actual particle event. Colocation means the particles have close numeric dimensional positions, not that they are actually close in a physical space.

If two particles are collocated an event is triggered and an actual mutual decoherence occurs to the values of the virtual one. An actual decoherence to exact dimensional values is necessary because events can only conserve exact particle component values. Thus events always require actual decoherences. The virtual decoherence is actualized in the event and the exact decohered dimensional values of all particles are then conserved in the event which begins to be computed as a new process by a new application of the processor and elemental program.

When a decoherence event occurs new trajectories for all resulting particles are created originating from their decohered values and begin to be computed together as coherent aspects of a single new process in a new dimensional fragment.

This model covers particle interactions but it's not complete. Particles interact with fields as well as other particles, so fields must be added to our decoherence model. In particular we need to include the interaction of particles with electromagnetic and gravitational fields in a manner that explains general relativity.

Fields are a mechanism by which particles interact with other particles at a distance. Charged particles attract or repel each other at a distance and the mass-energy charge of gravitation interacts with all other particles at a distance through gravitational fields. Fields are intrinsic aspects of charges. The charge itself includes its field.

To include the effects of fields charged particles are modeled as spherical fields of velocity density rather than points. As discussed earlier the charges of the four forces are all forms of mass-energy and thus different forms of relative motion whose velocity densities are their strengths. An alternate but largely equivalent model is to treat fields as exchanges of particles, which may be more useful for the strong and weak forces, which are more clearly mediated by their respective bosons. Recall that charges including their fields and space itself are both two aspects of the same phenomenon of spatial velocity.

Thus fields are dimensionalized as spherical clouds of intrinsic relative motion surrounding charged particles and traveling with them as they move. Everywhere a particle goes its field accompanies it because the field is an actual part of the particle. The charges of particles actually are intrinsic velocity density fields centered on their nominal particles. Charges are not points but velocity density fields with forms corresponding to the type of charge. These velocity density fields are equivalent to the dilation or curvature of spacetime in the general relativity model.

A particle located in a field experiences its intrinsic velocity and thus acquires an additional spatial velocity that reduces its temporal velocity in accordance with the STc Principle. Each point in the field can be characterized by an STc tilt that slows its clock proportional to its velocity in space so its total velocity through space and time is always equal to c. This produces the correct time dilation and other effects of general relativity.

Fields are part of charges so their dimensionalities oscillate with the dimensional oscillation of their central particle. The dimensionality of the whole field oscillates with its central particle at each P-time tick because it's computed relative to its center.

Particles interact with fields; so entangled particles computed as part of the same coherent process can be affected differently by fields because one particle can be deeper into the field than another. Thus the probability distributions of their dimensionality with respect to the field can be different even though they are parts of the same dimensional fragment and being computed as part of a single process.

In order to compute the movement of particles through fields including the 'null' field of 'empty' space virtual decoherence events must take place at every processor cycle. Every processor cycle virtually recomputes the dimensionality of all particles in the observable universe as aspects of a single universal event. All particles virtually decohere and are tested for field interactions and colocations with other particles that might trigger actual particle events.

At every processor cycle every wavefunction and dimensional fragment, including their fields, is tested against others to determine the probability of interaction and events are computed on this basis. If particles interact they decohere as before, but if a particle's virtual location intersects that of a field, then the effect on the particle is calculated and the trajectory of the particle is adjusted accordingly.

Since field effects are energetic they modify the trajectories of affected particles. The total conserved relative motion of the dimensional fragment is adjusted by the amount of the effect on the individual particle. There is an instantaneous virtual decoherence that produces an exact random position and velocity of the particle relative to the field so the effect can be computed. Then the total conserved relative motion of the dimensional fragment is adjusted by that amount.

The entire process repeats with every P-time tick for all particles. There is a continual series of virtual decoherences at every P-time tick of all the entangled particles of all dimensional fragments that modifies them by the field interactions of their individual particles. There is still an entanglement carried by the dimensional fragment, but it now carries the modified total mass-energy rather than just that of the event that created it. It's the original relative motion adjusted by the cumulative effects of field interactions on its member particles at any point in time.

To what extent are entanglements maintained when particles traverse fields of different strengths? Are they still being computed by a single process and its specific oscillations or are they now affected by the oscillations of the process computing the field? This apparently depends on the strength of the particle-field interaction.

The passage of a wavefunction through the null field of empty space obviously doesn't cause it to decohere or all particles would exist in a decohered state all the time, but there must be some process of virtual decoherence at every tick in case particles interact strongly enough to produce an actual decoherence.

A particle passing through a field is affected by the field but the passage doesn't affect the field. Of course if the particle carries its own field the fields will additively interact and each particle will affect the other but the mere fact of a particle passing through a field doesn't affect the field. There is no influence back to the central charge creating the field other than via the field of the passing particle. This is key because it means there can be no actualized mutual decoherence from field interactions alone. There is only an effect on the passing particle, not on the particle generating the field.

One could argue that if fields are relative motion then their effect on a particle is energetic and should conserve energy by affecting the field but we can replace the fields as relative motion model with the equivalent model of fields as STc tilts and the effect is now seen as inertial so this is not really a problem.

Thus particles in fields don't actually decohere with the fields. There is an instantaneous virtual decoherence as there is with everything at every P-time tick so that the instantaneous effect of the field on the particle can be computed but there is no actual decoherence of the particle and thus no mutual decoherence of it and its entangled particles in their dimensional fragment. However the effect on particles in fields does progressively affect their coupling with entangled particles and gradually weakens the entanglement.

This model enables particles to pass through fields that change their trajectories as well as null fields that maintain their trajectories without actually decohering them or their entangled particles, which is what we actually observe.

So actual versus virtual decoherence depends on the strength of the interaction a particle has with its environment. In general interactions with fields are not sufficiently strong to decohere particles though cases of scattering in which two particle trajectories are changed by each other's fields are probably best modeled as mutual decoherences. One needs to test the effect in particular circumstances to determine this. Our model covers both cases.

Another issue is that the quantum vacuum background is itself full of zero-point energy and its basic dimensional constituents fluctuate from tick to tick. This induces an inherent fuzziness of all dimensional values from tick to tick at the elemental level, which at the classical aggregate level averages out. This produces some level of fuzziness in the mutual positions and velocities of all particles though apparently not spin orientations.

This part of the model also explains the partial decoherences observed in some situations and those instances in which particles can be gently measured without decohering. This effect should be a measurable test of the model.

DIMENSIONAL FRAGMENTS

In our model each individual process is represented against its own oscillating plane of incipient dimensionality, each with its individual spacetime fuzziness at the quantum scale. However at the classical scale the overall forms of all planes merge to the local relativistic spacetime velocity density field. This makes it easy to computationally test for particle colocations as all particles virtually decohere with every tick.

When particles collocate in computational space particle events occur and begin to be computed as a single new process. The particles are transferred from their previous processes to the new process and their entanglement with their previous partners is broken and new entanglements are produced by the conservation of the particle components of the newly interacting particles.

Thus in the universe at large as particles continually interact their dimensionalities continually merge and are realigned, and in turn their relational connection with previously entangled particles is progressively diluted and previous dimensional fragments are gradually reduced as particles leave previous dimensional fragments and join new ones.

In this manner the entanglement network of all particles is continually shuffled and updated. At any point in P-time the current entanglement relationships among all particles is the data of the entire universe, both in its particle structure and dimensionality.

Each oscillating plane of a separate process can be considered a dimensional fragment, and observable dimensional values are produced when dimensional fragment planes transiently align as their individual particles decohere with those of other planes. In this manner the dimensionality of the observable universe is continually rebuilt from the observable actual mutual decoherences of particles from different dimensional fragments.

If we now take an orthogonal (non perspective) view from the top and collapse all the planes of separate processes back into a single surface of the present moment of the observable universe we arrive at a picture that is superficially much like a standard interpretation of quantum theory. The observable universe now consists of the current set of all decohered dimensional values, which together form a consistent logico-mathematical entanglement network in which the decoherence events producing the observable values are the nodes.

And the event nodes of this network are connected by links, which are dimensionally fuzzy and could well be interpreted as particle wavefunctions. Thus we have a theory that looks like quantum theory on the surface, but which is fundamentally richer and consistent both with general relativity and a non-paradoxical view of quantum events.

This visual model is not quite complete and needs to be tweaked because entanglement is not always either or, and the separate planes (dimensional fragments) making up the surface may not always be entirely independent. In general events dilute but don't completely destroy entanglements with previous partner(s) in the dimensional fragment. There always remains some relationship among all particle component values in the entire entanglement network including those of mass-energy across all particles. This underlying logico-numeric consistency among all components of the entanglement network is in fact the consistent structure of computational space with respect to which all observational dimensionality is ultimately relative to.

However the *coherence* of entanglements is lost in decoherences because the same single process as before is no longer computing the entangled particles. The decohered particle is now being computed by the same coherent process as its newly entangled partner(s) rather than that of its previous partner(s). Only a single application of the processor with its unique oscillation pattern can compute any one process. Thus if the newly entangled particle is now being computed by a new process it loses coherence with the previous process that is no longer computing it.

Thus there remains an underlying numeric relationship among the dimensional values of all particles in the entanglement network even though only the currently entangled sets are being computed as separate coherent processes with a single application and oscillations of processor cycles.

Thus all particles in the entanglement network will have consistent numeric relationships (within Uncertainty limits) among all their particle component values. They will not always be exactly one-to-one but they will never be inconsistent. And the set of all decohered dimensional values, which is the set of all *observable* dimensional relationships among all mass-energy structures, is the set of all dimensional values observers interpret as spacetime.

Thus observable spacetime consists of interrelated dimensional values (positions and velocities) sewn together by all decoherences from the dimensional fragments created by quantum events conserving and entangling the mass-energies of emitted particles.

Observable spacetime forms a consistent logico-mathematical structure because events exactly conserve mass-energy. While mass-energy is probabilistically manifested as dimensional observables at the quantum scale observable dimensionality seems exact at the classical relativistic scale. It's the logico-mathematical consistency of observable dimensional values that humans physicalize and interpret as an encompassing spacetime container within which events occur.

From an overall view particles and events consist of exactly conserved numeric particle component values computed by individual applications of the elemental program to each separate process. Because of the non-proportionality of particle rest masses conservation requires a dimensional spacetime in which positions and velocities can conserve excess mass-energy as linear velocity. However due to the intrinsic oscillations of processor cycles allocated to computing velocity in space versus velocity in time those observable positions and velocities are inherently fuzzy with respect to each other and the reference background.

The virtual template by which dimensionality can be constructed can be visualized as a plane x, y spatial surface rising at every recomputation in universal P-time. But because of the space versus time processor oscillations the surface of this reference template is oscillating differently at every point between space and time for each computational process.

Thus when the exactly conserved values of energy are mapped to positions and velocities against the oscillating background the resulting values appear intrinsically uncertain with space and time conflated at the quantum scale. These oscillations are also the source of the zero-point energy fluctuations of the quantum vacuum.

The oscillations are also the reason that the exact trajectories and time evolution of entangled particles are describable as wavefunctions. However each *set* of entangled particles is computed as a separate coherent process by an individual application of the elemental program and processor. As a result the oscillations of each entangled process are unique and must be visualized as a separately oscillating plane. These separate oscillation planes are effectively fragments of dimensionality, each a separate variant of the virtual dimensional template with respect to which observable dimensional values are computed.

Then as particles from separate planes interact and decohere in new events they begin to be computed by new applications of the processor and form new dimensional fragments. At this point the previous dimensional fragments from which the interacting particles decohered are reduced to their remaining particles and continue their independent evolutions. In this manner entanglements are in a process of continual reduction and recreation and this overall process produces the observable dimensional values of the entanglement network we interpret as a physical spacetime.

In this model all the quantum equations still work as they did before, it's only the interpretation that is new. A wavefunction no longer represents a particle smeared out in a single fixed space, but the uncertainty of how a particle's dimensionality relates to the oscillating background of dimensionality itself. Wavefunctions now become a description of the probabilities of how particles from separate dimensional fragments can join and decohere in an event common to both to produce a new dimensional fragment.

Thus the positions and velocities (momenta) of particles that are part of a single process and single dimensional fragment are computationally exact with respect to one another, but are not exact with respect to particles of other processes being computed with different processor oscillations. However if two particles from different processes are to interact their positions and velocities must become exact with respect to each other for their energies and momenta to be exactly conserved.

This requires the positions and velocities of interacting particles to become transiently exact with respect to each other. This process results in a decoherence in which the dimensional fragments of two particles become transiently exact with respect to each other and the decohered particles leave their previous dimensional fragments and begin to be computed as a new dimensional fragment with a new processor application with its own dimensional oscillations.

A decoherence is effectively a measurement of the position and momentum of each particle by the other and the results can be amplified to a lab instrument. This is the basis of all measurements, which are all decoherences of individual particle interactions amplified by measuring instruments or sensory systems.

So positions & speeds are exactly fixed among entangled particles exiting an event as a dimensional fragment, but not fixed with respect to particles exiting separate events. Thus if the positions and velocities of one particle in a two particle dimensional fragment was measured (within the limits of uncertainty) one would know the exact position and velocity of the other one at that same moment. Again within the limits of uncertainty, which is the underlying oscillation effect.

When decoherence occurs in an interaction between particles of two separate coherent systems it instantaneously aligns the two systems and all member particles. A decoherence of one entangled particle automatically decoheres the other. But as soon as the interaction event occurs it begins to be computed as a new independent process and the interacting particles lose their previous coherences with their parent dimensional fragments, which continue on in a reduced form on their own.

THE NATURE OF PROBABILITY FUNCTIONS

Ultimately a wavefunction is simply a probability function rather than anything 'real'. Contrary to the standard interpretation of quantum theory there is no physical wavefunction that particles are. Particles are exact numeric trajectories whose *observable* dimensionality over the exact trajectory varies probabilistically due to the oscillations of the dimensional background against which observable dimensionality is instantiated.

Thus there were never any other actual dimensional values than those actually chosen even though they appear randomly and weren't themselves actual until they occurred. The other possibilities in the wavefunction have the same status as the possibilities of a thrown dice landing on a particular side. As soon as the dice lands on one particular side the other possibilities 'collapse' and vanish erased by the reality of the side landed. So the probability distributions described by wavefunctions were never anything real, but only a *description* of how the particle's dimensionality *might* become real. Nothing physical or real has happened to the other probabilities; it's just the description that has changed.

If a man takes the train in the morning the probability that he didn't take the train vanishes to zero but there was never anything real or actual associated with that possibility. This is the essence of wavefunction 'collapse'. The wavefunction was never something physical or even numerically real that existed, it was simply a probabilistic description of what *might* exist, and as actual events occur the descriptions of what might be continually change with no actual consequences at all. They were never part of reality in the first place but only some observer's description of a possible future reality that never had any actual reality.

Millions upon millions of probability descriptions of practically anything can be imagined with greater or lesser accuracy. But making one up doesn't add anything to the real world, which continues on its merry way. Same with wavefunctions, they are merely descriptions of how things might become observable; they are not real in any other sense at all so there is nothing actually collapsing but the description of an unfulfilled possibility. Wavefunctions are useful in making predictions, but only in the same sense as anything else. It's not the prediction that is real but the actual behavior of what was being predicted. The whole notion that wavefunctions are somehow actual particles is entirely misguided and a wholly incorrect interpretation of quantum reality.

Note this is also true about every aspect of the past including the complete fine-tuning. No matter how we can imagine any part or parts of it could have been different these are all completely imaginary *descriptions*, which never had any reality whatsoever. And every one of those possibilities has now collapsed into non-existence by the actual events of the past. Thus there never was any actual possibility of the past being even one iota different than it actually was and this includes the complete fine-tuning itself. There was never any actual possibility of the complete fine-tuning being different than it was and continues to be. All other possibilities are completely and absolutely falsified by the actual data state of the observable universe in the present.

One might claim the difference between classical and quantum predictions is that we are dealing with a complex probability amplitude in the form of a wave that can interfere with itself (as in the double slit experiment) and other predictions. But this is not true. Classical level predictions are frequently interdependent on each other. And when any prediction is itself complex with multiple interconnected aspects, changes in one aspect can often affect the others. So wavefunctions are not nearly as mysterious as they are made out to be.

So wavefunctions are not real, they are not part of computational reality itself, but in our interpretation they do reveal something important about how processor cycles oscillate. They oscillate in a manner whose probabilities are accurately describable by wavefunctions and the Schrödinger equation.

THE SOURCE OF QUANTUM RANDOMNESS

The elemental program computes everything exactly to the extent it has algorithms that cover them. But whenever the laws of nature have no exact decision making rules decisions are made randomly among possible choices based on their probability distributions. This is a basic principle of reality and the source of all the randomness in the universe and the reason reality is not deterministic.

Random choices happen when events could validly emit alternate particle sets, or when particles can be emitted with alternate trajectories while still conserving energy and momentum. In such cases dimensional values are randomly chosen from available choices according to their probabilities, which generally have to do with minimizing energy levels though in the case of spin entanglement the probabilities of any equal and opposite orientation are the same because there are no energy differences.

And importantly it's also true of the space versus time oscillations of the processor that computes the observable universe. Or in the usual quantum interpretation the measurement of observables from wavefunctions where observable values are chosen randomly from their probability distributions in a process of decoherence.

It's also true of the spin orientation paradox where the decision is delayed. In the case of spin orientation the randomness occurs not as the entangled spins are created but when they are measured, as they must decohere to exact values to be conserved in events.

For example, the spin orientation of the first particle measured can have any possible random direction within 3-dimensional space. Whatever that random choice turns out to be the spin orientation of the other particle will automatically also decohere to be equal and opposite since that opposite spin relationship was already part of the two particle dimensional relationship, which is now aligned to the orientation of the laboratory dimensionality.

In general quantum events are random because there are no mathematical rules for how to exactly align the separate independent dimensional fragments produced by separate processes, so nature must choose alignments randomly within allowable constraints. This is effected by the randomness of the processor cycle oscillations.

Most randomness at the quantum level seems to occur with respect to dimensionality as it arises computationally in the merging of separate dimensional fragments as they are aligned and created by decoherence events. Thus prior to the emergence of dimensional interactions the computations of reality can be quite simple and exact as they contain none of the complexity necessary to describe events in relationship to intrinsically fuzzy dimensionality.

Ultimately the random oscillations of processor cycles as dimensionality is computed appears to be the source of most of the randomness of the universe, since all actual randomness occurs only at the quantum level. The apparently random processes of the classical level are either quantum level randomness amplified by supportive computational structures or simply processes that are too complex to be exactly computed. Most classical level non-predictability such as weather forecasts is a combination of the two.

This mechanism of random choice due to a lack of exact decision making rules at the quantum level is the ultimate source of all the randomness of the universe. Without this single mechanism the universe would be completely deterministic, the future would be fixed, and there would be no free will. Thus all the randomness of reality which gives meaning and frees the universe from complete determinism is an emergent manifestation of the manner in which dimensionality is computed by the processor cycles that compute the universe. It's the quantum scale oscillations of the processor that frees us and the entire universe from complete determinism.

RESOLVING QUANTUM PARADOX

The common characteristic of most quantum 'paradoxes' is that they are paradoxical only with respect to a mistakenly assumed fixed pre-existing spacetime background, but in Universal Reality this fixed pre-existing spacetime doesn't exist so the paradoxical nature of quantum events vanishes.

Universal Reality treats spacetime as a computational structure that emerges at the level of particle events as frames of dimensional relationships. Because these dimensional relationships are directly computed by the conservation laws we know they exist and we know what they are. But there is no reasonable source in traditional science for the single universal spacetime that it assumes. How does it get there? Where does it come from? Traditional science has no answer.

Combine this with the fact that the empty spacetime of traditional science is not even observable, that only dimensional relationships are actually observable. A general principle of science is that unless something is observable its reality is questionable at best.

Thus it's eminently reasonable to treat spacetime as independently oscillating frames of dimensional relationships computed by quantum events. Not only is this approach much more reasonable but it resolves the apparent paradoxes of quantum theory in a completely natural non-paradoxical manner as the alignments of separate dimensional frames as they are joined by decoherence events. An excellent example is how it resolved the spin entanglement paradox.

The model of entangled particles as dimensional fragments computed as independent processes neatly resolves the seemingly paradoxical nature of quantum phenomenon. All quantum paradox is now seen as an artifact of science's mistake of assuming an exact fixed preexisting spacetime within which events occur. This is due to a lack of understanding on the part of scientists of how our minds simulate a physical spacetime reality that is much different than actual reality.

We have already explained the source of quantum randomness, the wave-particle duality of nature, the Uncertainty Principle, and the spin entanglement 'paradox' in a non-paradoxical manner but it's worthwhile covering a few more aspects of quantum phenomena to really clinch the case.

For example in the case of the double slit paradox that is often taken to demonstrate the wave nature of material particles it's not particle waves that are interacting with the slits in a fixed space but oscillations of the dimensional background with respect to which the positions and velocities of the particles become observable. It just so happens that the wavefunction model properly describes the form of these oscillations, though the interpretation is reversed.

In this model the slit and screen apparatus itself consists of molecular structures composed of bound particle entanglements (see the upcoming section) so the dimensionality of the apparatus is more classically constrained.

In terms of the oscillating background model the dimensionality of the apparatus is oscillating relative to the exact trajectory of the particle, at least that's the way it appears to the particle. So the observational effect is the same. From the perspective of an observer conducting the experiment the particle appears wave-like, but from the perspective of the particle the observer and apparatus appear to be oscillating in waves of inverse form.

Then as the particle hits the screen it decoheres with the dimensionality of the apparatus and observer (within their own limits of uncertainty and bound entanglement) and becomes part of that system and begins to be computed as part of it.

This is consistent with our supposition that the supposed wave nature of particles is not an observable phenomenon and thus suspect. Only the particle impacts on the screen are actually observable. The wave nature of particles is never directly observed but only *inferred* from the *pattern* of impacts. But our interpretation in which processor cycle oscillations produce observable dimensionality explains these patterns perfectly well without wavefunctions.

When a particle is measured to determine which of the two slits it passes through the interference pattern on the back screen disappears. In our model the measurement decoheres the particle and it becomes part of the double slit system at that point and its coherent trajectory begins again from that location so the probability of its dimensionality oscillating through both slits has vanished to zero. All other apparent quantum paradoxes are explained in a similar manner by our model.

WAVEFUNCTIONS

Universal Reality can be considered a new interpretation of quantum theory that emerges naturally from the much wider perspective of its comprehensive Theory of Everything. But Universal Reality doesn't try to replace the equations that are the actual essence of quantum science. These equations are wonderfully accurate and comprehensive and are the cumulative work of a century of great genius.

Universal Reality accepts these equations because they work, but it completely revolutionizes our understanding of what they are telling us, and the overall context in which they have meaning.

For example wavefunctions are the primary mathematical description of quantum entities and they describe their behavior with great accuracy, yet even physicists admit they don't understand them.

Universal Reality provides a context in which wavefunctions naturally acquire a straightforward and non-paradoxical meaning.

In Universal Reality wavefunctions still work, but they aren't descriptions of particles with respect to a fixed pre-existing spacetime background, but descriptions of how individual particles in dimensional frames are formed, joined and aligned by quantum events. Thus wavefunctions don't describe particles that are smeared out in space and somehow everywhere at once, but how the separate dimensional frames associated with particles behave with respect to each other. The mathematics is the same but the interpretation is much more natural and reasonable and completely non-paradoxical as well.

In Universal Reality particles are exact computational data entities that always act as particles. This is true because only the particle nature of particles is ever directly measured or observed. Their apparent wave nature is always deduced from patterns of individual measurements *of particles acting as particles*. Because particle waves are never directly observed their existence must be considered questionable. Wavefunctions don't describe the wave nature of individual particles, but the randomness in the way their dimensional relationships can decohere with respect to the frame of an observer.

This resolves the fundamental quantum paradox of the apparent dual nature of particles as particles and wavefunctions. Particles are always particles with exact trajectories and time evolutions as their Schrödinger equations imply. Their apparent wave like nature has nothing to do with the particle itself but with how its dimensional observables are intrinsically uncertain due to the continual oscillations of the reference background with respect to which its dimensionality is computed. It's dimensionality itself that is fuzzy rather than particles fuzzy with respect to an exact dimensional spacetime.

QUANTUM TUNNELING

Quantum tunneling is a seemingly paradoxical phenomenon in which particles sometimes appear on the other side of an impenetrable barrier (Wikipedia, Quantum tunneling). The standard interpretation of tunneling in quantum theory is that the wavefunction of the particle is smeared out in space and that part of it extends to the other side of the barrier. And since the wavefunction is the probability distribution of where the particle might appear it can sometimes appear beyond the barrier.

Universal Reality reinterprets quantum tunneling in terms of its computational model. Here again it's not the particle that is smeared out in an exact fixed space, but dimensionality itself that is fundamentally oscillating as particle positions and velocities are computed and pinned against it to produce observable values.

So quantum tunneling is due to probability distributions not of particles but of the underlying dimensionality against which they are pinned. There is no physical spacetime in which objects have positions. The universe is all numeric, so there is no intrinsic problem of things appearing on the other side of things. It's just a matter of consistently generating observable positions and velocities.

The particle and the barrier are being computed as different processes and the processor cycles computing them have different oscillations with respect to each other. Thus if their oscillations overlap sufficiently the particle has some probability of appearing on the opposite side of the barrier.

The important point here is that each coherent process is computed by separate applications of the processor, each with its own oscillatory pattern. There is no single physical spacetime background each of whose points is characterized by a particular oscillation. The oscillations are in the separate processor applications within non-dimensional computational space. Points of the reference background can be visualized as oscillating differently in each process. The oscillations are different in each process because each is calculated by a separate application of the processor, and the oscillatory pattern depends partly on the individual details of what is being computed.

More massive mass-energy structures are computed with smaller dimensional oscillations. Quantum theory explains they have shorter de Broglie wavelengths, thus their oscillations are less than those of free particles (Wikipedia, Matter wave). As the mass and size of a structure increases it behaves more and more classically and impenetrable barriers are not penetrated.

HALF-LIVES

Half-lives are another good example of the seemingly paradoxical nature of quantum reality. There are two related cases of half-lives of interest (Wikipedia, Half-life).

Every radioactive isotope has a half-life that describes how long it takes for half of its nuclei to split and decay. The decay of any particular nucleus appears to occur completely randomly, but in aggregates of many atoms approximately half will decay in every half-life interval specific to that isotope.

In our model decay is a matter of certain dimensional adjacencies occurring. The internal elements of the nucleus are all oscillating randomly with respect to each other by a minute amount and when the overlap is sufficient to overcome the prevailing balance of forces holding the elements together a decay will result.

If a fissionable nucleus has internal components whose dimensionality jostles probabilistically with respect to each other they can come close enough to react on a random basis. They must be closer than some threshold distance from one another to interact for the nucleus to split. Thus it's the dimensional oscillations that bring them close enough for a decay to occur but with a predictable frequency in aggregate based on the particular details of the oscillations of the processor application computing them.

In nuclei this is due to the mutual jostling of the quarks and gluons in the nucleus, which are in a constant exchange that holds the nucleus and its protons and neutrons together. When this balance is randomly upset the nucleus splits. Everything jostles in just the right manner to hold the nucleus together until the jostling randomly exceeds one of its probability thresholds and the nucleus splits. The average statistical form of the jostling determines the aggregate half-life but its random nature determines the fission of any particular nucleus.

The other case of half-life decay is that of individual particles. For example free neutrons decay into protons, electrons and neutrinos with a half-life of around 10.2 minutes (Wikipedia, Neutrons). Many other particles including various mesons have much shorter half-lives and decay almost instantly. Basically only particles that are composed of multiple other particles in the sense their particle components can be validly redistributed into other particles are subject to decay and have half-lives. However free protons, which are composed of quarks like neutrons are not known to decay. Particle decay is essentially an event involving a single initial particle and is computed similarly to multi-particle nuclei.

Particle decay, like most processes in nature, is generally a mechanism to reach a lower energy state as some of the particle's internal energy is converted into the kinetic energy of its resulting particles.

Thus the probabilistic half-lives of composite particles and nuclei arise from the particular oscillations of the internal details of particles subject to decay relative to each other. Decay will result when particular overlapping oscillations occur and those occur probabilistically based on the oscillations computing the internal structure just as with radioactive nuclei.

Thus the mechanism responsible for half-lives is the same underlying mechanism of processor oscillations responsible for all quantum phenomenon and their seemingly paradoxical nature. Again the process is similar to that of quantum theory but the interpretation is reversed.

BOUND PARTICLES

Up till now we have been concerned with free particles but of course it is particles bound in atoms and molecules that make up most of the structure of the universe.

Bound particles are essentially bound entanglements of continuous interactions, and form a major part of the entanglement network. It appears that all the complexities of atomic and molecular structures emerge naturally from the simple rules that govern the interactions of free particles. Thus atomic and molecular matter is actually an emergent phenomenon that arises naturally from inter-particle interactions according to the rules of the complete fine-tuning.

This means that all of the very complex equations that describe atomic and molecular matter are not part of the programs that actually compute the universe. These equations are *descriptions* of the aggregate behavior of the elemental program that actually computes particle events.

All the emergent structure of the universe is simply the manifestation of individual particle interactions at the aggregate level.

There appear to be no higher level laws involved in actually computing the universe, though the programs of purposeful beings appear to act as such in the same manner that complex computer programs are directed towards computing various tasks but always only in terms of actual machine level operations. This is a complex subject that will be explored further in the chapter on Emergence.

The atoms and molecules that make up all the mass-energy structures of the universe are composed of bound particles, specifically the electrons that occupy orbitals around the protons and neutrons of nuclei, which in turn are composed of quarks bound by the strong force by gluons. The rules that govern particle binding are consequences of a few simple rules that govern how different types of free particles and their particle components interact.

In atoms, electrons are attracted to the opposite charge of protons in the nucleus and drawn towards them, however in most cases they don't have enough energy to react with a proton to produce a neutron. The mass of a neutron is significantly greater than that of a proton and electron combined. So for a reaction to occur there must be enough additional energy to be converted into the additional mass necessary to form a neutron. This can only occur with very high velocities or intense gravitational fields that provide enough intrinsic velocity energy to be converted into the necessary additional mass.

This does occur in extreme cases such as neutron stars where atoms are crushed by intense gravity and electrons do combine with protons to form neutrons and all the atoms of the star collapse to the size and density of their nuclei.

Thankfully in most cases electrons don't carry enough energy to react with protons to form neutrons, and electrons in atoms are unable to react with protons. Thus it's only the slight mass disparity between neutrons, and protons plus electrons, that prevents all atoms from collapsing into neutrons and all the ordinary matter in the universe from disappearing! An important example of how the universe is very finely tuned to maintain its structure.

Instead electrons are trapped in atoms by nuclear protons and continually bounce back and forth around them because they are unable to react with them or escape from them. The electrons become bound by the electrostatic attraction of protons and oscillate around the nucleus forming standing waves of the probability distribution of where the electron might be at any moment with respect to the nucleus.

The forms bound electron waves take around nuclei are called atomic orbitals. The basic principles underlying the forms of atomic orbital are fairly simple through the actual resulting forms in multi-electron atoms become quite complex due to electron-electron interactions and the imperfect spherical attraction of multi-proton nuclei (Wikipedia, Atomic orbital).

Because they are constrained by the attraction of nuclear protons bound electrons settle into harmonic standing waves centered on the nucleus. Harmonic waves are standing waves with an integer number of nodes that maintain their forms over time. They are analogous to a violin string fixed at both ends, which can only vibrate in a standing wave of one, two, three, or more integer numbers of nodes when plucked.

Because electrons form standing waves their energies become fixed. Each different standing wave has a specific energy and for an electron to jump between from one wave form to another it must absorb or emit a specific amount of energy in the form of a photon. The specific frequencies of the photons emitted or absorbed accounts for the distinctive spectral colors and lines of the various elements, and the fact that atomic orbitals have discrete energies is the origin of quantum theory.

The quantized energy levels result from the relation between a particle's energy and its wavelength. For a confined particle such as an electron in an atom, the wavefunction takes the forms of standing waves. Only stationary states with energies corresponding to integral numbers of wavelengths can exist; for other states the waves interfere destructively, resulting in zero probability density.

A more accurate analogy is that of a circular drum head whose circumference is fixed to the drum rim. Depending on how it's struck it vibrates as a standing wave with one or more nodes and the waves produced in the drumhead are nearly identical in form to those of electron orbitals. Wikipedia, Atomic orbital has some excellent animations.

So the secret to understanding electron orbitals is they are all the possible modes of dimensional oscillations of standing waves with increasing numbers of nodes in 3-space around a center that constrains them. This is the simple key to understanding electron orbitals and underlies the periodic table of elements.

The atomic orbitals form successive shells of increasing radius around the nucleus. Electrons in an atom are uniquely described by 4 quantum numbers so that no two electrons in an atom can have the same 4 quantum numbers. The first quantum number n denotes the shell and is simply the number of nodes electron waves in that shell have; 1, 2, 3, etc.

The second quantum number l is the azimuthal quantum number and ranges across all integer numbers such that $0 \leq l \leq n-1$. Thus for each quantum number n there is a set of $l+1$ quantum numbers corresponding to the number of possible harmonic wave forms with n nodes. The azimuthal quantum number basically describes the orientations of possible standing wave forms relative to the 3 spatial axes. There are 3 possible identical harmonic waveforms, one along each axis.

The 3rd quantum number m_l, the magnetic quantum number, describes the magnetic moment of an electron in an arbitrary orientation and is also an integer that varies within the subshell l_0 such that $-l_0 \leq m_l \leq l_0$. So for example for subshell $l=2$, m_l would take on the values -2, -1, 0, 1, 2 corresponding to the possible harmonic wave forms for that n shell and l subshell. All the m's within all l's for a given n correspond to all the possible harmonic wave forms with n nodes, and together for all n's these define all possible orbitals for an atom.

The number of possible waveforms a spherically centered standing wave of n nodes can have in 3-space is n^2. However two electrons can assume the same orbital waveform if their spins are oppositely oriented so the maximum number of electrons in a shell becomes $2n^2$. Spin, s, is the last of the 4 quantum numbers and is always plus or minus ½ (spin up or spin down) since the electron's spin is ½.

Thus the possible orbitals electrons can occupy in an atom is simply the number of possible harmonic waveforms the electron wave can symmetrically assume centered on a nucleus.

The orbitals are the waveforms that actual electrons must assume when they are filled. The actual number of electrons in a neutral atom is equal to the number of protons so the charges are balanced, and the number of protons determines the element.

In turn atoms form molecules primarily by sharing outer orbital electrons. When atoms combine in molecules outer atomic orbitals become distorted into molecular orbitals and the specifics of how these form and their complex properties determine the laws of chemistry and thus the structure and interaction rules of all matter. Thus chemistry and all mass-energy structures are the emergent results of the bound interactions of electrons, protons and neutrons.

A somewhat analogous situation occurs in the nucleus where the mutual repulsion of protons is overcome by the strong force, and proton and neutron quark waves vibrate in complex standing wave forms around each other. However, because the masses of protons and neutrons are much greater than the electron mass their de Broglie waves are much smaller, and thus nuclei are much smaller than electron orbitals.

The orbital forms shown in most illustrations are those of ideal individual harmonic waves. They are those a single electron would assume around a single proton as it increased or decreased its energy and jumped from orbital to orbital. However the presence of multiple electrons occupying orbitals in a single atom distorts their orbital forms due to the mutual repulsion of electrons and the necessary presence of multiple protons so the positive charge of the nucleus is not exactly spherically symmetric.

Thus though the principles underlying electron orbitals are fairly simple the forms they actually take in atoms become quite complex with increasing atomic number due to the mutual repulsion among electrons, including the screening of positive nuclear charges from electrons in further out shells, the uneven attraction of multiple protons in the nucleus, and even relativistic effects coming into play in larger orbitals.

In more technical terms the electron wavefunction oscillates around the nucleus according to a time independent (unchanging) Schrödinger wave equation, and orbitals are its standing waves. The standing wave frequency is proportional to the orbital's kinetic energy. The real part of the Schrödinger equation gives the form of the orbital, and the imaginary part gives the probability distribution of finding an electron at a particular location within it.

Thus emergence begins at the atomic level more or less. Atomic and molecular structures emerge automatically from the way that electrons, protons and neutrons interact, from the computational rules that actually compute their interactions.

Electron orbitals, and thus all chemistry and the structure of all matter emerges from the possible harmonic forms bound electron waves can take around protons, and electron waves themselves are the result of the space time oscillations in the processor cycles that compute reality.

Thus a small set of fairly simple particle interaction rules produces the atoms and molecules that compose all material structures:

1. The mutual attraction and repulsion of electrons and protons.
2. The inability of electrons to react with protons at normal energies.
3. The resulting possible orbital forms of standing harmonic electron waves centered on the nucleus.
4. The analogous strong force binding of proton and neutron quarks in nuclei.

Thus the structure of the observable universe from atoms on up is emergent rather than independently computed. All the actual computations of reality occur at the particle and particle component level, and larger scale structures automatically emerge from bound particles that manifest these elemental computations. This is how all the incredible complexity of the observable universe emerges from the finely tuned interplay of the simple elemental rules that actually compute them. This greatly simplifies the computational structure of reality. It's the incredibly amazing complete fine-tuning that is responsible for the wonderfully meaningful complexity of emergent structures that are simply its aggregate manifestations.

Because they exist as bound entanglements the dimensional relationships of electrons and protons are exact in terms of energy conservation, which is of course the basis of quantum theory. However when these conserved energy relationships are viewed in terms of positions and velocities of electrons relative to the nucleus they are subject to the processor cycle oscillations as viewed from an observer frame, which is why they appear as harmonic standing wavefunctions within which the actual position and velocity of the electron appears probabilistic.

MASS-ENERGY STRUCTURES

The emergent observable universe consists of the data of particles and their fields in a dynamic balance of the four fundamental forces. All the atomic and molecular structures of particles are balances of different types of forces. Just as all forms of mass-energy are different types of relative motion, emergent particle structures form when different forms of relative motion reach balances.

This is due to the attraction and repulsion of the various forces on the various charges. The charges of the forces manifest fields that attract and repel other charges. Due to the principle of differentiation, the universe consists of myriads of discrete particles with different types and signs of charges. Thus the fields of the various forces balance to form compound particle structures, and these balances define the atomic and molecular structures of the observable universe.

Atoms are balances of the repulsion and attraction of electric charges, and atomic nuclei the balance of repulsion and attraction of the strong force operating via gluons on the color charges of constituent quarks. And the gross structure of the cosmos is largely a balance of the gravitational attraction of mass charges with the linear relative motion of kinetic energy in combination with the repulsive gravitation of the Hubble expansion.

All the 4 forces can be viewed either as forms of relative motion or as exchanges of particles due to the equivalence of space and mass-energy as forms of spatial velocity. Basically the emergent structure of the observable universe is a balance of the relative velocities of the 4 forces with the relative velocity of linear motion. Everything is given its emergent structure by the interplay of various types of relative motion computed at the particle level.

Because spacetime and particle structures are computed together as an integrated structure, and because atomic and molecular structures are so pervasive, they are the main contributor to the overall structure of the observable universe. They are the primary constituents of the entanglement network at least in the vicinity of stars and planets.

IMMANENCE & THE QUANTUM VACUUM

To briefly relate the last several chapters to the overall theory, recall that the universe consists entirely of data being computed in the immanent reality of the quantum vacuum, which is the substrate or medium of existence. Because all the data and computations of the universe occur in the medium of existence they are all real actual things.

Immanent data is simply the fundamental nature of all the things in the observable universe including ourselves.

The immanence of things is also the source of consciousness. All things shine with the immanence of their existence and the immanence of their representations in our mind is what we call consciousness. Our consciousness of things is the immanence of their existence in which we all share.

However the structural details of things in consciousness is their information structures filtered through our individual perceptual and cognitive structures. The fact of consciousness itself, as opposed to the structure of its contents, is immanence; the structural details of things that appear in consciousness are logico-mathematical data structures. And how these are computed at the elemental level has been the subject of the last several chapters.

Universal Reality's model of the universe unifies quantum theory and general relativity in a single computational model in which relativistic spacetime emerges from quantum events. And it also explains consciousness, existence and the nature of the present moment.

Particles are still well described by the equations of quantum theory as are large-scale phenomenon by the equations of general relativity. Universal Reality just interprets them differently, and this new interpretation leads directly to their unification and numerous new insights into the ultimate nature of reality.

COSMOLOGY

THE REAL VIRTUAL REALITY

At the fundamental level the observable universe exists as logico-numeric data in the non-dimensional computational space of the quantum vacuum. There the data state of the universe is continually recomputed by individual processor applications of the elemental program operating in accordance with the virtual data of the complete fine-tuning. The quantum vacuum is the substrate or medium of existence thus the current present moment data state of the universe is the real and actual universe.

Though the computational space of the quantum vacuum is itself non-dimensional it contains the numeric data of dimensionality that's computed within it. Just as the numeric data of a simulated universe within a computer program can be graphically visualized on a computer screen so the purely numeric dimensional data computed in the quantum vacuum is displayed to observers as the actual observable universe.

The data of the observable universe is the integrated data of the mass-energy structures and their dimensional relationships produced by quantum events. The data of the observable universe corresponds to Level 2 data in the 4 level computational model explained above. In this model both mass-energy structures and the observable spacetime they appear to exist within are computed together by quantum events.

The resulting observable universe can be accurately visualized as an immersive virtual reality which each individual observer experiences centered on himself. However this is the real actual universe programmed by the evolution of myriads of quantum events over the life of the universe rather than by any human or alien programmer. And it's the only real existing universe and not subject to reprogramming or reruns. Though we and other beings act freely and purposefully within it to some degree our powers are quite limited and on the whole it evolves according to its own immutable laws carrying us along.

At the quantum level some events, especially those involving the computation of dimensional relationships, occur stochastically, but this randomness is constrained by the general form of the complete fine-tuning so that on the whole the universe evolves statistically around general trends.

There is an important additional complication. The actual data form of the observable universe is quite different than how observers view and experience it. Just as the headset of a virtual reality display consists of speakers with vibrating electronics and two separate flat screens emitting photons in the visible spectrum, so the actual observable universe consists of myriads of raw sensory data inputs that our brains must organize into the semblance of the physical universe in which we believe we exist but which is not at all like the actual universe. Our simulation of reality is explored in detail in the chapter on 'The Simulation'.

So when we look at the bright world around us we know that at its finest level it consists of countless myriads of logico-numeric data in a continual process of recomputation. And this is computationally projected by our minds into the moving images that make up our familiar world including ourselves. Everything is the classical level projection of the elemental particle events that actually compute reality, and what we see are the emergent patterns manifested by huge aggregates of those quantum scale events as they play out before us, around us and within us.

The observable universe can also be thought of as a massive cellular automaton whose surface is composed of innumerable cells or pixels too small to be individually seen. The data state or color of each cell is being continually recomputed at each P-time tick from the computational interaction of adjacent cells (Wikipedia, Cellular automaton).

At the classical level the elemental interactions of vast myriads of these cells produce emergent patterns that take the form of all the things and events that populate our world. All the vast complexity of aggregate scale processes emerges from these simple computations like the patterns of cellular automata emerge from aggregates of simple cell-to-cell interactions.

Because the elemental program that computes the individual cell-to-cell interactions runs according to the rules and constants of the complete fine-tuning the observable universe it computes is the one we observe around us.

Everything in the universe is data, interpreted by our minds as a physical universe within an encompassing spacetime. Though this is largely an illusion it's a beautiful and very convincing illusion. It's the ultimate virtual reality and we live our entire lives within it as part of it. It's the only actually real virtual reality and the ultimately interesting show. Because it runs in the computer of actual existence we and the world around us have actual existence. Thus the universe it computes is the real living universe and we are real living beings.

THE COSMOLOGICAL HYPERSPHERE

In the previous chapter we outlined a useful visual model of an arbitrary section of the observable universe. This consisted of a sequence of rising surfaces, each surface representing a section of the 3-dimensional universe with one dimension suppressed for clarity. In each P-time tick all the data of the surface is recomputed to produce another surface on top of the stack of previous surfaces. The entire stack represents the past computational history of the observable universe and the top surface corresponds to the current actual observable universe in the universal present moment in which everything is recomputed.

This surface was flat and not curved by the presence of mass. Instead masses add surrounding fields of intrinsic velocity density across the surface. Relativistic effects are due to the sum of linear spatial velocity and the intrinsic spatial velocity at any point. The total spatial velocity reduces the velocity in time by the STc Principle and is responsible for all relativistic effects.

The STc Principle arises from the fixed number of processor cycles allocated to computing the total spacetime velocity of every process. If some of these cycles are used to compute velocity in space fewer are available to compute velocity in time and velocity in time slows as spatial velocity increases as relativity predicts.

At the quantum level events conserve and redistribute the particle components of interacting particles among new particles. These newly emitted particles are entangled separately on each of their particle component types. Each set of entangled particles constitutes a dimensional fragment, which is computed as a single coherent process by an ongoing single application of the universal processor.

All the dimensional indeterminacy of quantum processes (such as wavefunctions and the Uncertainty Principle) is due to random *oscillations* between the space versus time allocation of the processor cycles that compute them. These oscillations result in a conflation of space and time at the quantum level that can be described by probability distributions of space versus time positions and velocities

Each separate dimensional fragment is computed by a separate processor application, which has its own random oscillations. Since entangled particles are computed together as part of a single process a measurement of one automatically fixes the dimensionality of the other as in the spin entanglement 'paradox' as explained in the previous chapter.

This visual model gives us a good understanding of an arbitrary section of the observable universe at the local level but we need to generalize the model to the cosmological scale to incorporate the entire observable universe into the picture. We can expand the P-time surface model to the whole universe to provide a picture of its overall geometry in terms of the elementary computations that create it and the continuous extension of its P-time surface.

It turns out there is only a single possible geometry of the observable universe consistent with Universal Reality's two kinds of time and the STc Principle. This is a 4-dimensional hypersphere whose surface is the 3 dimensions of space, and whose radial dimension is historic P-time. Imagine the 3 spatial dimensions of our universe as the surface of a 4-dimensional balloon being inflated by P-time. We, and the entire universe around us, occupy the current present moment in 3-dimensional space on the surface of the hypersphere. The entire observable universe is the surface of this hypersphere. The no longer existent interior of the hypersphere is the past history of the observable universe tracing back to the center point of the big bang.

This is a straightforward model based on the concept of a universal P-time in which the data of the entire observable universe is recomputed in each successive moment. However modern cosmology has been unable to discover this simple elegant geometry due to its inability to recognize the concept of a universal present moment.

Modern scientists are positively schizophrenic when it comes to the notion of a universal time. First physicists adamantly deny the concept of a present moment and the notion of a common universal time, and in the next breath they tell us the universe is 13.8 billion years old and that's true for every observer in the universe. Then they engage in all sorts of genuflections to try to reconcile these clearly contradictory views.

This lack of progress is due to the inability to recognize that clock time and the time of the present moment are two separate kinds of time. The age of the universe in clock time clearly depends on how fast your clock is running but clock time runs at different rates in a single universal P-time, the time of the common universal present moment. The unfortunate result is science has no clear picture of the overall geometry of the universe and tends to fall back on clearly inaccurate expanding tube like images with flat surfaces and edges (Wikipedia, Metric expansion of space, #Topology of expanding space).

Only a small volume of the cosmic hypersphere is visible from any location within it since its spatial surface is uniformly expanding at a rate that exceeds the speed of light beyond a distance called the particle horizon. Because space itself is expanding away from us faster than the speed of light beyond the particle horizon, light can never reach us from there and that area of the universe is not visible to us. Likewise we are not visible from points beyond our particle horizon because we are beyond the particle horizons of those points.

However the entire current P-time surface of the hypersphere including all regions beyond the particle horizon is the whole actual universe since it's the current present moment in which the entire universe exists. The entire surface of the hypersphere, the entire universe, is in the same P-time present moment all around its surface, irrespective of particle horizons, and irrespective of the various local rates of clock time.

Note that the particle horizon is an observational as opposed to an actual relativistic effect. Beyond the particle horizon nothing is actually moving faster than the speed of light. Processes evolve normally just as they do in our area of the observable universe. The absolute dimensional background of computational space in which all processes are computed extends around the whole surface of the hypersphere with no interruptions or anomalies. It's only when processes near the particle horizon are observed that anomalies appear to exist.

All the past onion-like P-time surface layers represent the universe as it was at previous present moments that no longer exist. The interior corresponds to the past states of the universe's history back to the big bang at the center when the actual computations of the observable universe began. These previous layers no longer exist and can only be inferred from the state of the universe in the current present moment.

Thus the universe is a closed finite hyperspherical surface with positive curvature and no edges. It cannot be infinite because nothing actual can be infinite because infinity is not an actual state or fixed number but a never-ending *process* of continual addition. Infinity is a useful mathematical concept but nothing actual can be infinite.

Nor is there any reason to suppose the universe is not a closed continuous surface and has edges. How could the point universe of the big bang develop edges as it inflated? That seems nonsensical. Traveling in a straight line in any direction across the universe one would theoretically eventually end up at approximately the same place ignoring any local curvatures of space just as one does by circumnavigating the earth ignoring the mountains and valleys.

The evolution of the universe through time consists of the ongoing extension of the radial P-time dimension of the hypersphere, which carries along the 3-dimensional surface space as its current information state is continuously recomputed. This extension is not in clock time as computations generate different clock time rates within the current 3-dimensional present moment surface depending on the spatial velocities of local relativistic processes.

Now there is a very obvious apparent problem with this model. If P-time is extending the radius of the hypersphere uniformly that doesn't seem consistent with the apparently accelerating Hubble expansion of its spatial surface. This is an important point that will be addressed shortly in the section on The Hubble Expansion.

If P-time is the radial time dimension of a hypersphere then the circumference of the spatial surface of the universe should be a function of its P-time radius, and measurements of the curvature of space should provide a measure of its radius and the P-time age of the universe. Current measurements suggest that 3-dimensional space is fairly flat within its observable volume but a hypersphere is not ruled out. A hypersphere also makes sense from the perspective of general relativity, as the mass-energy content of the universe should curve it in on itself at the largest scales.

This hyperspherical geometry should be subject to experimental confirmation since the curvature of space is measurable (Wikipedia, Shape of the universe). It should turn out to have a very small positive curvature. But even if it doesn't that raises doubts but doesn't necessarily falsify the theory since if the hypersphere is not perfect it could be closed and finite and still contain some areas with greater or lesser or even negative curvature.

Though the hypersphere model is useful in visualizing the overall geometry of the universe it should not be taken for a physical structure. The hypersphere is not a physical structure but a logico-mathematical structure that emerges with the dimensionalization of spacetime at the largest scale. It's the overall structure of emergent dimensional relationships at the cosmological scale of the entire universe. It's the aggregate result of the precise complete fine-tuning of the quantum vacuum manifested by the elemental program as it computes dimensionality in the form of individual dimensional relationships at the particle level.

Recall that in Universal Reality spacetime doesn't exist as a physical or even a geometric structure within which events occur. All that actually exists is a vast interconnected entanglement network of dimensional relationships produced by particle interactions. This network doesn't exist in any physical space. Any notion of a physical space is an *interpretation* of this network of dimensional relationships. Dimensional relationships simply have a logico-mathematical consistency that observers interpret as a physical spacetime.

Events are not actually occurring within a pre-existing physical spacetime so the universe doesn't actually exist as a cosmic hypersphere within which events occur except in a logico-mathematical sense. Events are computed in the abstract computational space of the quantum vacuum, and the dimensionality that emerges assumes the logico-mathematical structure of a hypersphere at its largest scale. Nevertheless this is a real and actual emergent structure that can be taken as defining the geometry of the universe if we remember it's an abstract logico-mathematical structure rather than a physical structure. As we have seen all apparent physical structures are actually information structures.

Thus the hypersphere is an interpretation at the largest scale of the overall geometry of the network of dimensional relationships. The universe and everything in it ultimately exists only as consistent logico-mathematical data structures within the abstract computational space of the quantum vacuum.

So it's misleading to think of P-time inflating the hypersphere of the universe as if it were a balloon. The happening of P-time processor cycles computes the evolution of this structure but it remains fundamentally a data structure analogous to a data structure in a computer program. The hyperspherical model is a *visualization* of this data structure in our simulations of reality just as dimensional computer data can be visualized as a spacetime on a computer monitor. The evolution of the happening of the universe is not really an extending radial P-time dimension of the hypersphere producing a corresponding geometric expansion of its surface though this is a useful conceptual model.

Thus the computational surface in which everything is computed doesn't actually rise or go anywhere at all since it's completely non-dimensional. The actual logico-mathematical surface of the cosmic hypersphere can be extending, contracting or remaining in the same place depending on its overall dimensional structure as it's computed. Thus any rate of Hubble expansion including a collapse into a universal black hole is consistent with this model.

So the actual radius of the hypersphere is not strictly its P-time radius but the current radius of its overall computed dimensionality. This is the consistent dimensionality that emerges from the combination of all the individual dimensional relationships among particles as they are computed by quantum events. So the P-time surface and radius of the hypersphere is a useful but imperfect visualization and should not be carried too far.

Thus there need be no equation of a hypersphere in the quantum vacuum used to compute the geometry of the universe. The code of the elemental computations of particle interactions in the fine-tuning just results in its emergence at cosmic scales. The conservation and other laws that govern the interaction of particle components results in a dilated dimensionality around mass-energy concentrations, and this resulting curvature of space manifests as a logico-mathematical hyperspherical structure when considered in its entirety.

This is true of all emergent laws of nature. They are aggregate expressions of the elemental particle component computations due to the way the complete fine-tuning works rather than higher-level laws that actually compute the evolution of classical processes, unless the fine-tuning itself is somehow being tuned by hidden high-level processes.

CURVED LIGHT CONES

Though we can only infer the past state of the universe since it no longer exists, we can directly observe a slice of it into the past by looking down our 4-dimensional light cone into the past.

Due to the finite speed of light it takes time for the light of distant objects to reach us and thus we see the universe as it was at times in the past depending on its spatial distance. So we do have some 'direct' knowledge of the past even though that knowledge occurs only in the present moment.

An interesting and largely unmentioned aspect of light cones is that they curve inward as they extend back in time. Though they are invariable portrayed with straight sides in scientific articles they obviously must curve inward since they must all eventually meet at the point of the big bang (Wikipedia, Light cone).

The observational effect of this curvature is that the farther we see back in time along our light cone there comes a point when things begin to get closer and closer together rather than farther and farther apart. Effectively our light cone becomes a giant magnifying glass where at great distances back in time everything was much closer together than it appears to be. Our light cone acts like a giant lens that curves light beams from the distant past just as a glass lens magnifies by curving light beams.

If we could actually see all the way back to the big bang or very near to it, we could actually see elementary particles interacting with our naked eyes if only in terms of the now greatly red shifted radiation they emitted.

In fact our view of the CMB (cosmic microwave background) is a greatly magnified view since the universe was considerably smaller when the CMB was emitted only ~380,000 years after the big bang (Wikipedia, Cosmic microwave background).

COSMIC INFLATION

The inflationary period of the universe was an apparently enormous exponential expansion of the volume of the universe in the first slight fraction of a second after the big bang (Wikipedia, Inflation (cosmology). There is considerable evidence for inflation and the theory is widely accepted. In Universal Reality this enormous near instantaneous expansion of the size of the universe immediately after the big bang has a straightforward explanation.

In Universal Reality the universe consists of data and doesn't exist in a physical space with any actual dimensional extent. However this data does include dimensional relationships among the data of elementary particles and their particle components. It's the network of these dimensional relations that observers, and their science, interpret as spacetime.

Thus when particles first arose immediately after the big bang they furiously began interacting and producing networks of dimensional relationships among them. Dimensional entanglement networks manifest as dimensional spacetime, and these spacetimes are automatically 'large' enough to contain all the dimensional relationships generated in a consistent framework.

Prior to the creation of particles there could be no dimensional relationships among them and thus no spatial extent to spacetime. But as soon as particles and their dimensional relationships formed their interactions generated a dimensional entanglement network large enough to accommodate them in a logically consistent framework.

Thus from our current look back viewpoint it appears there was an enormous near instantaneous inflation of spacetime, but actually there was just an enormous network of newly computed dimensional relationships that suddenly manifested as a spacetime large enough to contain them.

Thus inflation produced enough logico-mathematical space to consistently contain all the dimensional relationships generated by the interactions of all the particles that had just been created. The dimensional relationships created automatically manifest as the space necessary to contain them, and this accounts for the initial exponential inflation of the universe. There was no actual physical space that inflated. There were just an enormous number of dimensional relationships computed that only make sense to modern scientists in terms of an exponentially expanding physical space large enough to contain them.

This explanation also resolves the apparent problem with the hypersphere model of the universe mentioned above. If P-time is the radius of the hypersphere and has a constant rate of extension, then the Hubble expansion of the spatial surface of the universe should also be constant and proportional. The surface of a sphere obviously can't be expanding faster or at a different rate than the radius is increasing, and it certainly can't be expanding at an accelerating rate if the rate of increase of the radius is constant.

However it must be remembered that we measure the radius of the universe in *clock time* as 13.8 billion years. This is not its actual *P-time dimension*, which has no intrinsic measure and can only be measured in terms of the clock time it produces. Thus if P-time and clock time run at different relative rates there isn't necessarily a problem.

For example if clock time just ran much slower during the inflationary period that would explain why inflation occurred in almost no time at all from our perspective. It could have taken millions of years of our clock time with only a minute fraction of a second passing on clocks back then. And of course there is a very good reason why this should be true.

If inflation were an enormous expansion of the space of the entire universe it would automatically produce a proportionally enormous time dilation (slowing of clock time rates) across the entire universe due to the STc Principle. With the enormous spatial velocities of all particles generated by inflation clock time across the entire universe would have slowed uniformly across the universe to almost nothing while the P-time rate continued unchanged as before. So even the enormous expansion of inflation could seem to take place in almost no time at all because time was running so slow.

So relative to the rates of today's clocks inflation could have taken eons, but because all the clocks of the universe were running at almost no rate at all it would seem to have expanded in almost no time at all from our current perspective. Thus the extraordinary inflation of the universe in a minute fraction of a second may not have been so extraordinary after all. It could just seem that way compared to the rates of our current clocks.

The takeaway lesson from this is that P-time and clock time rates are not necessarily proportional. P-time computes all clock time rates, and these can vary widely even today so it's quite likely that the overall clock time rates we use to date past events in the history of the universe could vary over the age of the universe as well.

Thus we may think of the overall evolution of the universe

progressing smoothly through clock time but that could be an illusion as clock time itself could have been running at different rates in different eras in the history of the universe even as the P-time radius of the universal hypersphere extended to produce the correct Hubble expansion of that period.

If the overall clock time rate of the universe did vary from age to age it's not clear how this would be distinguished from the apparent changing rates of the Hubble expansion. They might be observationally identical. Thus the apparent changes in the Hubble expansion rate could simply be an artifact of changing rates of cosmic clock time.

THE HUBBLE EXPANSION

After the initial period of exponential inflation, the expansion of the 3-dimensional space of the universe seems to have gradually decelerated to a much slower rate before beginning to gradually accelerate again. The expansion appears to be still accelerating. This expansion of the universe is called the Hubble expansion after its discoverer, Edwin Hubble (Wikipedia, Hubble's law).

Even though the expansion is fairly slow on local scales, at sufficient distances it does add up to produce a particle horizon equidistant from every point beyond which the expansion exceeds the speed of light and nothing is visible.

The Hubble expansion is an expansion of empty space itself relative to the dimensionality of particulate structures within it such as galaxies and galaxy clusters. This means that all material structures including gravitationally bound areas of space are not expanding while the distances between them becomes greater in non-gravitationally bound intergalactic areas.

As previously mentioned if the average clock time of the universe is passing uniformly and clock time was the measure of the radius of the universe then its spatial surface must also expand uniformly and proportionately. This is the only way the universe could remain a hypersphere. Thus the accelerating Hubble expansion seems to be a problem assuming the rate of P-time remains constant.

But if we simply take P-time as the radius of the hypersphere and have the clock time it computes vary to account for the changing rates of the Hubble expansion then the hypersphere is preserved and there is no problem. P-time just computes varying rates of clock time throughout the history of the universe proportional to the Hubble expansion it also computes. We have just seen this is likely to have occurred during inflation but why would it still be occurring and at an apparently increasing rate?

First the Hubble expansion of the universe and its currently accelerating rate are not increases in the size of a physical universe, but an increase in the relative *scale* of the dimensional entanglement network produced by elementary particle interactions.

There are several current theories for the expansion and its current acceleration whose uncertain source is called dark energy. General relativity suggests dark energy could be the presence of a cosmological constant consisting of the predicted negative gravitational effect of the energy density of the quantum vacuum. This becomes more important as the universe becomes less dense due to its ongoing expansion. Other proposed explanations involve new types of particles (Wikipedia, Dark energy).

Whichever is true, Universal Reality would simple take that as additional dimensional relationships being produced that automatically manifest to observers as the necessary expansion of intergalactic space necessary to accommodate them in a consistent logico-mathematical framework.

The hypersphere is not a physical structure but a logico-mathematical one so it's expansion need follow only logico-mathematical rules rather than physical ones. There is no intrinsic reason to consider the measure of the P-time radius in the same units of measurement as the surface. It seems more reasonable to assign units of measurement to the radius in terms of a cosmic clock time derived from running the spatial expansion backward. But if the average rate of clock time has been much slower in the past as inflation suggests, the actual age of the universe could be considerably older and its radius much larger.

A measure of the P-time radius could also be derived from the curvature of the spatial surface but that would need to be known with considerably better accuracy. Current evidence suggests that the universe is very nearly flat. Thus if it's a hypersphere its circumference must be very much larger than one with only a 13.8 billion year radius.

Thus the evidence does suggest that the universe is a hypersphere with a very large circumference and that the true clock time age of its radial dimension must be much older than 13.8 billion years in terms of current clock time rates. Thus clock time must have run much slower in the early universe. This effect may have occurred mostly during inflation but there still could be some effect during the subsequent slower Hubble expansion.

Part of the difficulty in figuring this out is our very limited knowledge of past expansion rates. Due to the finite speed of light we observe the past states of the universe only at specific distances, and we observe the states at different distances only at specific past times. Therefore we have direct observational evidence of the past expansion of the universe only as surfaces of specific combinations of distance and time. We must assume that the entire universe was expanding at the same rate as we observe it expanding at only a single distance for each past time. So if the universe was/is actually expanding at different rates depending on distance, or even if it appeared to be because of some other spacetime effect that varied with distance, the accelerating universe could be an illusion.

So it's unclear how accurately we know the rate of Hubble expansion through time. We are actually completely ignorant of the actual current state of the entire universe except in our immediate vicinity because the present state of the entire rest of the universe is hidden from us by the finite speed of light. The rest of the universe could completely vanish and we wouldn't know it until its light reached us, so it's really a quite a stretch to think we know what the current expansion rates of the universe are anywhere at all.

So there is no necessary dependence of the apparent expansion rates of the surface space to the rate of extension of its radial time dimension. The hypersphere is not an independent entity that is expanding but an extrapolation of the *current dimensionality* of the entanglement network. This frees us from any troubling dependency of the non-uniform Hubble expansion rates of the spatial surface on the putative uniform extension rate of the P-time radius and is consistent with our treatment of inflation.

Thus any notion that each of the past onion layers of the universe represents a common past *clock time* is misleading as clock rates run at different rates within the universe and have throughout past time. Even the notion of a fixed age of the universe depends on ignoring this fact and using a universal standard time science calls 'cosmic time'. Relativistic observers don't even agree on the times of their clocks in the present moment, much less the age of the universe on their clocks running at different rates (Wikipedia, Chronology of the universe).

The emergent *space* of the universe clearly has a meaningful metric in that all measurements of distances between events produce results consistent with the overall network of dimensional relationships. So we can meaningfully assign a metric on this basis to the radial time dimension of the universe to get its clock time age. This must be a standard clock time metric since P-time has no observable metric as it is what computes the very existence of clock times.

Thus it's meaningful to consider the length of the radius in terms of a universal standard time clock time age, and its past onion-like surfaces as the past states of the universe as science traditionally does if we are careful and don't read too much into it.

This enables us to preserve the hypersphere model and, with the caveats above, use the accepted age from the big bang of 13.8 billion years even though the radius is actually a P-time radius with no intrinsic metric. The STc Principle suggests we should use a universal standard time, namely the time of a theoretical motionless clock in empty intergalactic space, though astronomy's cosmic clock is slightly different. In this view the 13.8 billion year age of the universe would be the *proper time* reading of this clock, but in terms of our current clocks the universe is likely very much older.

It seems natural to assume that the passage of P-time and thus the lengthening of the radial dimension would be uniform. But it's not clear what this would even mean since P-time has no intrinsic metric and can only be measured in terms of the clock times it produces and those vary from location to location. P-time could be computing events at different rates in different areas of the universe, which might produce different values for the speed of light but there is currently no evidence of that. And if the average density of the universe changes as it expands that might well affect the clock rate of a stationary clock in empty space.

So we have assumed that the radial P-time dimension extends more or less uniformly in all directions even as the clock time rates of processes on the surface progress at different relativistic rates within it. However it's possible this isn't true and the surface of the universe is not exactly spherical but wobbly like the surface of a giant soap bubble. This is another possibility that could account for the currently observed accelerating expansion of the universe, which could conceivably be the result of an unseen wobble on the other side of its surface but there appears to be no current evidence for this.

In a computational universe the Hubble expansion is just an expansion of the dimensionality of the entanglement network in areas of low mass-energy density. Recall that this and the hyperspherical geometry of the universe is not a physical process but the emergent dimensionality of a computational process. Thus whatever explains it computationally is reasonable to consider.

There are several possibilities. It could be due to a conversion of some of the concentrated mass-energy velocity of the universe to that of empty spacetime, the injection of additional zero-point energy into the flat spacetime of the quantum vacuum, some process injecting negative entropy into the universe, or some variation in the intrinsic P-time rate that calculates the expansion of the universe and becomes its radial dimension. It could also be some gradual change in the hidden templates of scale that define dimensionality in the quantum vacuum. All these are possible scenarios.

The question is related to whether the quantum vacuum is an inexhaustible fluid that can be stretched indefinitely without becoming somehow a less dense reservoir of virtual particle components or whether as the universe expands particles are less likely to actualize from the quantum vacuum.

REDSHIFTS

The redshifts of distant objects are not due to their relative recession velocities as light is emitted as is often mistakenly thought. They are actually due to the cumulative stretching of space over the entire path of the light from emission to reception (Wikipedia, Redshift). Therefore a redshift automatically reflects all the different expansion rates of the intervening space the light traveled through from the time it was emitted to the time it was received.

Thus by comparing the redshifts of objects at various times in the past we can determine the changes in the expansion rate of the intervening space over time. And by measuring objects at different times in all different directions we can confirm that all of space was expanding pretty uniformly at any given time in the past at least within the visible universe.

An interesting consequence of the continuous expansion of space is that things we see in distant space were actually considerably closer to us than we might imagine. For example the light from the CMB was emitted nearly 13.8 billion years ago so we might imagine the CMB we observe was nearly 13.8 billion light years away when its light was emitted but this is not true. Its light took nearly 13.8 billion years to reach us but most of that time was due to overcoming the stretching of the distance through the expanding space between us during that time. Because the intervening space stretched by a factor of over a thousand the CMB light took over a thousand times longer to reach us.

What this means is that the CMB that we observe was actually only around 12.7 *million* light years away from us when its light was emitted rather than the 13.8 *billion* light years we might expect. So we observe the CMB over a thousand times closer than its age might suggest. This is consistent with the magnifying effect of the curvature of our light cones as we look back at the CMB.

This actually leads to an interesting paradox in which the CMB we see *beyond* other distant objects such as quasars is actually *closer* to us than some of those objects in front of it, or at least we see it so because it was when its light was emitted.

Only if the universe hadn't expanded at all would distances in space be consistent with distances back in time. In a uniformly expanding universe our light cones would be straight but uniformly slanted inward from 90° so that the time to some distant object was increasing faster than the distance to it.

In our universe there was an initial exponential increase in the rate of expansion during the inflationary period, which then slowed and now is increasing again. Thus in our universe our light cones are curved proportional to the changes in expansion rates over time.

For example the distance between points in space an arc minute apart increases proportional to the distance from us as we go deeper in space but as redshift increases the distance between the points increases more slowly and eventually begins to decrease as we near the earliest universe. At first our light cones extend outward but eventually they begin to curve inward and would finally converge through inflation to the point of the big bang if we could see back that far in time.

So at the farthest distances, the farther we look back in time the closer we are seeing to us, and if we could see all the way back to the putative single point of the big bang we would see not the farthest possible point from us but the closest to us. We would see the big bang where we are right now because that's where it actually occurred because everything in the universe was collocated at that same single point.

This can be confusing and science must be careful not to let it mislead our interpretations of our understanding of how the Hubble expansion rate has varied over the age of the universe.

A NEW DARK MATTER THEORY

The existence of an invisible form of matter called dark matter was first proposed to explain observational anomalies in the motion of galaxies. For example observations suggest that galaxies rotate as if they had halos of invisible mass around them because they are rotating faster than would be expected based on their apparent masses. The amount of dark matter necessary to explain the movements of galaxies is huge, about 5 times the amount of visible matter in the universe (Wikipedia, Dark matter).

Dark matter has been sought in the form of various types of new particles but so far none have been found. However Universal Reality suggests another possible explanation for the dark matter effect, which so far as we know is original to the author's 2013 book 'Reality'. This proposal is a simple and rather obvious consequence of the Hubble expansion.

The Hubble expansion is an expansion of the relatively empty space *between* galaxies and galaxy clusters which makes up most of the universe. By contrast the space *within* galaxies isn't expanding because it's gravitationally bound by their mass (Misner, et al, 1973, p. 718). Thus the earth, the solar system, our galaxy, and we are not expanding but the space between galaxies is expanding.

The result is an uneven Hubble expansion that warps space around the boundaries of galaxies; precisely in the area that dark matter is expected to be found! And from general relativity we know that any warping of space must manifest as a gravitational effect. Thus we have a natural explanatory mechanism for the dark matter effect that involves only the expected warping of space from the uneven Hubble expansion and doesn't require the existence of any new particles.

This warping may or may not be the cause of the entire dark matter effect, but it certainly should be producing a large gravitational effect, since the uneven expansion over the lifetime of the universe should produce a very large warping of space.

Distributions of dark matter can be mapped by tracing gravitational deviations of the expected paths of light beams from sources beyond them. These maps indicate a distribution of dark matter generally around galaxies but sometimes offset as well. However there is nothing to prevent these Hubble space warps, once they are created, to have a life and movement of their own. Thus dark matter distributions should initially form as halos around galaxies and galaxy clusters but then be able to move as massive objects on their own.

Once Hubble warps are formed they are effectively just additional areas of gravitational mass that can move through space just as galactic masses do. The continued existence of a dark matter mass is not dependent on the original galaxy it was created from. There will be a continuous creation of new dark matter warps around galaxies, but once created these can trail away and should leave detectible plumes of warping behind that indicate how galaxies moved over time.

Over the course of the expansion of the universe the actual effects will be extremely complex because the distribution of galactic matter with time is extremely complex. It should be fairly easy to test at least the viability of this theory by comparing the current distributions of dark and visible matter and inferring their relative motions over time and making a calculation of whether the expected warping would account for the gravitational effects of known dark matter concentrations.

This is one possible explanation of the dark matter effect, but not necessarily the only one. Nevertheless there should be a very substantial warping due to the uneven Hubble expansion, and that warping should be producing quite a large gravitational effect. Where is that effect if it isn't the dark matter effect? It must be somewhere. The evidence seems quite strong and it certainly simplifies things by not requiring any new unknown types of particles.

This theory of dark matter also neatly explains why dark matter is dark. Not being an actual form of particulate matter it obviously doesn't emit light. Thus it's invisible and interacts with regular matter only via the gravitational force.

THE BIG BANG

The entanglement network traces back to its origin in the original event of the big bang. The big bang was an actualization event in which particles in the form of valid particle component sets were computationally instantiated and given relative motion in the quantum vacuum. This formed the observable universe within the medium of existence of the quantum vacuum and set it into motion.

The big bang injected relative motion into the quantum vacuum in the form of valid associations of particle components. Something in the form of particle component associations must exist for it to have relative motion with respect to the background.

In the beginning there was no effective dimensionality but because there was relative motion the newly created particles began interacting and these interactions created the dimensionality of the observable universe.

Because everything was originally at a single non-dimensional point an enormous cascade of events instantly occurred. It was this cascade that computationally created enough space and clock time to contain them. The injection of relative motion into the quantum vacuum instantly creates a computational dimensionality large enough to contain it.

This is the computational source of cosmic inflation. Since dimensionality is created by events it is instantly created large enough for those events to occur within it. In a sense events open up a spacetime big enough for those events to occur within it. The events themselves create the spacetime that contains them.

When the observable universe opens itself in the big bang on one end of its scale are all the elemental forms of relative motion and on the other end the spacetime that contains them. Relative motion and spacetime are both aspects of the same phenomenon of happening. They are equal and opposite aspects of the opening of the quantum vacuum into actuality. They are the smallest and largest aspects of the same thing (spatial velocity) and are reflections of each other. Relative motion is what is poured into spacetime to open it up. And the dimensionality of individual particles is a reflection of the templates of dimensional attributes of the computational background in which they are computed.

The big bang injected relative motion into the quantum vacuum in the form of particles. This relative motion can go into the vibrational relative motion of mass, or the charges of the other forces, or into the wave motion of electromagnetic energy, or into linear relative motion, or into the zero-point energy of the quantum vacuum.

Thus space and time were created so relative motion could be expressed. In particular linear relative motion exists because particle events can't occur without it. The total relative motion of mass-energy in events can't be conserved without dimensional space because the rest masses of particles are not proportional and they need positions and velocities in dimensional spacetime to manifest. Thus when the Big bang injected relative motion into the universe it had to create a dimensional spacetime and open it up so it could manifest linear relative motion so that total relative motion could be conserved.

Originally perhaps all relative motion was distributed evenly in the form of zero-point energy. In this state of maximum entropy then the dimensionality of spacetime wouldn't exist because events would have an improbably low chance of occurring.

For some reason a vast amount of this relative motion was converted into other forms and distributed among valid particle component sets in the form of particles. One can speculate on how this might have occurred possibly by the conversion of the relative motion of a previous universe collapsing into a universal black hole. Since particles immediately begin to interact in events that conserve and entangle their different forms of relative motion dimensionality is immediately created in the observable form of velocities and positions. The observable universe is just particle components in relative motion and the particle components have to have something to manifest their relative motion in so they create spacetime. Spacetime is created by events to manifest relative motion so it can fully conserve the total relative motion of their mass-energy.

Total spacetime velocity is conserved. The total vector velocities of space and time of all processes always equal the speed of light c. The vector velocities of space and time can be converted into one another as long as their total remains equal to c.

And relative motion, or more accurately relative velocity, is always conserved because all forms of relative velocity correspond to different forms of mass-energy. All forms of mass and energy can be converted into one another by changing the form of their relative velocities. For example the vibrational relative velocity of mass, the wave frequency relative velocity of electromagnetic energy, and the linear relative velocity of kinetic energy and heat are all inter convertible.

But relative velocity actually is velocity in space, so what is fundamentally conserved is mass-energy and time velocities. The total mass-energy velocity and time velocity of all elemental processes is always equal to c.

So the conservation of spacetime velocity and the velocity of mass-energy are unified in the more fundamental METc Principle of the conservation of mass-energy and velocity in time. The total velocity of mass-energy and time is always conserved at every point to the value of the speed of light. And this is true at all points in space and time. It is true everywhere and at every time.

So we could speculate that originally the quantum vacuum was pure time (P-time). It was the pure living happening of existence in an unobservable form because there was no mass-energy and thus no observers to observe it. Then the big bang converts some of that unlimited time velocity to mass-energy, which is relative velocity in space, and as a result the maximum possible velocity value slows to c based on the amount of time converted. The amount of relative motion created in the big bang sets the value of c and the allocation of a fixed number of processor cycles to compute velocities in space and in time.

The total relative spatial velocity of mass-energy at any point reduces the velocity of time at that point and this is origin of both the STc Principle and the conservation of mass-energy. They are both aspects of a more fundamental principle, the METc Principle of the conservation of mass-energy and time, which together with particle component conservation covers everything in the observable universe.

Together everything in the observable universe is conserved in a manner that explains all the effects of both general relativity and quantum reality. The computational conservation of mass-energy and time produces the effects of general relativity and the computational conservation of particle components produces the effects of the quantum world.

A BIG BOUNCE?

The dependence of entropy states on gravitation suggests a possible big bang theory that could possibly solve the question of how the initial entropy state of the big bang seems to have been the most unlikely possible minimal entropy state.

The standard discussion of cosmological entropy presumes that an inexplicable minimal entropy state in which all matter was originally clumped together existed at the big bang (Penrose, 2005), but this is based on the unrecognized assumption that gravitation was attractive. If it were actually repulsive as is likely given the very nature of the big bang and the exponential expansion of the universe with its immediate inflation then entropy would have actually been at the most statistically likely maximal state instead.

It is almost certain that during the big bang inflationary period gravitation was in fact repulsive. How else could the immensely concentrated energy of the initial very small universe not instantly collapse back into a black hole if a very strong repulsive gravitation was not expanding it to cosmological dimensions? The hypothetical inflaton field proposed to explain inflation just adds an unnecessary additional force of nature that has since vanished into nowhere (Wikipedia, Cosmological inflation).

It also seems to me unclear how entropy could even be measured in a universe with essentially zero volume since everything packed together tightly in a constrained space seems both minimal and maximal at the same time. But assuming it to be minimal following Penrose we can offer a reasonable theory to explain why.

Assume a big bounce universe in which the final previous state was a runaway attractive gravitational collapse into a universal black hole. Now assume that when this universal black hole collapses through its singularity that we get a white hole big bang in which gravitation instantly reverses from attractive to repulsive. We then automatically get both the required minimal entropy state to start with (the maximal entropy state instantly reverses to a minimal entropy state as gravitation reverses) and we get a gravitationally repulsive big bang and instant inflation of the early universe as well. Then over time the mix of positive and negative gravitation changes to what it is today.

So a good explanation for the Big Bang is simply the reversal of positive gravitation in a universal black hole with everything clumped in the same place (now the only actual place) in a state of maximal entropy to a state of gravitational repulsion where everything starts flying apart with expanding space in a big bang from a state of minimal entropy. This minimal entropy then enables meaningful form to evolve into a new universe with stars, galaxies and life. This theory, which so far as I know is original, explains both the big bang and the supposed statistical anomaly of an originally minimal entropy state.

This assumes that the previous universe collapsed into a universal black hole, but there is no indication this is the expected fate of the current universe. But since there is no clear explanation for the currently accelerating expansion of space the jury is still out. The ultimate entropy state depends entirely on what gravity does. If gravitation turns entirely attractive and the universe stops expanding then all matter, and space itself along with it will eventually collapse into a universal black hole which could conceivably rebound in a big bang into a new universe with initially repulsive gravitation.

The currently accelerating expansion of the universe does seem to indicate that the mix of attractive and repulsive gravitation is changing which at least lends credence to the theory.

If the universe does eventually begin to collapse there is no reason to think clock time would then reverse its arrow as is sometimes assumed. Processes would not run backwards and clock time should still run in the same direction. And time doesn't necessarily reverse through the black hole to white hole transition of a big bounce. Clock time is presumably just reset and starts again from the beginning while P-time continues eternally uninterrupted with happening driving the universal processor computing the whole process.

Of course this is a speculative theory that requires some good evidence but it does seem reasonable. What would be fairly convincing would be some indication that the force of gravity would actually reverse through a black hole white hole transition.

Perhaps spacetime would turn itself inside out? As spacetime turns inside out clock time might reverse as well but this might not be observable since even clocks reversed in time always seem to go forward because all the processes of observers are inevitably going in the same direction. Space turning inside out might be observable only as a reversal of gravitation.

Note that from the point of view of an observer in empty space objects that have fallen through the event horizon of a black hole are traveling faster than light and this is also true of objects beyond the particle horizon. By the STc Principle this means that the clock times of these objects will be the square root of a negative number. In other words their clock time velocities will become imaginary numbers.

Stephen Hawking and others have explored the concept of imaginary time as way of avoiding the singularity at the center of black holes though our model in which spacetime is actually computed by events seems preferable (Hawking, 1998, p. 157). In any case it's still unclear exactly what happens to time, space, and mass velocities inside black holes from an internal view. Thus it makes sense to consider the possibility that mass and time might turn negative or imaginary and this might initiate a white hole inflation by repulsive gravitation.

However individual black holes don't contain enough particles to create entire new universes. We'd need a universal black hole white hole bounce to accomplish that. So the jury is still out on this one.

TUNING THE FINE-TUNING

Another purely speculative but quite intriguing hypothesis concerns the origin of the complete fine-tuning and what might happen to it in a big bounce. If the universe effectively turns inside out in a big bounce perhaps the cosmological becomes the small and the small the cosmological. In this case perhaps the evolutionary results of the previous complete fine-tuning might somehow become or at least inform the complete fine-tuning of the next universe, effectively retuning the complete fine-tuning hopefully to produce a more effective universe in the next go round.

If the more evolutionarily successful emergent programs of the previous universe could somehow reprogram the complete fine-tuning of the new universe to more efficiently produce them or even better results in the next go round this might involve a slight tweaking of the fundamental constants such as c, the zero-point energy, the strengths and balance of the forces, the masses of particles, the numbers and valid associations of particle components and or other fundamental aspects of the complete fine-tuning. Though it's not clear how this might work or what kind of hopefully new and improved universe would result it's always possible.

It's as if the universe might tell itself, "Hey intelligent life was a neat result of the last bounce so let's adjust the complete fine-tuning this time to make it a little easier for intelligent life to evolve and flourish in the next universe. Perhaps tweaking things so brains could be a lot bigger and more efficient and easier to evolve might work?"

As a result the new universe would stochastically produce a new and improved set of emergent high-level programs, which in turn would tweak the complete fine-tuning of the subsequent universe. In this way the overall universe of bounces could evolve through successive incarnations towards an ultimate end that can only be imagined.

If the universe does turn itself inside out in successive big bounces this may well involve spacetime parity reversal. Currently antiparticles exist so that many types of particle events are able to occur that couldn't occur without them. However the skewed particle antiparticle ratio is not understood and could well involve the way spacetime parity reversal might work in a big bounce.

In the author's view the immensely rich and improbably effective complete fine-tuning of our universe through all levels of emergence is the biggest mystery of all. There is much more to this mystery yet to be discovered, and within it may lie the greatest insight of all into the true deep nature of reality. The robustness of the DNA structures that emerge naturally from the complete fine-tuning are only one of a seemingly improbably coherent set and are a wonderful example of this mystery.

INFORMATION COSMOLOGY

THE INFORMATION UNIVERSE

In Universal Reality the universe consists entirely of data being computed in the medium of existence, which is identified with the quantum vacuum. This data consists of the actualized data of the observable universe and the virtual data of the complete fine-tuning and elemental program that computes it. In this chapter we examine the universe from the perspective of its information structure.

We use the terms data, information, and forms more or less interchangeably though the context should make the emphasis clear. Information is simply meaningful data. However all data is meaningful in a fundamental sense and it may have different meanings to different observers. Forms and information are identical in our usage. Information places emphasis on the meaning, and form on the structure but these are essentially two perspectives on the same thing.

Data is the raw elemental differentiable units of reality. Data is non-physical and abstract in the same sense as the data of a computer program. However the data that makes up the universe is various forms of the underlying medium of existence itself within the common medium of existence in the same sense that waves, currents, and ripples are different forms of a common medium of water. All the myriad forms of data that make up the universe are different forms of existence within the universal sea of existence. All the different forms of data of the universe have no self-substances other than that of their common existence. Everything that exists is a data form of and within existence.

The actualized data that makes up the observable universe consists of innumerable identifiable interacting data processes in a continual state of recomputation by the universal processor of happening. Because these processes are in a continual state of recomputation they act as emergent programs that together compute the data state of the entire observable universe at each P-time tick.

The information of the observable universe consists entirely of abstract data and exists in the computational space of existence rather than a physical space. This computational space has no dimensional characteristics; it's pure formless existence, within which all the data of existence exists as forms in a single vast universal computational nexus that takes the form of the entanglement network produced by conservation events among elementary particles. However this data does contain the *numeric* dimensional data from which a dimensional spacetime can be computed.

The data of the observable universe exists in the computational space of the quantum vacuum just as the data of a computer program defines a computational rather than a dimensional space. However just as astronomical data in a computer program consists only of numbers but can be displayed as a physical space on a computer screen, so the real data of the actual observable universe exists only as numbers in the computational space of the quantum vacuum but can be interpreted as physical processes within a dimensional spacetime by observers.

INFORMATION DOMAINS & OBSERVERS

The universal program that computes the entire universe thankfully doesn't produce a completely homogenous amorphous universe. Due to the complete fine-tuning manifesting in particular the binding energies of particle structures it computes a universe with an extremely rich and complex emergent structure.

Though all the information of the universe is computationally connected as aspects of a single process there are great differences in the degree and types of connection among different areas of information. These differences produce information *domains*.

Domains are meaningful subsystems within the universal program and in particular among different areas of data. They are generally characterized on the basis of boundaries that are less computationally dense or of different types than their interiors. However as aspects of a single universal program domains are never completely independent information structures as they are always in continuous interaction with computationally adjacent areas.

There are different measures of computational density corresponding to areas of computational difference. For example computational areas that move relative to their background (for example the programs and data of living beings), areas of similar types of data structures (the data of organized cellular structures), or rates of computational change (eddies and waves in a river) demark domains.

Computational density tends to conform to how observers identify material boundaries, dimensional relationships, or different types of processes, and in general to what observers identify as individual things and processes. However the individual things that observers discriminate on the basis of domains are generally parts of *ad hoc* and overlapping hierarchies.

Thus individual things and programs are not fixed discrete well-defined areas of the universal program as they tend to be in observer simulations. Individual trees, limbs and leaves are all meaningful programs or things that can be conceptually isolated even though they reference overlapping domains which are all parts of larger domains we can identify as species, forests, ecosystems, continents etc. Surfers, oceanographers, and smelt selectively isolate waves, currents and tides from the larger domain of an ocean to facilitate their actions.

So though domains are the actual data structures that emerge as computational manifestations, the things and programs that observers conceive the world in terms of are not innate emergent structures in reality but data constructs in observer simulations of reality. They are invariably based on actual emergent domains, but observers map domains to things in ways meaningful to individual organisms and actions.

Biological observers naturally tend to isolate discrete individual things on the basis of domains though there is much flexibility in how this occurs in changing situations. Thinking in terms of individual things greatly simplifies observer simulations of reality by reducing the enormous actual complexity of computational reality to conveniently defined individual entities, actions and relationships. This makes mental computations much simpler and more efficient. It also explains how the concepts of individual things can change as needed in a meaningful way as exhibited in the great flexibility of human language.

Observers also tend to take mental snapshots of dynamic ongoing processes as individual things though things are more accurately the running programs that are continually updating that data.

Thus everything that exists within the universe, including us, is almost certainly an enormously complex hierarchy of often overlapping information domains, computational structures that are emergent aspects of the universal entanglement network. Because every element of a domain such as that of a human being is continually happening the domain itself acts as a complex hierarchical program in continual computational interaction with its computational neighbors.

The actual forms of things and programs are not merely the names or simplistic descriptions that observers often denote them as. Rather they are the complete enormously complex hierarchical aggregates and associations of all their many sub and super forms down to the elementary particle component level. All this data in a continual state of recomputation is what any individual thing actually is.

Theoretically forms can be defined in every way possible across the total information of the entire universe. For example whole categories of forms or processes can be defined as programs themselves. Any categorization that has meaning to any observer is meaningful and thus valid in the simulation of that observer.

All possible sets and subsets of forms can be considered forms in their own right. There is virtually no limit to how forms can be defined. The form of a biological being can be considered either as a single object, or the enormous aggregate form of every one of its elementary particles, cells, organs, and body parts. Any one of these can be considered as a separate form or as parts of another form. Take all the information of the universe and any grouping whatsoever that is meaningful to some observer can be considered a form.

Forms are any and all possible sets or groupings of the information of the universe, though of course they will have different relevance to different observers. The important thing is to understand a form is an observer specific mapping of a subset of all the sufficient information of a domain that gives it its identity to the observer in the current context. In actual reality the domain is the totality of all the hierarchies of data of that thing, only some of which will be relevant to any observer at any particular time.

The only data that exists in the sense of actually being stored in computational space is the data of elemental particles, particle components and their entanglement relationships in the entanglement network. All the emergent programs and data can't be stored as separate data entities because their multiple possible definitions and overlaps would inevitably lead to contradictions that would tear a computational universe apart. Thus it's important to understand that all emergent programs are dynamic *manifestations* of aggregate operations of individual applications of the elemental program. Emergent programs and data are mainly meaningful only to observers because only observers are able to recognize them. Emergent data structures are not stored as independent entities apart from their elemental data; they are always aggregate manifestations *of* their elemental data in the same sense as cellular automata are.

The patterns that emerge in cellular automata from individual cell interactions aren't computed or stored as separate data structures but are entirely dynamic *manifestations* of those individual cellular interactions in aggregate. They have meaning and are observable only at the emergent level, and only to observers able to recognize them and form simulation models of them.

Thus domains are the actual emergent information structure of the entanglement network, and individual forms, things and processes defined along domain boundaries are observer views of domains; vastly simplified observer views of the enormously complex emergent structure of the entanglement network meaningful to individual observers.

THINGS ARE THEIR INFORMATION

All the individual things of the universe are meaningful observer mappings from emergent domains in the entanglement network. Emergent things are meaningful only to observers because only observers are able to detect emergent relationships among aggregate particle data. Individual things are observable and meaningful only to the extent observers are able to model them in their simulations of reality. They are invisible to the elemental subroutines that compute them and to observers unable to recognize them.

When we look at the world around us we tend to see it in terms of material things and physical processes in an encompassing space. However this is our mind's simplistic interpretation of the complex reality of the data of the running programs that underlies it.

The actual reality of all things is the complete data or information of those things in a continual process of interactive computation. Everything in the observable universe consists entirely of it information in a continual process of computational evolution. Thus everything in the observable universe is the complete information of what it is and that is all that it is.

This insight was already being expressed in ancient Indian and Zen philosophy as the concept of the emptiness of forms. In other words all the forms of the world (the data or information) consist only of their forms. All forms are only forms and are empty of any self-substance other than the common immanent existence (śūnyatā) all things share (Wikipedia, Heart Sutra).

Thus all the individual things we experience as humans including us actually consist only of their dynamic empty information forms given immanence by existence. And this can actually be verified by analyzing anything at all into all its individual data constituents and discovering there is nothing else there.

If we carefully examine all the conceptual components of any thing whatsoever we find that every one of those components ultimately reduces only to the information of what it is. What we think of as a thing is actually the association of all the various types of information of all its various aspects. Ultimately things are only the complete information of all the aspects that make them seem to be physical objects in our mental simulations of them. This applies to everything in the universe without exception.

This is also confirmed by the obvious fact that only information can ever be observed. If all the information of anything is subtracted from it there can be nothing else left of it other than the pure substrate of existence common to all things within which its forms appeared. Thus everything is its information only and the immanence of its existence that it shares with all other things. Thus there is nothing physical to anything at all. It's all information *interpreted* as physical in our human simulations of reality.

Everything in existence reduces to the information of what it is, and that is the actuality of what it is. We can analyze everything into all the information that composes it including the information that appears to make it a physical object. Thus there is no need to add anything at all other than information to anything to make it appear to be physical. It's apparent physicality in our simulation of it consists only of the combination of all the specific information that makes it seem to be physical and gives it its seemingly physical appearance.

This clearly applies to the representations of things we experience in the world around us which exist only as information structures in the neural circuits of our brains, but it's also true of the actual existence of things in the external world independent of observers. This is further confirmed in the chapter on Computing Reality by showing how mass-energy and spacetime, the last bastions of physicality in modern science, also reduce entirely to information. Thus everything we see in the world around us including ourselves is actually its information only, its running program only.

We all are only our data states in the present moment in a continual process of recomputation by the processor of happening that computes the evolution of the entire observable universe and all the information within it. We are all our running programs computing our data in continual interaction with the other programs of the observable universe and that is all we are.

This insight changes nothing about the world, we just notice that everything in the world around us actually consists of computational processes or programs whose current information states in the present moment are the things we interpret them as.

Humans and other biological organisms are among the programs that can be meaningfully discriminated from the universal program. All biological organisms are clearly enormously complex integrated computational systems from the computations that govern the chemistry of individual particles and cells up the complete hierarchy of their organs and control systems, to the computations taking place in their neural circuits that constitute their simulations of their environments.

Because humans and other biological organisms are computational systems, the rapidly developing science of intelligent robotic systems provides a rich and useful model to understand how we function in our environments as programs.

THINGS ARE THEIR COMPUTATIONAL HISTORY

Everything in the universe is its program computing its current data state. But there is also a much deeper secret here because things are not just the information of what they are in the present moment, but the current computational result of their entire information history as well. Things are the information of themselves, but the information of themselves is the current result of their entire computational history. The information of any thing at all is the current computational result of its entire computational history all the way back to the big bang through all the evolving forms and interactions that resulted in its current form.

For example a dead leaf lying on the lawn consists entirely of the information of what it currently is, but the information of what it currently is contains the information of its species, the information of its DNA and cellular structures and information about the processes of biological life itself, even the fundamental information of elementary particle attributes and processes including information about the original fine-tuning. Not to mention the position of the leaf lying on the lawn carries the information of the coming of Autumn, its changing temperature and sunlight, and even every precise detail of the breezes that brought it to its exact current position from the twig from which it fell.

All the past information history of the leaf is contained to one degree or another in the current information of the leaf, and the leaf is this information of its current state which in turn is the exact computational result of its every possible aspect down to the finest detail of its entire past information history. The leaf would not be in the exact position it is and condition it is if it had not undergone every one of those exact interactions. The information of the total results of all these interactions is the current information of what the leaf is, and that information is all that the leaf is and exactly what that leaf is.

All individual forms are transient. From the total information form of the universe individual forms are continuously emerging, evolving, and fading back into the total form of the universe. From the single running program of the universe, individual programs continuously emerge as identifiable individual programs, run and evolve for a while, and then computationally dissolve into other programs

Thus everything that exists in the present is in essence a recording of the past. The entire information of the present states of things is a recording of the entire past states of things. The entire present is a recording of the past and that is all it is. But this information is distributed through the network of computational interactions. We just need the proper technologies to play it back and eventually perhaps we may be able to actually play back past events and watch them occur. It's certainly theoretically possible.

Consider an event that occurs in a closed container. All the information of that event is redistributed among the contents of the container. Thus the information of the event remains in the container but redistributed among its contents. Since the past-present is a completely determined data structure we could theoretically play the information in the container backwards, at least in a simulation, to recreate any past event within the container. In general it should be possible to play back any past event from all its redistributed information.

This is of course done to a limited extent in forensics, science, and even the application of common sense. So it certainly does work. Its accuracy is only limited by the amount of redistributed information available and the logical capacity of the investigator.

In effect the exact deterministic consistency of past and present and the rich redistribution of information by events gives us present moment mirrors in time through which we can logically recreate and visualize events from the past and even predict future events to some degree.

THE CONSERVATION OF INFORMATION

This raises the question as to whether and to what extent information is conserved. Assuming for the moment the universe is composed of a fixed number of particle components there is a fixed number of valid arrangements of those particle components, each corresponding to a discrete information state. From this perspective it appears the universe would contain a fixed amount of information or at least a fixed amount of data.

However we must be careful in measuring quantities of information. Information is usually considered to be meaningful data, and that implies meaningful to some observer, but we can also use it the absolute sense of the amount of possible variation in data. In this sense the amount of information in a million digit random number is equal to the amount of information in any other million-digit sequence including that of a mathematical treatise or work by Shakespeare. The million digits of the random number are the information required to express that random number just as the million digits of any other number are exactly what is required to express that number. They both contain the identical amount of information by this definition.

A million digit expression of the number pi is apparently random, yet many would argue it's inherently more meaningful in its information content than a random number generated by some other process that didn't represent a meaningful geometric 3-space ratio.

But if events continually redistribute the information of the past into the information of the present, and the information of the present contains the information of the past, doesn't that mean that the information of the present is continually being added to that of the past and that the total information content of the universe is exponentially expanding with time? So is any information ever lost or does it keep accumulating?

The answer is that the universe continually keeps adding information as it redistributes the data elements of its information but the information of the past gradually sinks below the threshold of retrievability.

But it also depends on how information is defined. Information is generally defined as the meaningful content of data, but this is ambiguous because the amount of meaning depends on the observer as well as the data. Data only has meaning in the concepts of individual observers. A mathematical paper has different information content to a mathematician than it does to a paper-eating insect.

Take an initial state consisting of newly created elementary particles that have had no interactions. The amount of information of the state is a function of the number of particles. It's true there is a sense in which the lack of all possible interactions is information but this is not properly part of the information content of the universe but of some observer model of the universe.

Now as the particles begin interacting information is added consisting of the information of the particle events and their resulting entanglement relationships. This adds information to the information of the particles and the amount of information in the universe increases.

Though only the actual state of particles, events, and relationships in the present moment is properly part of the current information content of the universe, that information also contains the redistributed information of past states. So the total information content of the universe continues to increase as time passes. There is a sense in which no information is ever lost even as it slips below the threshold of retrievability.

The information contained in all the elemental data of particles, particle components, and their stored relationships is constant but the amount of *emergent* information manifested by this data in aggregate varies because it is depends on observers able to recognize it.

And take the analogy of the past information history of the Pacific Ocean. It's exact current information state, the exact form of every ripple, wave and current, is the exact computational result of every past interaction of every ripple, wave and current, thus the information of its entire past history is hidden within its current data structure which is the exact computational result of its entire past history.

However all past individual waveforms have now interacted and merged and been redistributed into the forms of the current ones. The information of prior individual waveforms is still there but is now redistributed among the current ones. Thus the actual forms of past waves are still present in the current waves, which would be different if they hadn't existed but widely redistributed among them.

Thus instead of being able to directly observe the forms of past waves they must be reconstructed from the current forms of all the waves they have influenced. With ocean waves this is an impossible task because we also have to know the forms of all of the other waves they interacted with to produce the current waves. It becomes effectively impossible even over short periods of time. The information of all past waves and their interactions is there but it becomes impossible to retrieve it.

However the sea of existence is not a simple homogeneous fluid medium like water but the basic problem is the same. The observable universe is analogous to the forms of all the waves, ripples and currents of an ocean but its elements are much more complex so the information forms that populate it are also much more complex. In particular some forms are vastly more persistent than other types of forms, and some forms like DNA continually retain their information identity by copying themselves.

This means that some past information forms are easy to retrieve from current forms while others are nearly impossible. It all depends on the forms and how they are computed from previous forms and redistributed.

Nevertheless the current information state of the observable universe is always the current present moment result of all the individual computations of the entire universe all the way back to the big bang. Thus the information of every one of the past interactions of the universe is distributed within the current information state of the universe.

In a sense none of the past information of the history of the universe is ever lost. It's all there redistributed through the current information state of the universe, which would be different otherwise. The current information state of the universe actually is the redistribution of all of its past information states.

Everything in the universe is information and the evolution of the universe is just the continual reshuffling of that information. The old information is interactively reshuffled and becomes the information of the new. Everything in the universe is its information only, and that information is the information of its complete computational history, and that is all anything is including even ourselves.

THE SHERLOCK HOLMES PRINCIPLE

Through the history of the universe uncountable myriads of forms computationally interacted within the single evolving universal form. Within the universal form the forms of individual things continuously appear, transform and disappear into new forms. And in this grand cosmological process the information that is individual forms is continuously fragmented and redistributed throughout the universal form continuously coalescing into new individual information forms before those too dissolve into other forms.

Thus every individual form we choose to examine does actually consist entirely of the information of what it is in the present moment, but that information of what it is now always consists of varying amounts of the information of all the forms it has interacted with through its entire computational history back to the very beginning. When forms interact with other forms the forms change and these changes reflect the other forms. Thus all forms contain some of the information of the other forms with which they have interacted.

This is a general principle of the interaction of information forms. For example even in our perceptual interaction with the forms of external things the forms of our retinas and neural data structures change to register them. When the forms of two stones collide, both of their information forms will be deformed in ways that reflect the information forms of the other and the programs that compute them. Thus by examining the information form of a stone one can gather information about what other forms it may have interacted with in its computational history to produce its current form.

The entanglement network whose current state is the observable universe is the computational result of countless particle interactions. At the classical level it's the computational result of countless interactions of running emergent programs. And when these programs interact they invariably exchange information about each other. By definition programs interact by exchanging information, by modifying each other's information. An event or interaction is by definition an exchange and modification of information.

The result is that the information of things is continually distributed among the other things of the universe, and this has been true throughout the entire history of the universe. Thus the information of everything in existence today will contain information about the complete fine-tuning, the basic particle structure of reality, as well as all the emergent processes that lead to its computation. Everything in existence is its complete information history and that is all it is.

Because the information of a thing is the current state of its entire information history that information contains information about the things and programs that led to its being here right now throughout the entire history of the universe. It contains varying amounts of information about many other things and programs now separate from it in both time and space.

Because all things contain information about other things and processes we can extract information about those other things and processes that we can not directly observe by analyzing the information of things and in particular related sets of things.

Universal Reality calls this fundamental principle of reality The Sherlock Holmes Principle. It states that the information of any individual thing or combination of things is composed of retrievable information about the other things and programs with which it has interacted in the past. The Sherlock Holmes Principle is the fundamental principle of science, forensics and of most knowledge.

The fact that the universe consists of the computational interactions of programs is what makes knowledge possible. And the fact that the information structure of the universe is so computationally rich is why science has revealed so much of its incredible awesome wonder. By analyzing a number of forms that have interacted with past forms we can often build very accurate pictures of those other forms.

In a real sense every form other than the most elemental is composed entirely of other forms because of the myriads of interactions that have contributed to its current form. And those other forms are the redistributed information of innumerable other forms and carry information about all the other forms with which any particular form has interacted through its entire history.

The Sherlock Holmes Principle continually redistributes the information of the past into the present information state of the observable universe. The evolution of the universe consists of the redistribution of its information just as it consists of the continual redistribution of its particle components because its particle components are information and thankfully much of that information is retrievable.

FROM CAUSALITY TO CONSISTENCY

Causality is not actually a concept of science but a rather a confusing metaphysical *interpretation* of science. Essentially it's a hang over from the old idea of a physical universe in which physical things seemed to push other physical things around to produce effects. But in Universal Reality where everything consists only of data being computed the usual interpretation of causality makes no sense and must be abandoned.

Because the universe isn't a physical structure causality loses its relevance. And this is actually confirmed by science itself since there isn't a single variable of causality in any equation of science whatsoever. So there is no loss to science at all in completely abandoning the concept of causality. In an information universe the data states that are the actual reality of the universe are *computed* from prior data states. Nothing physical happens in a universe that consists of data. The elemental program that produces particle interactions computes them but doesn't cause them in any physical sense.

So causality is not really even part of science but a meta-principle or interpretation that just refers to the fact that events predictably precede other events in logical sequences. But this is because they are computed according to fixed rules, not because material objects somehow push other material objects around in predictable ways to cause them as was originally thought.

In a computational universe it's enough to know that programs predictably compute subsequent data states according to consistent logical rules. There is no necessity of any physical mechanism. Saying that computations cause events is meaningless. Adding 2 + 2 doesn't *cause* 4, though it does predictably *compute* a result of 4.

When causality is abandoned important new insights into reality and the nature of knowledge emerge. The main implication is that the history of the universe is not a causal history but a consistent history. This enables it to be understood as a single integrated structure that is logically consistent in both temporal directions.

HOW THE PRESENT DETERMINES THE PAST

In Universal Reality the past is the entire computational process beginning at the big bang that has resulted in the current present moment data state of the observable universe exactly as it is in every precise detail. The problem is that the past doesn't exist. It has no reality whatsoever since the entire universe exists only in the present moment. Only the present moment is real and actual.

Thus our only possible knowledge of the past comes from its computational traces in the present. The past exists only in memories and its computational effects in other present moment information forms including the arrival of light from distant astronomical objects. The past exists only in the current data states of all present moment forms.

Knowledge is based on observation and unlike the present there is no way of directly observing the past to confirm either its existence or its information structure. Thus to make sense of current reality we must create a past that would logically result in the observable present. We create this past to the best of our ability, as the computational process we think would result in the present. Thus the past exists only as our best current information model of what would have computed the present.

We can only deduce the past from our current understanding of the current present moment information state of the universe. Since the past doesn't actually exist we create it as part of our simulation of reality and come to believe it is something real. This evolutionary adaptation helps us make sense of our current existence and learn from what we remember of our experience so as to function more effectively.

Thus to the extent the past does exist it exists only as an abstract logical structure in our collective and individual consciousnesses with no actual corresponding existence in reality. The only reality of the past is in our simulations of it. This reality is not that of an actual existent past, but a conceptual model of a past that exists only in our simulation of reality rather than as an actual state of reality.

Thus the entire concept of a past exists only in our heads, and isn't a part of the actual universe. The only reality of the past is a mental construct deduced backwards from the current information state of the universe based on what we think we know about how processes work which is also part of our simulation of reality. The entire reality of the past is our best current understanding of what would have resulted in the present information state of the universe, and of course this understanding is continually being refined. The model of the deep past that was current a hundred years ago contained none of the insights of the modern evolution of the universe from the big bang.

Thus the past is a theoretical computational progression of information states though past times that we think would result in the exact actual information state of the present. It's the best theoretical model of computational evolution through time that from moment to moment consistently results in the subsequent data state up through the present moment. The only actual existence of the past is our collective theoretical model.

Thus from the perspective of the present moment, which is the only possible perspective, it's clearly the present that now determines the past. While it certainly does make sense to say that the past determined the present this is ultimately a completely unverifiable statement. It's certainly consistent with our usual model of reality but unverifiable by direct observation because we can never visit the past to conduct an observation.

In contrast we continually directly confirm the existence of the present through our experience. The present exists independent of our knowledge of it. However our concept and knowledge of the past continuously changes with our knowledge of what would be necessary to produce the present information state of reality. And since an actual past doesn't exist we must accept that the past itself continuously changes as our theoretical model is continually revised since the past exists only in our collective simulation of it.

The only past that actually exists is a theoretical model that is different from person to person and certainly from species to species. Every individual person and organism will have its own individual model of the past and collectively all these models can be said to define the past as well as it can be known.

One could of course argue that while our *knowledge* of the past is created from the present there must have been an actual past that actually existed that did result in the present. And yes, this certainly makes sense from a theoretical standpoint, but the point is there now is no such actual past, and at any point in time there never was an actual past. At every point in the past only that current present moment existed and the previous past didn't exist. So there is no way to ever directly verify what the past was or even that it ever existed. All we have is a theoretical model in the present moment that makes good sense.

Logically we can assume that there must have been an actual past that evolved into the present information state of the universe. But we must be very careful in understanding what this really means. There are subtleties with respect to past and future tenses difficult to accurately convey in language, especially when it comes to speaking about different possible pasts. Nothing expressible using the past tense actually exists. It's always a backward inference from our present conception of things.

We can however define the past as a logical structure even if our knowledge of it will always be imperfect and incomplete. The actual past is the complete computational evolution that actually did result in the actual complete current state of the universe. But again this is a theoretical description rather than the actual past which remains ever elusive. The only actual past that can ever actually exist remains a mental model in the present deduced from the present.

Since the only past that actually exists is a current present moment information structure we must conclude that the present actually creates the past, specifically the past that would have computationally created the present to the best of our current understanding.

This doesn't mean that the past doesn't create the present. It certainly does according to all our consistent mental models of reality. And these must be accurate at least to the extent they enable us to function and survive in the real world. So the real point is that past and present must now be seen as a single bidirectional computational system completely logically consistent in both temporal directions.

The present creates a past that creates that exact same present. The past creates a present that creates that exact same past. The entire past up to the point of the present stands as a single consistent bidirectional unity exact and unchangeable in every last detail, but this is entirely a theoretical structure that exists in its most complete and accurate form in the collective information model of current science in the present. Only the current present moment of this vast structure has a real and actual existence apart from its abstract mental model.

This novel epistemological understanding of what the past really is has important cosmological consequences. Only the present moment exists, and it exists exactly and absolutely as it is in every minute detail. Therefore only one possible past could have existed, and it must also be exact in every minute detail throughout its complete evolutionary history from the big bang on. Only one exact possible past could have existed because only this single unique past could have computationally produced every exact detail of the present.

Thus the complete past - present information state of the universe forms a completely deterministic fixed and unalterable theoretical structure totally consistent in both temporal directions. And nothing about this entire structure, not even the minutest detail, could have possibly been different than it was. Not even a single one of the uncountable myriads of random quantum events over the entire history of the universe could have possibly been different than it was.

The actual exact current information state of the universe conclusively and absolutely falsifies any other possible past than what resulted in the present. There is only one single possible past that results in the present in every exact detail of the entire universe. There are no other possible alternative pasts in even the slightest detail. The entire past is completely and deterministically fixed in every minute detail.

In common parlance we often speak of the past as if it might have been different, and this is a useful tool in learning to predict future events. We imagine different possibilities in the past and consider the differences they could have made in the present to better understand the workings of reality and the effects of our choices in the future. But what does it actually mean to say something existed (past tense) in the past? All this can ever mean is that this present concept is part of an internally consistent mental model of reality in the *present*.

Being able to imagine different possible past events doesn't mean the actual past could have been any different than it actually was. It couldn't have been different because the actual past is quite clearly impossible to change even though our knowledge of it improves. What we really mean when we speak of alternative possible past events is that we can devise a *similar* event in the *present*. But of course we can never actually change any past event. It's completely impossible to exactly repeat any past event because all information states are connected with all the information of the entire universe as it was at the time the original event occurred and by definition those are now different.

Thus the entire past back to the big bang is completely and exactly determined by the current information state of the universe in every last detail and simply could not have been any different whatsoever and that includes the actual results of every random quantum computation.

The entire notion of different possible past states is an illusion based on our ability to repeat conditions *similar* to past conditions in the present on a *local* scale. Thinking in this way is convenient but applies only to experiments that are actually possible. It's certainly not possible to go back to the past and actually change anything and it's certainly not possible to recreate the origin of the universe. Thus alternative possibilities have no meaningful application to the actual cosmological past. Alternative possibilities in the past are meaningless nonsense.

BEYOND THE ANTHROPIC PRINCIPLE

The ultimate consequence of this understanding is that the complete fine-tuning of our universe is the only possible fine-tuning. There is no possibility of any other fine-tuning because the one that existed at the big bang and still exists is precisely what resulted in the present moment state of the universe exactly as it is. The existence of the observable universe exactly as it is in every last detail completely and absolutely falsifies not just the existence but also even the possible existence of any other complete fine-tuning whatsoever.

The present moment information state of the universe is all that exists and this completely determines what the original complete fine-tuning of the quantum vacuum that resulted in it had to have been and still is. There is simply no other possible alternative for the complete fine-tuning given the absolute unchangeable reality of the present moment universe it produced.

Thus the entire computational history of the universe back to the big bang is an exactly fixed and determined logical structure with no possible alternative from the viewpoint of the present, and since only the present actually exists this is the only valid viewpoint.

This is true all the way back to the original fine-tuning. Given the actual state of the present there is not the slightest possibility the original fine-tuning could have been different than it was in any detail whatsoever because only it led exactly to the present information state of the universe as it actually is. Thus beyond the Anthropic Principle (Wikipedia, Anthropic principle) we find the reason for our original complete fine-tuning being exactly as it was because there is simply no other fine-tuning possible. The actual existence of the current information state of the universe conclusively falsifies any other possible fine-tuning, and even the possibility of any other fine-tuning.

Of course we can *imagine* other possible fine-tunings and the universes they might have produced, but the possibility of their existences is completely falsified by the actual existence of our current universe exactly as it is.

Though initially counter intuitive this argument boils down to the key epistemological issue of how we know what we know. Carefully considered it turns out that the whole realm of past and future tense in human language and thought, though quite useful, is epistemologically unsound because it describes aspects of our internal mental simulation of reality rather than aspects of reality itself. In a very real sense it's completely meaningless to speak of the past or future except with respect to our mental models of the past or future. Statements about what happened in the past or will happen in the future are not so much statements about the universe but about our simulation of it

Ultimately it's meaningless nonsense to assume there could have been other possible complete fine-tunings when there is no evidence for that whatsoever. The only universe for which there is any evidence whatsoever is ours, so there is only one complete fine-tuning that exists. We can certainly imagine other possible ones, but there is not the slightest evidence that any of them could ever exist. Just because we can imagine something doesn't mean we need to give it any credence.

It's the apparent unexplainable randomness of the fine-tuning of the fundamental constants that was the rationale for the supposed existence of multiverses so that each could have one of all possible fine-tunings (Vilenkin, 2006). But it's now clear there are hidden relationships among these constants waiting to be discovered and it's also clear that much of the complete fine-tuning must be as it is to produce a consistent universe.

So it appears the rationale for the uncountable number of universes some physicists imagine is fading. In any case even if the fine-tuning of our universe is not further reducible it's still an enormously unwarranted leap to think that not being able to explain the fine-tuning somehow implies the existence of a Google of new universes! Much more reasonable and parsimonious to assume only the single universe that we know exists and try to figure out why its fine-tuning is as it is.

As a result the rationale for most of the multiverse and bubble universe theories goes by the wayside. The reason most of these theories were developed was because it seemed like the original fine-tuning of our universe must have been random because there was no known reason it had to be as it was. This has led some cosmologists to unreasonably and unparsimoniously assume that there must have been another universe for each of the other possible fine-tunings they could imagine, a very large number indeed. But this is a completely unwarranted assumption for which there is no evidence whatsoever.

When we are faced with something for which there is no known explanation the rational approach is to try to discover an explanation, not to imagine an entire universe for each of all the other myriad possible variations of that something that can be imagined.

In light of this common sense epistemological approach I think it's now safe to discard the idea of myriads of other possible fine-tunings on which most of the multiple universe theories are based. And I see no reason to assume the existence of any other universes absent any direct observational evidence.

Thus it appears that the apparent problem of why the complete fine-tuning is as it is actually a pseudo problem. The complete fine-tuning is as it is because only that exact complete fine-tuning results in the actual universe, and the existence of only the single universe that does exist for which there is any evidence at all completely falsifies even the possibility that the complete fine-tuning could have been any different that it was and is.

In addition, as previously explained, when all its aspects are considered it seems fairly likely that only the actual complete fine-tuning results in a logically complete and consistent computational universe. Certainly much of it is required in the exact form that exists.

THE PROBABILISTIC FUTURE

Though the information structure of the entire past to the present is completely determined and could not be otherwise, the future is probabilistic and subject to constrained quantum randomness. We can claim that the past could have been different due to quantum randomness, but once random events have occurred the results are completely determined and with them the entire past bi-directional computational history of the universe up to the current present.

But the future doesn't exist, as it hasn't been computed. However the present continuously advances computationally through time, and its future states are not completely determined due to the randomness of quantum choices and the oscillation of processor cycles between space and time. Many aspects of the future are certainly predictable with varying degrees of certainty, they have to be for life to be possible, but the future doesn't exist as a completely fixed information structure like the past does. It can't because it has not yet been computed and thus is subject to the constrained randomness of quantum events.

The present can be thought of as the process of converting the possibilities of the present into the exact actualities of the past. The universe is a program that converts stochastic randomness into exactly fixed deterministic reality.

EMERGENCE

THE EMERGENT UNIVERSE

The elemental program that constitutes the observable universe operates in terms of a relatively simple set of computations that are part of the intrinsic virtual nature of the quantum vacuum. Due to the specifics of the complete fine-tuning the aggregate effects of these computations produce all the enormous interconnected complexity and profound beauty of the observable universe including us. Thus the large-scale aggregate design of the universe is implicit in the design of the complete fine-tuning.

The evolution of the universe is nondeterministic due to the inherent randomness of quantum processes, but is strongly constrained by the complete fine-tuning. Thus the observable universe is one of many possible variations on a single grand design.

The details of how the universal program self organizes into the individual programs that constitute all the things and processes of the observable universe and their interactions is the subject of all the sciences, a vast realm far beyond the purview of any individual book. Thus only the most general principles of emergence will be outlined here from the computational perspective of Universal Reality.

The elemental computations of the quantum vacuum simultaneously compute all the mass-energy structures of the universe and the dimensional spacetime in which they seem to exist. This takes the form of the entanglement network of relationships among all elementary particles and their particle components including their dimensional relationships. The entanglement network is a single universal data structure incorporating every aspect of the observable universe. Simply put the current state of the entanglement network *is* the observable universe and it consists entirely of data in a continual process of recomputation.

All structures and processes above the level of the actual elemental computations themselves are emergent. Thus the emergent universe is all the aggregate structures produced by the elemental computations. Properly speaking the emergent universe refers to everything above the particle interaction level because all the computations that actually compute the universe appear to occur at the particle and particle component levels.

However most of the laws of science are descriptions of things and processes at the emergent aggregate level. It's important to understand that these laws of nature *describe* but don't actually *compute* the structures and interactions of emergent things and processes. All the actual computations occur only at the elemental level of particles and their particle components.

Due to the exquisitely precise design of the complete fine-tuning of the computational system emergent structures and laws automatically emerge from the aggregate operations of the elemental program. It produces the emergent universe and follows the emergent laws of nature observers use to describe it. These are both implicit in the precision of the complete fine-tuning.

The complete fine-tuning specifies what elementary particles can exist and the constants and forces that determine their binding energies. The binding energies and balances of charges in turn determine the rules of particle interactions that build atoms and molecules. The structures of atoms and molecules in turn determine the laws of chemistry and the conditions under which chemical compounds form. And chemistry in turn determines what types of materials exist and how they interact. And these in turn determine the large-scale structures of the observable universe and what types of biological organisms can exist and evolve within it and the rules by which they function.

So the emergent universe that exists today is the cumulative large-scale result of a single elemental program computing uncountable numbers of elementary particle events. From myriads of these deceptively simple computations the entire emergent complexity of our amazing universe has blossomed. And how and why this beautiful and profound complexity is hidden within the complete fine-tuning of the quantum vacuum is perhaps the greatest mystery of all.

EMERGENT PROGRAMS

Though all the computations of the universe occur at the particle component level the entire universe can be thought of as a single universal running program. And this single universal program can be thought of the interactions of innumerable individual programs that make up all the individual things and processes of the universe.

Though everything is actually computed at the elemental level, all the aggregate things and processes of the emergent universe appear to act as individual programs themselves, and even in the case of living organisms as purposeful programs.

This is analogous to ordinary computer programs, which fulfill specific high-level functions even though all their computations are actually taking place at the level of machine language operations. It's the overall structural organization of individual machine language operations that gives a computer program a meaningful independent function at the emergent level.

This is also true of the running programs of the universe. They too are emergent manifestations of organized aggregates of elemental operations. However these reality programs have been programmed not by human programmers but by self-organizing evolutionary processes selecting among the 'fittest' aggregate groupings of elemental computations.

There is another fundamental difference between computer programs and reality programs. Computer programs are sequences of *pre-programmed* code executed sequentially by discrete processors that make calls on externally stored data. Only one operation in an existing sequence is being executed at a time by a processor. But reality programs consist only of present moment data states every one of which is being simultaneously recomputed in interaction with its neighbors within the ubiquitous processor of the quantum vacuum in which they all exist.

Thus reality programs are not linear sequences of pre-programmed code, but organized data aggregates every element of which is continually recomputed in its interaction with others. The entire observable universe is a single structured nexus of particle and particle component data whose interactions are being continually recomputed by the elemental operations of the quantum vacuum in which they all exist. Thus the entire observable universe acts as a single program running in the quantum vacuum and evolving through time.

Thus everything we see around us, including ourselves, is not just its data but also every bit of its data in an active process of computational evolution. Everything that exists, including ourselves, exists as data within the quantum vacuum of existence where we are all continually recomputed together into continuing existence in the present moment of existence.

The interactions of programs with other programs are all computational and they all consist of information only. For example the program of a human being hit by a bus exists only as the information generated by the computational interaction of their programs. But this information representing the breaking of bones and loss of life is the reality of the actual event in the real world. Everything is computational, everything is information, but this information is always the real things and events of the world.

The entire universe is a living system in the sense that it continually happens with no external cause or motivating force. Thus everything in the universe, all the individual programs that run as parts of the universal program, are also alive in the sense they are all direct manifestations of the actively self-motivating presence of the universal program running within them.

We, and everything around us, are the active living presence of the quantum vacuum happening within us continually recreating us into continuing existence. We are the elemental computational subroutines eternally active within our data and that of everything in the universe. We exist not just as living biological entities in the observable universe of science, but also as living computational entities within the active quantum vacuum that underlies it and gives it existence.

We are all running programs, computational domains of the universal program that continually computes the unified mass-energy and dimensional spacetime structure of the observable universe. We are purposeful processes at the emergent level because all the elements of our data structure make up an emergent structure that is self aware and purposeful.

Thus we are not just our complete data states, but also the active manifestation of the fundamental computational processes of the quantum vacuum operating within our data. We glow with the life of the universe continually happening within us. The universe lives within us, and we exist as parts of a living universe. The universe lives us from within by continually recomputing our data into the universal present moment of existence.

The data of all things acts as running programs because it exists in a continual state of recomputation within the ubiquitous processor of the quantum vacuum. Everything is data in a continual state of happening and thus every entity acts as a program in computational interaction with the other running programs that constitute its computational environment.

Because aggregate data has such a rich emergent structure thanks to the complete fine-tuning, things emerge as discernible running programs that can be named and described in terms of their overall function. The data of what things are continually interacts to compute the function of the whole. And the function of the whole is determined by its continual computational interactions with the elements of all the other programs that constitute its environment.

EMERGENT LAWS

Emergent laws are laws that *describe* emergent processes but don't actually *compute* them. Emergent processes and programs are programs that produce computational results as the ordered aggregate results of innumerable individual applications of the elemental program that performs the actual computations of the universe. All the laws and structure of the universe save for the fixed routines that perform the actual elemental computations are emergent.

The emergent laws that describe the states and processes of the universe are those of the experimentally verified laws of science insofar as they are known. Universal Reality accepts all the established theories of science, always subject to revision based on new evidence. In this way our theory maintains complete consistency with the actual equations and logical structure of science even though our *interpretations* of what the equations really mean, are often completely novel.

The emergent laws of the observable universe form the basis of the *logic of things* that humans and other organisms use to understand and function within our environments, and the laws of science based upon them. The logic of things is the fundamental logical rules that underlie the interactions of everyday things at the level of our simulation of reality.

These fundamental rules emerge naturally with the processes they describe and are extracted and codified in our simulations to make sense of our environments and our interactions with them. They are the basic logical rules that both organisms and robotic control systems use to make sense of their surroundings and function effectively.

The logic of things includes simple classical rules like single things can't be at more than one place at the same time. To get from one place to another things must move along the distance between them. Things don't arbitrarily appear and disappear. Existent things must be somewhere. Events have causes, and so forth. The logic of things is the whole set of these simple fundamental rules of the world as we simulate it in our minds.

These emergent laws are in general statistical descriptions of emergent programs that derive from rules intrinsic to the fine-tuning. They tend to have exceptions and inherent limits to their accuracy especially with respect to smaller and smaller aggregates. Random processes tend to converge on exact results in aggregate as clearly demonstrated by half-lives and the laws of gases for reasons explained in the chapter on Quantum Reality.

It is in the nature of the *super-consistency* of the universe that the emergent laws of nature are largely consistent at whatever levels they apply. Emergent laws emerge along with the aggregate processes they describe and are discovered by humans who model them as best they can in their simulations of emergent processes. Certainly the overall logic of things of emergent processes emerges with the processes themselves though it also strongly incorporates the structures of the simulations that model it.

Thus the logic of things describes emergent reality as it is simulated in the simulations of particular organisms however its basic rules are identical among all organisms because it must accurately map the actual logic of reality for organisms to successfully function within their environments.

THE GENERAL PRINCIPLE OF EVOLUTION

Evolution refers to the manner in which emergent information forms change over time, to how the emergent programs of the universe develop as they computationally interact. Evolution describes how the data that constitutes the universe has reorganized throughout its history. Though evolutionary changes are the large-scale results of fixed elemental computations they follow general laws in a stochastic manner in accordance with the complete fine-tuning.

The data forms that compose the universe remain the same unless they are changed computationally. This is effectively an evolutionary law of inertia. All things stay the same unless they are computationally changed. When things do change they change only as the emergent results of interactive computations among their elementary particle components.

The evolution of all the individual programs of the universe is determined by their computational interactions with their environments, which consist entirely of other programs. Thus programs that survive are selected and modified through computational interactions with the programs that constitute their environments. In the aggregate in any particular environment of programs some survive, others vanish, and in general all programs continually transform in interaction with the other programs that constitute their environments.

Thus the evolution of the observable universe consists of the continual computational interaction of all of its individual programs. This continual process continually selects the current mix of programs that constitutes the observable universe at every present moment. Thus the complete fine-tuning of the quantum vacuum manifests its hidden design through time through the interactive evolution of all its emergent processes.

The General Principle of Evolution states that all things in the observable universe are data forms in a process of continual computational interaction in which some have greater survival rates than others. Those programs or data forms with greater survival rates can be said to have greater fitness within their computational environments. In turn it's these programs that tend to increase and spread at the expense of others and to become the progenitors of subsequent programs and data forms.

Thus the mix of programs that exists in the observable universe automatically converges towards the most successful and adaptive at any given time. In this way the observable universe automatically converges towards expressing itself in the optimal possible manner through an intrinsic evolutionary process. The observable universe automatically evolves towards and converges on a data state composed of the fittest running programs.

However due to the lags in computational results spreading through the system this state continually changes over time. As computational changes spread through the system the programs that constitute the environments of other programs change and individual programs must adapt to their new environments or be superseded by fitter programs. Thus it's the finite spread of computational changes across the whole system that allows the universe to continue to evolve through time rather than immediately reaching a permanent end state.

Darwinian evolution is simply a special case of the General Principle of Evolution that applies to biological programs that reproduce their kind. The survival and procreation of individuals of all species is selected via their computational interactions with their local environments including individuals of the same and other species. Species whose individuals prosper relative to others are said to have increased fitness within their environments, and populations of these species tend to increase over time. In this way the mix of individuals of all species becomes better adapted to their common environment.

Through the individual interactive selection of programs the entire mix of all programs, both biological and inanimate, becomes better adapted to the common universal environment of programs. However this is a never-ending process due to the time lag of computational changes propagating through the entanglement network that underlies the observable universe of emergent forms and programs.

INANIMATE PROGRAMS

Since every particle component interaction of all the elementary particles that make up aggregate structures is being actively computed simultaneously at every moment, all emergent structures can be meaningfully considered as independent programs that compute their own existence and actions. The myriads of individual computational interactions of all the elementary particles within emergent forms are actually being computed with the result that aggregate structures act as if they were computing their own emergent interactions.

The non-biological programs of the universe compose a vast hierarchy of systems from the largest scale cosmological processes, down through the geological processes of individual planets, to the building, interaction and erosion of the smallest grains of minerals, and all the flows of energy and materials that connect them. These are all integral aspects of the universal program that can be identified and studied on an individual basis. They are all computational processes on the data that composes them.

These processes are described in wonderful detail by the physical sciences and new discoveries are continually adding to our knowledge of their forms, processes and evolution into the world we observe today. All that Universal Reality adds to this picture is the perspective that these are all emergent computational processes that can be viewed as individual programs. They are all emergent manifestations of vast numbers of elemental computations that have been selected over the history of the universe through their computational interactions.

The inanimate programs of the universe are characterized by being non-intentional or non-purposeful. They operate on the basis of immutable stochastic laws of nature. Those laws determine which programs tend to emerge within the total environment of programs through countless generations of elemental computations. They are the current results of vast numbers of quasi-random computational interactions over the history of the universe. They are the current manifestation of the complete fine-tuning as it plays out in the observable universe it has designed.

When it comes to specifying the particular programs that constitute the inanimate universe observers have considerable latitude. They can think in terms of the individual chemical reactions that form rocks and minerals, the deposition and erosion of geological processes, the plate tectonics that drives the building of mountains and the movement of continents, or the cosmic scale processes that create stars, planets and galaxies, or anything in between.

All these views are legitimate and all identify computational processes that can be studied as the actions and interactions of individual programs within the complex hierarchy of the whole. This is true of all the hierarchies of all the processes of the universe. All are parts of a single universal program but all can be meaningfully viewed as individual programs at all levels of the universal overlapping hierarchy of computational domains.

In general the individual programs of interest to observers tend to be identified on the basis of the hierarchies of the overlapping computational domains that naturally emerge as the aggregate results of myriads of elemental computations.

Domains are simply areas of computational density or computational similarity. They are defined by factors such as having borders of different computational type or distinct computational forms relative to their backgrounds. What we identify as twigs, leaves, branches, trees and forests are all examples of computational domains within a single hierarchy. They can all be viewed as independent programs as appropriate to specific needs of the viewing organism.

BIOLOGICAL PROGRAMS

The programs of the observable universe operate as described by the standard physical and biological sciences and evolution but are now viewed from a computational perspective. In this view all the things and processes of the observable universe are running programs that together in interaction make up the single universal program that continually recomputes the current data state of the observable universe.

All individual programs are emergent processes that are aggregate manifestations of their individual particle interactions. Inanimate programs are the direct emergent manifestations of their elemental particle interactions but biological programs have evolved to purposefully compute their interactions with the other programs that constitute their environments to some degree.

As with inanimate programs the programs that constitute biological life forms can be considered either in terms of individual organisms, social groups, or even in terms of the historical processes of families, nations or cultures. All can be viewed as the operation and evolution of individual programs within the context of the universal program.

At the emergent level living programs form interactive simulation models of themselves within their environments and compute their actions on this basis in furtherance of the instinctual imperatives transmitted in their DNA.

In order for a program to act purposefully it must be able to encode the elements of its purpose, the elements of its environment, and how to effect that purpose within its environment. This is accomplished in an organism's simulation of itself within its environment.

All individual things are the programs of themselves, and they are all subroutines of the single universal program that computes the observable universe. This includes the living programs of biological organisms including human beings such as you and I and all the individuals of other biological species.

Even though all computations are computed at the elemental level, biological organisms appear to act as independent programs that act purposefully in accordance with instinctual imperatives such as the basic imperatives of survival, reproduction, and preference for pleasure and satisfaction over pain and discomfort, and their many species-specific behaviors. These instinctual imperatives are part of the basic software of biological organisms encoded and passed through the generations encoded in their DNA.

If all the data of a human was constructed piecewise from its cells or particles and loaded with the appropriate software stored in its brain, it would begin to function as a normal human being and after a learning process one would not be able to tell the difference from a naturally born and raised person.

The fundamental nature of all programs is code. We are the complete program code of ourselves, continually running in the present moment as happening occurs. But our code is the elemental data of ourselves as it's organized within us, as every last bit of that data is being continually computed by individual applications of the elemental program on each of our elemental data processes.

Our individual programs continually interact with the other running programs that make up our environments. And the results of those computational interactions generate the facts of our lives and our effects on our world. It's all a computational process in accordance with the laws of nature and science and human behavior.

Thus we are the biological robots of ourselves. Our programs are living, intelligent to varying degrees, sentient, feeling and self-modifying and we move fairly effectively through our environments. Our programs are a wonder of design, and are even capable of reproducing their kind.

We are the fully human conscious robots of ourselves, and our true nature is that we are the running programs of ourselves rather than physical beings. As beings consisting solely of the abstract information of our data being continually computed we actually inhabit an information universe rather than the illusory physical universe of our simulation.

Our programs are enormously complex hierarchical systems that include the information processes of every elementary particle, cell, and organ of our selves in a wonderfully effective integrated whole with all sorts of internal feedback, regulatory and repair systems. Working in concert these systems are the subroutines of every part of our bodies operating in harmony as the single program of our selves.

Our program also includes the information systems encoded in our DNA which transmit the hardware and software design of our programs from our parents down through countless generations, and which control the programs of our growth, self-repair, and even the encoded deteriorations of our deaths.

And of course part of this program is our nervous system and brain with its internal self-modifying simulation model of our whole program and its modeling of and interactions with the programs of the surrounding world. All in all an incredible masterpiece of program design!

This total program is the true nature of what we really are. We are not just a consciousness carried along by a physical body, we are the entire program of our selves down to its most elemental operations, and all this is ultimately only running code and data.

Though initially counter intuitive this is really not so difficult to understand and can even be directly experienced if we put our mind to it. All we have to understand is that the *information* of all the processes of our bodies and being down to the finest level is all that is actually observable in any way whatsoever. Even our apparent physicality is ultimately entirely the *information* of that apparent physicality. So just put all this information together leaving nothing out and this is all that we actually are. It's always what we have been and it's just a matter of recognizing this verifiable fact. Our awareness of the internal processes of our selves is simply our experience of our program running even as it reads and comprehends this sentence.

So the recognition that we are our information, the information of our complete running program, in no way diminishes us as humans whatsoever. It doesn't change us in the least; we remain exactly as we always were. We think, feel, and act exactly as we did before. There is now just a deeper understanding of what we really are compatible with the deeper understanding of the entire universe of immanent information revealed by Universal Reality.

Thus all the programs of reality including those of our selves consist entirely of data or information in the same sense that the code and data of ordinary computer programs is information only. Computer programs run in the medium or substrate of microchips so they manifest as computer programs. The programs of reality run in the substrate or medium of existence, so they manifest as the real actual things of the world. The running programs of real things actually *are* those things; running programs are the true fundamental natures of all things in the universe.

The primary difference between ordinary computer programs and the programs of Universal Reality is how they run. All the information that makes up the universe exists in the present moment driven by the processor of the happening of existence that computes the entire data state of the next moment simultaneously. All the computations of the universe execute simultaneously rather than sequentially and individually as they do in even multi-processor super computers. All the information of the universe exists together in the universal substrate of existence that is also the single ubiquitous processor that computes it.

And unlike ordinary computer programs, the programs of reality don't consist of long pre-coded sequences of operations that are executed sequentially. That would inevitably lead to logical contradictions among different programs as each would have to predict the future computations of the others, and it's also impossible because as yet unexecuted code would in effect deterministically predict a non-existent future, and how could that code ever be written and for how far into the future? It's simply an untenable model.

In contrast the programs of reality consist only of their present moment data states in the act of continual recomputation, and the outputs of those recomputations are their next present moment states in the act of recomputation. Only in this way can all the computations of the universe occur together in a logically consistent manner.

Thus the universe doesn't consist of a sequence of static data states, it consists only of continual dynamic computations. The current data state of the universe or any of the programs running within it exists only as observer snapshots of the data state of that program in some simulation of reality. Observer simulations of reality tend to consist of sequences of static data states for easy comprehension, while reality itself consists of a continual flowing process of running programs.

FREE WILL

Biological programs exhibit some degree of free will because the complexities of their programing exhibit quantum randomness at the elemental level that are magnified up through the levels of decision making.

Even though the universal program is computational it's not completely deterministic because its computations incorporate a constrained degree of stochastic randomness at the quantum level. Therefore the universal program, and certainly the individual programs that run as parts of it, exhibit varying degrees of freedom from deterministic causation depending on their individual information structures. In living organisms this is the basis of free will.

Many but not all processes of the universe incorporate quantum randomness. For example the fundamental conservation laws are exact, at least to the level of granularity of elemental reality, but computations involving dimensional spacetime generally exhibit some degree of quantum randomness due to the random oscillations in the processor as it computes space and time velocities as explained in the chapter on Quantum Reality.

Since all the emergent programs of the universe incorporate quantum randomness they all exhibit varying degrees of freedom from determinism. While all the processes of nature obey the laws of nature, they do so in a non-deterministic manner so their evolutions are never predictable below certain levels of detail depending on their individual structures. We know mountains will erode over time in a generally predictable manner but it's impossible to know exactly how this process will play out down to the individual grains of stone.

The amount of freedom the programs of various systems exhibit is highly dependent on their internal data structures and whether quantum randomness tends to be concentrated or damped out at higher levels. For example human mechanisms such as digital clocks or industrial robots are designed to almost completely eliminate the expression of randomness at the level of their design operations, while others such as dice or lottery drawings are designed to maximize it. The number of quantum processes may be the same in each but whether their randomness is magnified or damped out depends on their design.

Living organisms including humans are designed to act purposefully in response to external stimuli but to exhibit a significant degree of freedom in doing so. We respond meaningfully to external stimuli but our actions are not completely determined by the external world. Fundamentally this free will is due to the presence of quantum randomness within the elemental processes of our programs. We are designed to make individual decisions based on relative evaluations and weightings of masses of information bubbling up the hierarchies of our programs from below. Thus minor changes either in the emerging information, or computational choices among information streams, can effectively magnify the degree of effective free will exhibited.

Since the actual computational interactions of complex programs like humans with the programs of their environments take place at the level of elementary particles, our emergent level processes are considerably insulated from the emergent level processes of our environments. So it's our great hierarchical complexity that allows humans to exhibit considerable free will from environmental influences. And of course greater complexity means greater internal randomness as well. This randomness is damped in some areas to maintain structure and function but enhanced in other areas to allow more effective adaptation to environmental situations.

So the actions of living programs such as humans are not entirely determined by their external computational environments. In biological organisms, this manifests as the ability to generate actions in response to, but not exactly determined by, external influences. The randomness in the elemental computations of a biological organism is the source of its free will whose effect is magnified to a considerable degree by the hierarchical structures of its decision-making processes.

Biological organisms have some additional free will in the sense that they are programs with minds designed to generate intelligent and purposeful action. Most decisions are computed with a modicum of free will at the unconscious level. Consciousness, in its quality control function, allows some small additional degree of free will to modify or override unconscious decisions.

As most humans mistakenly identify with only their conscious selves, the relatively minor conscious decision making capacity is what most people think of as 'their' free will because they consciously experience it as such. This explains why people tend to think of their free will as the freedom of their conscious self to override their *own* instinctual imperatives, but that's a very minor part of the whole story.

All true randomness is quantum randomness and occurs only at the elemental level. All the apparent randomness and freedom of classical level events and actions is a structural amplification of quantum randomness up to the classical level, or in many cases simply the non-computability of extreme complexity which is not actual randomness. The non-predictability of turbulent flows, the weather and other extremely complex phenomena are combinations of both effects.

LARGE SCALE EMERGENCE

Emergence is a multiply hierarchical phenomenon with many nested and overlapping layers of meaningful computational processes, all ultimately deriving from the design of the complete fine-tuning as it constructs emergent domains. It's not clear there is any necessary limit to this hierarchy and there are no doubt many programmatic processes at work at levels far beyond human comprehension. It's the goal of science to discover these hidden large-scale processes that describe the operation and evolution of the universe at all levels of complexity.

All individual organisms are enormously complex hierarchical programs, but the much more complex computational *interactions* of these individual programs determine all the great processes of history. Social interactions among biological programs are themselves programs as well, and are part of the running program of the evolution of the earth and all its individual life forms, natural cycles, and geological history and the enormously larger cosmological programs that have created and maintain our universe.

All these individual programs are intertwined computational processes that are part of the single running program of the observable universe Each can be teased apart into individual programs running in the context of all the others at any level. All these programs are the progressive emergent manifestations of the innumerable ongoing computations of all the particle components in existence.

The operation of the elemental program that computes these interactions creates the particle component entanglement network that forms the integrated mass-energy and spacetime structure of the observable universe. The universe is the complete integrated entanglement network in the continual process of recomputation by the elemental subroutines of the quantum vacuum.

Due to the incredible design of the complete fine-tuning that informs this universal program, the resulting entanglement network exhibits super-consistent structures at all levels of emergence. The entanglement network manifests as a vast super-consistent hierarchy of emergent programs each consisting of aggregates of lower level programs.

The entire structure is super-consistent in the sense that every individual process is self-consistent, every level is internally self-consistent, and every level is logically consistent with every other level. The entire information structure of the observable universe is a computationally super-consistent entity. The descriptive laws of nature that emerge at every level are themselves consistent with the laws of all other levels. The entire hierarchical unity of the universal program is a single multiply self-consistent computational process at every level.

At every level there are certainly as yet undiscovered laws in operation describing as yet undiscovered programs running hidden within those levels. And the continuing discovery of these hidden programs will certainly lead to a much greater understanding of how the universe and life on earth operates, and a greatly increased ability to predict and benefit from this knowledge that will hopefully guide future developments in a beneficial manner.

By adopting a systems approach to the understanding of the great processes that guide our universe, the evolution of our earth, and our individual lives, we improve our ability to accurately simulate these processes and thereby guide our progress into the future in a more effective manner. Whether this will be done only for the benefit of those who control such knowledge and technology or the broader benefit of civilization and the earth as a whole remains to be seen.

Though it seems likely that all the emergent meaningfulness of the observable universe emerges blindly from the operation of the simple computations of the quantum vacuum, the fact that such remarkably complex and meaningful programs emerge naturally from seemingly blind inanimate processes suggests there could be something more going on here. That somehow there may be some sort of feedback mechanism from the overwhelmingly meaningful higher level laws that somehow has tuned or is tuning the complete fine-tuning that produces it. What this might be is the ultimate mystery, and the relation between the complete fine-tuning and emergent meaningfulness at the universal scale will be examined in more detail at another point.

There is no apparent upper limit to emergence. The flows of individual lives from birth to death, the great flows of civilizations and history and evolution, the interactions of all the programs that constituted the biosphere (Lovelock, 1995), and no doubt even greater as yet undiscovered processes, can all be considered as interacting programs of the universe emerging at every level from the interactions of their constituent programs, and all ultimately from structured sequences of elemental operations. Systems analysis of big data will progressively reveal many of these hidden programs.

Thus considered all together over the life of the universe we observe its evolution towards locally higher and higher levels of emergence and complexity exhibiting an intrinsic though often unrecognized evolution based intelligence implicit in the complete fine-tuning of our universe. There is certainly much more to be discovered here though the limits of human intelligence may ultimately limit their understanding at least by our species.

CONVERGENT EMERGENCE

The precise specifics of the complete fine-tuning determine the general direction of the evolution of the observable universe. They deterministically produce the laws of chemistry and other elemental laws but they also stochastically determine the evolution of the large-scale structure of the universe towards ends already implicitly present in the virtual structure of the quantum vacuum as the observable universe was created.

Thus the general design of the universe and the laws that describe it were predetermined at the big bang, but the myriads of individual details of that design are the result of quantum randomness deriving largely from the creation of dimensionality by the entanglement network.

Along with the observable large-scale structure of the cosmos this evolutionary process seems to naturally produce some form or forms of intelligent life where conditions are favorable. Intelligent life forms tend to be more successful over long periods of development where conditions are favorable and will almost certainly acquire certain characteristics.

They will likely have appendages able to manipulate their environments to create supportive technologies. They will also tend to develop means to record and share collective knowledge. And they will likely be aggressively competitive and prone to ruthless violence to promote their success both individually and as a species.

To what extent this initial stage of highly intelligence technological life can be superseded by a more benign, wiser and compassionate intelligence is unclear. Ultimately that would be adaptive as it would foster the longer-term survival of civilizations but that could be at the expense of thousands of 'intelligent' civilizations that destroyed themselves and their planets. The jury is still out based on the current human evidence.

In any case it seems likely the universe inevitably tends to evolve towards becoming aware of itself through the intelligence and sensory organs of the life forms it naturally produces. One could say the evolution of the universe is an evolution from unconsciousness to self-consciousness, a process of producing intelligence, consciousness, and sentient knowledge and sense organs through which the universe becomes self aware.

This evolution has occurred in profusion on earth in the form of the vast interconnected network of the individual sensory organs and intelligences of humans and all other species that continually exchange incredible volumes of information about the world. And it's greatly extended in the collective intelligence and knowledge we humans store outside our brains in various media. And an enormous range of scientific technologies from the invention of magnifying lenses to remote sensing and supercomputer visualization has exponentially enhanced our sensory abilities.

This great explosion of sentience, consciousness and intelligence on earth effectively makes our small planet the brain of the universe, the only one of which we are currently aware. And this brain enables the universe to become consciously aware of itself in exceptional detail. Every thought we think, every feeling we feel, is the universe thinking and feeling itself through us, and this is equally true of the sensations and intelligences of all the living creatures on our planet. Every one of them is the universe experiencing itself and knowing itself through the life it has evolved and the technologies that life has produced.

Thus evolution can be seen as the process of the universe awakening to its own existence and to some extent gaining conscious control of its destiny. While previously it was the presumably blind complete fine-tuning that programmed the universal program through a long slow evolutionary process, the universe now suddenly gains the ability to begin to consciously program itself. Who knows to what extent this capability may develop? In any case we humans now act as the main repository of self-conscious intelligence of the universe as we look into the future and begin to guide it towards its ultimate destiny.

The ultimate destination of convergent emergence is of course speculative but it's very likely there is one that may already have been implicit when the universe began. The entire life of the universe may be only the process of making this original implicit virtual design actual. Whether it's an eternal dark entropy death in which nothing more ever happens, or some sort of truly cosmological intelligence that ultimately fills the entire universe with awakened consciousness is unclear. But it certainly seems possible that the entire emergent system produced by all the elemental computations in the universe could already be manifesting as some sort of cosmic mind in the act of awakening.

If the aggregate effect of all the individual firings of the neurons in the human brain emergently manifests as the intelligent consciousness of a human mind, then it could well be possible that the aggregate effect of all the elemental quantum computations in the universe manifests as some sort of cosmic mind, or at the very least something far beyond our current understanding. It certainly manifests as the observable universe, and at the topmost emergent levels there is likely much going on far beyond the comprehension of the human mind.

A bacterium might well be aware of the firing of a single proximate neuron but would have no possible comprehension of the workings of the human mind that neuron was part of. Likewise we humans are aware of many of the individual processes of our universe, but likely have no idea at all of the hidden deep scale processes they manifest at the level of universal emergence.

It is the continual computations of the quantum vacuum that give us our own individual lives. The universe is not a biological organism but it is a computational organism, and each of us is part of it in our own way according to our own forms. As explained in the chapter on Consciousness all interactions of all the forms of the universe can be considered as experience and the observable universe can be thought of as consisting entirely of xperience. To what extent the ultimate destiny of that xperience is to become conscious experience is unclear but certainly intriguing.

OBSERVERS AND EMERGENCE

Because emergent structures are not stored as separate individual data entities they are only apparent to observers able to recognize and infer them by comparing their individual components in the entanglement network and storing models of them in their simulations. Thus all emergent structures including all material structures and spacetime are apparent only to properly configured observers. These will be observers that construct internal simulations of their environments at some level. There is certainly much beyond the human ability to simulate reality waiting for higher intelligences to discover.

Thus it's observers that extract and model explicit emergent structures from the computational domains produced by individual particle component computations in aggregate. This includes all biological organisms to varying degrees since they all compute their functioning on the basis of their recognition of emergent structures in their environments and the logic of things that describes those interactions.

However emergent structures are essentially invisible to the elemental computations, which compute only the relationships among individual particles at the particle component level. Nevertheless, thanks to the complete fine-tuning these elemental computations in aggregate produce all the immensely rich emergent structures of the universe.

The only way the emergent information of the entanglement network is observable is through comparing aggregate relationships of particle sets to determine their relationships. Only biological observers can recognize them because only they have memories in which emergent aggregates can be stored and compared over time. Some level of mental model of reality is required to enable this as the observer must store information in a standard form and compare forms to discover their relationships. These emergent relationships aren't apparent at the level of the elemental computations though they become apparent at the aggregate level to properly configured observers.

Thus all levels of emergence require observers to make them explicit. In particular the individual things of the world are all observer concepts that tend to be based on the detection of computational domains in the entanglement network.

Thus to a great extent the discrimination of the universal program into individual programs is arbitrary and observer dependent. Observers tend to discriminate things and processes of particular relevance or interest and these often overlap the views of other observers, both of other species and among individuals of the same species.

One observer may think of an individual stone as a program in interaction with its environment, while another may want to consider the whole riverbed full of stones as the program of interest. All such views are valid; they are each based on meaningful *domains* within a single universal computational process. What observers identify as individual programs from the single universal program are generally based on natural computational domains defined by areas of computational density and the perceptual filters they employ to extract and identify processes of interest. These naturally lend themselves to being viewed as individual programs by observers.

THE INTELLIGENCE OF DESIGN

The total intelligence of design encoded in the universe is awesome and immense. Though certainly the product of evolutionary processes based in the design of the complete fine-tuning rather than the product of any designer god or alien programmer it's many orders of magnitude greater than any possible human intelligence.

After all it has produced and continues to produce the enormously intelligent design of every last entity in the universe as part of a single intelligently integrated computational system. By contrast we humans are incapable of designing the simplest component of even the simplest living organism from scratch, and even if we could design it we would still face the impossible task of constructing it. All that we are able to design we accomplish only by tinkering with what the universe has already designed.

Thus the total intelligence incorporated into the entire universe is far beyond the comprehension of human intelligence and no doubt will always remain so. Even though the intelligence of design of the universe is far beyond even that of any god we humans could imagine the important point is that it does raise the universe itself to god-like status. Thus if we want a god, the universe itself is the only viable candidate.

LIFE

THE PRECURSORS OF LIFE

The following several sections offer a somewhat speculative but quite reasonable scenario for the origin of life in terms of the evolution of natural physicochemical processes. The origin of life is one of the major unsolved mysteries of biology but it's quite clear it must be the natural outcome of underlying chemical processes resulting from the original complete fine-tuning of the universe (Wikipedia, Abiogenesis). As usual Universal Reality views this process from a computational systems perspective.

Due to the original fine-tuning it was most likely inevitable under favorable environmental conditions, that certain non-biological programs first gained the ability to incorporate information about their interactions with their environments into the information of themselves that already characterized them. Incorporating information about the environment in a form that enables programs to act with better than random efficiency to improve their fitness is the key determinant of biological programs. It provides them with the ability to begin to match their actions to their environments and so to act purposefully. It's especially necessary to maintain the function of the very complex data systems able to act purposefully.

From an information system's perspective life involves several essential elements: an enclosing boundary that defines an individual unit called a cell, an environmentally isolated medium protected within the cell's membrane that contains chemistry that maintains the function of the cell, and a mechanism to copy itself.

Though the specifics of the origin of life are still uncertain, it seems to have been the natural outcome of at least one particular set of conditions on the early earth. In fact it is quite possible these conditions may still exist and be producing the precursors to life even today, only to have them be consumed by currently existing life forms before they complete the process of producing new life.

Many of the precursor organic chemicals to life, such as the lipids that are the main component of cell membranes, and some of the amino acids that combine to form proteins and DNA, are known to form naturally in a wide variety of conditions and have even been detected in space. So the question becomes how do these chemicals combine to form life.

Lipid molecules are hydrophobic and like other fats naturally tend to form membrane like films on the surface of liquid water. When agitated by physical processes it would be natural for lipid films to form bubbles enclosing water-based interiors within their surface membranes. This simple physicochemical mechanism is the probable initial precursor to life as it automatically produces the initial form of the cells that are the foundation of all life.

These proto-cellular bubbles would trap drops of whatever water based chemical medium they formed within inside a protected environment where its chemistry could begin to evolve towards that of fully living cells.

All that would be initially needed would be lipid bubbles with an interior chemistry able to maintain the bubbles over significant periods of time. The necessary chemistry can likely be determined by experiment.

There is likely even a natural physicochemical mechanism that leads to the splitting of bubbles above a certain size so that each resulting bubble shares the existing internal chemistry. Thus if internal chemistry tended to increase the size of these bubbles they would automatically begin to duplicate themselves in a favorable environment.

The swelling and division of lipid bubbles occurs since lipid membranes are porous and chemicals can pass through simply by osmotic pressure. This allows the interior volume to increase depending on chemical concentration gradients across the membrane. Initially additional lipid molecules would be attracted to the membrane and its size would increase up to the point it split into two bubbles based on the surface tension. Thus there is a natural physicochemical mechanism that produces generations of lipid bubbles of approximately the same size before life even begins.

A regular process of agitation of the liquid medium would naturally promote the process of proto cell division just as the agitation of a soap solution produces hundreds of little bubbles of foam many of approximately the same size. This could easily occur in tidal pools via wave action or other processes. In fact there are many situations in which we see dense accumulations of organic foams appearing on the surface of water today.

This entire process doesn't require an energetic or information based interior medium but just the proper physicochemical conditions to form reasonably stable lipid based bubbles and allow them to split to produce additional copies. Yet it duplicates the basic mechanisms of single celled life and only requires the addition of more advanced internal chemistry to make the transition.

In proper conditions this model could provide a population of stable self-perpetuating proto-cells within which the chemistry of life has time to evolve. It certainly seems the most reasonable theory to explain the initial stages of life because it neatly solves the chicken and egg problem of cell and DNA.

Up to this point no information in the form of RNA or DNA is necessary to support this process. The traditional big sticking point in explaining the origin of life is how cells could first arise without DNA to create them, and how functional DNA could arise without precursor cells to produce it. In this new model neither of these steps is necessary.

This initial step probably occurred rather quickly so it seems likely it could be duplicated today in the laboratory in a proper water based chemical medium. Of course the process would have to take place in sterile conditions so no current microbes would be able to dine on the bubbles produced. Outside the laboratory the bubbles may still be being produced only to be eaten by already existent microbes, thus effectively covering the tracks of life's origins.

THE ORIGINS OF LIFE

To complete the transition from these self-duplicating lipid proto-cells to the earliest life all that is needed is time to allow their chemistry to evolve towards fitter more stable forms. This involves the incorporation of increasingly complex internal information mechanisms that differentiate lineages of proto-cells and the automatic selection of the fittest (those that survive best) among them.

As soon as we get more or less uniform lipid based proto cells that incorporate more or less stable internal chemical soups passed from generation to generation by simple bubble division we immediately have the possibility of distinct lineages of cells arising with variations of internal chemistry that evolution will select among.

The result of this evolutionary selection will favor lineages with better survival rates. Thus evolution automatically produces populations of fitter proto cells by selecting among naturally arising variations in internal chemistry. This process clearly tends towards the evolution of more robust cellular life. Through what is likely a very protracted step-by-step process of selection between alternatives the fine-tuning automatically tends to evolve the chemistry necessary for advanced cellular life, and the lipid proto cells provide the protected self-perpetuating environment within which this chemical evolution can occur.

The critical step is for the internal medium to incorporate or produce amino acids within its protected environment. Amino acids will tend to self-organize into many different possible stable arrangements in the supportive conditions of proto-cells. Different amino acid structures could originally be assembled from their constituents within the mediums of proto-cell lineages and be stable enough to be passed from generation to generation as their parent cells divided.

In this way distinct lineages of free-floating amino acid structures could arise prior to any functional DNA. They would not encode genetic information but their varied forms would act as enzymes to facilitate advantageous chemical reactions. The fittest would convey survival advantages on some lineages of proto-cells and be preferentially selected.

By this evolutionary process amino acid structures would begin to function as internal factories producing chemistry to increase cell fitness. This improved chemistry would convey considerable advantage over obtaining chemicals from the environment and enable cells to begin to function in a wider range of environments.

Thus we have a natural process in which the tendency for amino acids to self organize in protected environments leads to an evolutionary improvement of the lineages of proto-cells prior to the existence of any DNA or other genetic information.

Even so these pre-DNA amino acid structures incorporate specific information that identifies cell lineages and their functioning that is selected along with the lineages. So in a real sense these amino acid structures already are a form of genetic material. They encode some of the function of cell lineages though not information necessary to reproduce the whole cell. However that information is not necessary as proto cells continue to divide on their own as before in the absence of genetic information. And their amino acid structures continually tend to evolve towards more specific RNA/DNA structures.

Eventually this process of amino acid self-organization and evolutionary selection leads to early types of RNA based life evolved to survive in the environments that selected them simply because RNA molecules are able to carry greater amounts of functional information in smaller packages and convey improved evolutionary fitness because of their natural tendency to assume compacted forms.

Thus the function of the original amino acid information structures in proto cells would not be to carry the design of the cell but to enable the proto cell to function with increased fitness. This is the only way it's possible to evolve over time into DNA.

By tuning internal chemistry to external environments these original amino acid structures automatically carry information about the design of the cell and its chemistry and its interaction with its environment and in this is the origin of the genetic information that cells carry today.

As these amino acid structures stochastically assemble in different forms more and more complex variants of existing structures will appear by chance and be selected to the extent they convey survival advantages under prevailing environmental conditions.

These more complex versions naturally tend to produce stable molecules converging towards RNA/DNA forms based purely on natural chemical laws under evolutionary selection. And additional survival advantages will accrue from the increasing tuning of all aspects of the proto-cell to its environment.

Thus the origin of life is clearly a natural evolutionary process based in fairly simple natural physicochemical processes tuned by evolutionary selection. It is in the mysterious nature of the complete fine-tuning that this eventually leads to complex single celled life entirely via natural chemical processes under selection by evolution over many generations of proto cells.

In this manner complex single cellular organisms naturally evolve over time with internal DNA structures that carry the information of their functioning within the environment that selected it.

These early amino acid based information structures are the probable origin of life's key ingredient, the ability to store information about its functioning in its environment that permits improved functioning within that environment. This leads eventually to the minds of humans, which gain ever-increasing fitness in their ability to explicitly simulate their environments and their functioning within them.

NUCLEATED CELLS

The original cells with nuclei probably developed prior to the existence of DNA as some types of proto cells incorporated others within them. The standard theory is that these first eukaryotic cells were originally combinations of simpler archaea and bacteria (Wikipedia, Eukaryote). This merger would depend on the prior development of different cell lineages of appropriate types to stably isolate and maintain the inner nuclear proto cell within the larger one. Once lineages of these nucleated cells arise the chemistry of inner and outer cells can evolve somewhat independently so long as the integrity and function of the whole system is maintained.

Evidence for this model lies in the fact that DNA content is not directly involved in the original construction of its own cell since a cell must exist first for it to have DNA inside it. Existing cells simply divide via their own non-DNA mechanisms and in doing so copy their DNA content. Since DNA is in the process of being pulled apart and copied it is unlikely at that point to have any direct function in the process of cell division itself. There must be pre-existing non-DNA cell processes largely responsible for cell division. Thus cell function and division mechanisms are separate from and prior to the development of DNA in cell nuclei. However modern cell division may possibly be initiated and aided by information coded in DNA.

The original precursors to DNA must have begun as relatively simple stable sequences of amino acids. Through enzymatic actions these original amino acid sequences could have produced the original histone protein structures to which amino acid bases attach in stable ordered readable sequences to form DNA (Wikipedia, DNA). In modern cells histone structures provide the framework to which the amino acid bases of DNA attach to form chromosomes. The beauty of these structures is that they stabilize amino acid sequences in a linear easily read order that naturally curls into a tight helix able to fit within the minute volume of a nucleus.

The original function of these early amino acid sequences would not be genetic but to serve as a master template for the internal chemistry of the proto cell. Once enzymatic amino acid templates arise to control internal chemistry the fitness advantage of a single master template is clearly to avoid clashes among conflicting templates. Much more efficient and problem free to avoid potential conflicts and use a single master template to manufacture the proven chemistry necessary for proto cell function, and to pack it away in a doubly protected nucleus.

Once this basic mechanism exists more complex amino acid sequences can develop on the same histone framework and serve as templates for more complex chemistry to improve cell function.

Proto cells that have templates that produce chemistry for improved cell maintenance and even repair will have greater fitness and tend to increase their lineages. These maintenance and repair templates are the likely origin of the genetic functions of modern DNA. The same mechanisms able to repair cells are likely to be involved in creating cells as well as altering the design of existing cells and differentiating the cells of multicellular organisms.

The first proto genetic functions would likely have been to alter the functioning of cells in response to environmental changes. This is seen today in the ability of various single celled organisms to assume different forms in different environments. Examples are spore formation and reversal, enabling cells to persist through harsh conditions in suspended animation and reemerge when conditions improve (Wikipedia, Endospore). Other examples are the diverse forms assumed by some single celled organisms such as coral at successive stages in their development, and the colonial reorganizations of other single celled organisms.

All these functions require information within cells that serves as quasi-genetic templates to modify form and function *within* generations of cells that can then evolve further to control the differentiation of cells within multicellular organisms.

Thus proto cells already exist as living programs able to effectively function and divide prior to the existence of full DNA. This is increasingly done on the basis of multiple amino acid templates in their internal mediums. But eventually to avoid chemical conflicts master templates arise that are protected within nuclei in stable forms that are copied by messenger RNAs that facilitate cell chemistry.

Eventually these master templates encode information useful in maintaining and repairing the cell, and this in turn evolves into the ability to differentiate the form and function of the cell to survive through more drastic environmental changes.

So at this point proto cells already exist as robust little programs able to survive and adapt to changing environmental conditions, which enables them to proliferate and colonize many additional environments. And this all occurs as the stepwise evolution of natural physicochemical reactions according to the amazing design of the complete fine-tuning of the quantum vacuum.

MULTICELLULAR ORGANISMS

Single celled organisms are discrete little programs adapted to their environments. When cells of the same lineage come into contact they become part of each other's environment. From a computational perspective they can either compete for resources or act together to increase their mutual fitness. In conditions of plentiful resources but increased external threats there will be a tendency for cells of the same, and sometimes diverse lineages to form colonies under selection through membrane adhesion.

Cells of the same lineage will be more likely to adhere to each other in colonies due to the inherent chemical compatibility of their cell membranes. Thus colonies of cells with the same internal amino acid sequences will preferentially form, and be selected due to the compatibility of their internal chemistries as well.

In some situations colonies of such cells will convey superior fitness to the whole group and stable colonial behaviors can arise. There are many known examples of colonial single celled organisms in the seas and on land as well, and many of these have evolved to behave sometimes as independent single celled organisms, and sometimes as colonial multi-cellular organisms.

These organisms hold the key to the development of true multicellular organisms with differentiated cellular functions. This is a fairly well studied field of biology whose complexities are far beyond the scope of this book, but the essential element is the exchange of information among cells about their internal states and interactions, and systems that recognize and optimize the function of individual cells in such assemblages.

As colonies of cells form it becomes more efficient for individual cells or groups of cells to differentiate to perform more specialized chemical functions for the collective benefit of the whole so that the whole organism receives a survival advantage.

Colonies of cells in multicellular organisms naturally lead to the evolution of cell specialization for greater efficiency. It is intrinsically more energy efficient for cells to specialize so long as others support them rather than for all cells to duplicate all functions. This in turn leads to the evolution of specialized tissues and bodily structures using the existing mechanisms that enabled individual cells to alter their functioning in response to their environment. In particular internal tracts with external orifices to efficiently distribute nutrients to all cells of the organism, and remove waste products will naturally tend to evolve. Thus the digestive tracts of advanced multicellular organisms evolve by necessity as the separation of cells around internal tracts facilitates improved gradients of nutrient ingestion and waste elimination.

Likewise blood and lymph systems, oxygenating systems, hormonal systems, and nervous systems will evolve stochastically under natural selection to facilitate the efficient functioning of multicellular organisms through the selective encoding of cell functioning in more complex amino acid structures of DNA.

Multicellular organisms are all the emergent programs of the organized interactions of their individual cellular programs. They all interact by fairly simple rules on the basis of feedback from their neighbors. There has been considerable success in simulating the behavior of swarms of animals on the basis of small sets of simple interaction rules among neighbors, and this is likely also true of the genesis of the coordinated functioning of individual cells in multicellular organisms though these evolve to much greater complexity as the interaction of diverse individual types of cell. Each cell functions in the environment of the other cells of the organism and this determines its function.

In multicellular organisms the complexity of cells' amino acid sequences will tend to increase to encode the increasing complexity of differentiation of cell function within the new organism and RNA and DNA structures will appear that encode the increased information complexity.

Each cell now acts on the basis of a *subset* of the encoded DNA common to them all, which now begins to include information not just of the function of individual cells but must also encode the information of the design and operation of the entire multicellular organism and how individual cells must differentiate to form and maintain it.

The proximate environment of each cell in a multicellular organism is the surrounding cells it shares identical copies of DNA with. Thus through evolutionary adaptation the cells modify the expression of their genes to take on differentiated individual functions beneficial to the whole organism and to the extent this confers fitness on the whole organism these mechanisms tend to be selected and the DNA that encodes them perpetuated.

BRAINS & NERVOUS SYSTEMS

For a multicellular organism to function effectively it must have some means of communication among its individual cells and groups of cells. Initially this is due to chemical gradient flows as it still is in unorganized colonial animals and even in larger organisms without backbones or nervous systems such as plants.

However chemical gradients tend to be accompanied by electrical gradients so chemical gradients are generally electrochemical in nature. Thus it's reasonable to assume that organisms with developed internal channels to promote chemical signaling would have naturally evolved towards more efficient electrical signaling to communicate among their internal systems. The resulting channels further evolved into nervous systems, which still today work in terms of electrochemical signaling though at this point the chemical gradients primarily communicate the electrical signals across the synapses between nerve cells.

All higher animals today including humans still control the integration of function by both extensive chemical signaling with hormones and other chemical gradients carried by the blood and other fluid flows as well as the electrochemical signaling of their nervous systems. So these are both natural evolutionary developments of the differentiation of cells into diverse functional systems within multicellular organisms and their need to communicate to work in harmony in the whole organism.

When the density of electrical signals reaches a certain point it becomes necessary to establish control systems to properly organize and route them to prevent overload and properly route the signals. At this point nerves begin to form ganglia in which signals are not just transmitted but recursively monitored, sampled, analyzed and controlled. Some nerves begin monitoring and controlling the states of other nerve signals rather than just those of somatic structures.

It is this recursive enhancement of nervous systems that eventually leads to primitive central brain structures and eventually to the more abstract computational system of the cerebrum (Wikipedia, Brain). A brain provides an organism the means to monitor and become aware of its functioning and to exercise central control over it. And it's this computational recursion that eventually leads to self-awareness and the internal structures that enables consciousness. And this central control in turn provides a greatly enhanced evolutionary advantage.

There is an associated question of how the collective operations of seemingly blind elemental computations at the particle level can manifest as purposeful intelligent behavior at the emergent level of brains.

The fundamental reason is the exquisite design of the complete fine-tuning that automatically leads to emergent processes that are self-aware, purposeful and intelligent. The complete fine-tuning automatically tends to generate the evolution of life forms with self-aware brains and even fosters it where conditions are favorable.

We can get a glimpse of how intelligent behavior emerges by considering the collective intelligence of social insects, and swarms and schools of various species, and the behaviors of colonial single celled organisms. These phenomena are generally pretty well explained by attributing a simple set of behaviors to each individual member in its interactions with others. The aggregate interactions of individuals, acting in accordance with simple behavioral responses, tends to automatically manifest as considerably more complex and purposefully intelligent action in the aggregate.

This basic mechanism also applies to eusocial colonial organisms where useful information and necessities are exchanged for the good of the whole group by individual organisms such as ants, termites and bees (Wikipedia, Eusociality). Other social animals and especially humans with their very complex divisions of labor are another more loosely organized variation on this basic model.

Thus it seems reasonable to assume that the individual neurons of the brain each operate by a fairly simple set of interaction rules that emergently manifest as human intelligence in the aggregate structures of the brain. This is functionally equivalent to describing the programs of living beings and their simulations as consisting of specialized *associations* of the same elemental particle computations as everything else in the universe.

REPRODUCTION

The complete information of a multicellular organism must be encoded in every cell for cells to differentiate and perform individual functions within the organism. And this is also necessary in multicellular organisms that reproduce via individual germ cells, either sexually or asexually.

The simplest method of reproduction is to bud off individual cells that carry the complete genetic code of an organism. The genetic mechanisms that enable single germ cells to develop into new multicellular organisms of the same species is a complex process beyond the scope of this book but the entire growth and reproductive process must be coded in the germ cell's DNA and function through a process of cell division and differentiation according to a common overall pattern with variations corresponding to individual species.

The basic precursor to reproduction depends on incorporating the entire encoding of the design and function of all types of cells of a multicellular organism into the DNA of every cell. We have already seen that the likely precursor to genetic DNA was amino acid sequences encoding the abilities of single celled organisms to undergo phase changes in response to abrupt environmental changes, such as bacterial endospores, which necessitate the encoding of plans for the transitions to these differentiated forms even when not being expressed.

A similar mechanism must account for the origin of DNA in multicellular organisms. Cells with the ability to differentiate into different forms in an environment of multiple cells are similar from a computational perspective. So cells with the ability to differentiate into a few types of cells in simple multicellular organisms would be selected for their increased fitness just as individual cells with the ability to alter their phases would be.

Under an iterative evolutionary process of increasingly complex multicellular organisms this would be accompanied by increasingly complex amino acid sequences in cells which would gradually develop into the DNA sequences we see today.

It is worth mentioning that much of the DNA of current organisms appears to consist of fragments of other species many of which may be non-functional in the organisms that contain them. These are left over artifacts from the evolutionary histories of organisms and the ability of cells to incorporate significant sequences of DNA from foreign cells due to the natural ability of amino acids to form complex stable structures.

Thus we have a fairly clear natural evolutionary sequence through which complex multicellular organisms are produced that requires only selection among the results of fairly straightforward physicochemical processes firmly based in the amazing richness of the complete fine-tuning of the fundamental computations of the universe.

There are many complex steps in the development of eukaryotic cells and multicellular organisms that are best left to other discussions. However it seems clear that a sequence of natural inanimate physicochemical processes anchored in the complete fine-tuning is sufficient to explain them all. The process of evolutionary selection among physicochemical processes naturally leads to the broad spectrum of living organisms throughout the history of the biosphere. Though the precise life forms that have existed have not evolved deterministically the general types of organisms possible is strongly constrained by these basic laws of chemistry and physics.

FUNCTIONAL AUTONOMY

Biological organisms including humans are emergent level programs existences are computed in terms of elemental operations yet behave as purposeful, intelligent and sentient programs on their own. The design of biological programs has evolved through the selection of system designs that emerge naturally from the computational richness of the complete fine-tuning.

Just as computer programs of essentially any degree of complexity can be programmed from sequences of a small set of machine language operations so biological programs are hierarchically structured computational systems ultimately composed of aggregates of elemental particle computations. It is the entanglement structure evolved by the elemental subroutines that enables biological programs to function as seemingly independent and intentional programs.

More important than the form structure of living organisms to our discussion is their functional structure as programs. There is a general programmatic design common to all organisms that tends to maximize fitness across all environments.

The essence of this design is the ability to modify action in response to information about the environment. This enables even simple single celled organisms to respond with greater than random intelligence to their dynamic environments, and in this is the genesis of the intentional actions of all higher organisms.

What makes this possible is the incorporation of information about an organism's environment in its own information. Everything is the complete information of itself, but biological programs begin to incorporate information about their relationship with their environment into the information of themselves.

In a sense even the information of inanimate things necessarily incorporates information about their external environments through their interactions with it. After all the underlying principle of science, the Sherlock Holmes Principle, is that all things contain information about the other things they have interacted with. However biological programs begin to incorporate and systematize this information in a more explicit form, and in doing so they develop the ability to improve their fitness by better adapting their responses to their environments.

All biological programs are characterized by what can be characterized as purposeful or intentional action, action directed towards particular goals, in particular towards the goals of survival and procreation. Though it manifests as purposeful action directed towards the fulfillment of instinctual imperatives these imperatives are the result of blind evolutionary selection. This mode of intentional action has been selected simply because those programs that act according to it have improved the survival of their species and thus they are the ones that preferentially survive.

Thus purposeful action directed towards the instinctual imperatives of survival and reproduction automatically evolves through the process of evolutionary selection of those species that survive and procreate more successfully. There is a natural convergence of emergence towards this design among the range of possible programs that arise out of inanimate physicochemical interactions.

Intentional action is a direct evolutionary development of the basic physicochemical processes that furthered the survival of even the earliest proto cells. The ability to of a proto cell to modify its form in response to environmental changes can be seen as an early example of computational choice based on internal encodings of those changes. There is a natural evolutionary pathway from such basic mechanisms of choice to the complex choice mechanisms of higher organisms.

From a programmatic perspective the defining characteristic of life is the incorporation of information about the environment in an organism's own information and the ability to modify behavior based on this information. This results in the evolution of what we call the instinctual imperatives of survival and procreation through the process of selection of fitter individuals who implement them more successfully. But these instinctual imperatives are ultimately descriptions of emergent systems as much as they are actual programs that compute the actions of biological organisms.

The programs of biological organisms are code that has been programmed by evolution, and that code must have been designed by evolution to successfully function to survive and procreate or it would have vanished long ago. Thus all existent biological organisms now contain the code that enabled them to survive. A very simple and elegant and immutable systems development plan.

The incorporation of information about the environment in the information of self, which occurs to a rudimentary degree even in single celled organisms, is the evolutionary origin of the memory, learning and intelligence common to all organisms in varying degrees. The responses of all living organisms to their environments are not merely automatic in accordance with the laws of chemistry and physics like those of inanimate programs, but exhibit some degree of intelligent choice among possible alternatives.

The more possible functional responses a biological program develops, the greater its array of choices and the more freedom of action it exhibits. This is the essence of the free will that characterizes all living organisms. Free will is the ability of organisms to evaluate and select among their possible responses to environmental conditions.

Though free will characterizes living programs, it evolves naturally from basic physicochemical origins. All that is initially necessary is a rudimentary chemical response to incoming environmental information. To the extent that conveys fitness on any program it will tend to be selected for and passed on to its successors.

FUNCTIONAL DESIGN

The basic systems design of biological programs is characterized by a number of integrated systems:

1. An identifiable functionally discrete program distinct from its environment; a self within a not-self environment.
2. The ability to extract essential nutrients from the environment and excrete metabolic waste so as to maintain the program.
3. The ability to input and store information about self within environment in sensory and cognitive systems.
4. A range of basic instinctual imperatives centered on survival and reproduction.
5. The ability to formulate and select among possible actions to further instinctual imperatives on the basis of information.
6. A feedback system that returns information on the effectiveness of responses; the ability to learn.
7. Some ability for maintenance and self-repair.
8. The ability to reproduce its kind.

All these subsystems are necessary to maintain viable biological programs and propagate their species, and variations on this plan functionally define the programs of all individual life forms. A vast spectrum of biological programs corresponding to the different species of life exists, but all forms share these basic functional subsystems. From the most primitive single celled organisms to human beings, they are all variations on the same functional plan. And this fundamental design is the natural evolutionary result of the same basic physicochemical mechanisms that made the first proto cells successful. There is a natural evolutionary path through all of them and beyond.

Every one of these subsystems is an enormously complex programmatic system in itself tightly integrated into the whole. There is a huge range of species of biological programs each of which is another variation on this common underlying pattern.

Biological programs are usually studied in terms of biochemical structures and processes and in terms of individual responses to environmental stimuli. But to really understand them in depth they must be modeled as whole beings in terms of their overall programmatic structure. Simulating their entire *functional* structures as sentient intelligent biological programs will provide a much deeper unified understanding of what organisms really are and how they function as total beings. This will require a unified effort of biological and robotic artificial intelligence sciences that will also lead to useful designs for intelligent sentient automata. But as yet there has been very little work done to really understand and simulate the total systems designs of individual living species.

LEARNING AND CULTURE

One of the things that distinguish the human species of biological programs is the extent of human culture, but culture is not unique to humans but a natural evolutionary outgrowth of the general cultural systems of other species.

Basically culture is the storing of meaningful information across the members of a group so that it's potentially accessible to all and to the group as a whole, and the transmission of that information from member to member and generation to generation. It's rooted in the ability of parents to transmit information to their infants and for individuals to learn from conspecifics beginning with their parents.

Culture conveys a significant survival advantage in that it enables individuals to acquire useful information directly rather from often risky personal experience. Passing on information directly from individual to individual is much more efficient and less risky than having to learn everything through experience and likely failing critical lessons in the process.

In this sense culture is pervasive throughout at least the avian and mammalian kingdoms, and it's quite certain that individuals of other kingdoms learn from observing the behaviors of their conspecifics as well.

In most social species culture consists primarily of collective information about the environment, specifically knowledge of the seasonal availability of nutrient sources, the presence and dangers of various predators and other dangers, and appropriate behaviors related to social rank and individual interactions. All these define local social cultures that distinguish one group of a species from others in terms of how it and its members interact with their local environments.

Thus social groups as well as individual beings function as programs as they consist of the computational interactions of individual biological programs as they continually exchange information.

Nearly all multicellular species, and to some extent even single celled organisms have some ability to learn in the sense of being able to adjust their responses to their environments. This is intrinsic to the ability to exchange information with their environments in an ordered manner. This provides better than random fitness within their environments. The ability to select among alternatives involves learning and leads to the ability to weight alternative future 'what if' scenarios.

Even insects learn from experience and are able to preplan and select among multiple future options as is seen in ants and tarantula hawks dropping prey and scouting multiple routes ahead toward their burrows. And of course the ability of ants and spiders to adjust their constructions to local conditions requires considerable intelligence as well.

Birds and mammals, and to a lesser extent other species, also have brains that construct environmental simulations that are able to learn both in response to their environments but also to be taught by parents and other group members.

This learning that is passed from generation to generation through teaching and observation constitutes culture, which is pervasive among social organisms and to a lesser extent among asocial species that learn from their parents. Members of the group will learn what foods are good when and where, what predators and other dangers are about when and where, what behaviors are appropriate and other information important to functioning within their environments. This generational knowledge is an obvious source of fitness in addition to their individual instinctual imperatives and enables them to hone their intelligence and survival instincts.

Unfortunately in times of rapidly changing conditions such as those due to human interference much of this cultural legacy can be lost diminishing the survivability of affected species.

TECHNOLOGY

Based on their ability to simulate their environment many organisms have also developed the ability not just to adjust their response to their environments, but also the ability to manipulate their environments in ways that improve their fitness within them. Fitness within an environment can be improved either through better responses or by making aspects of the environment more favorable.

This technological ability is seen in countless species of living programs from hermit crabs and coral, to nest and burrow building arachnids and insects, to nest building birds and mammals and of course humans. These behaviors, at least in higher species, depend on the ability to plan functionally intelligent actions on the basis of simulations of themselves within their local realities.

Human specific technology begins in the ability to control fire and fashion tools and weapons, clothes and crude shelters. These are all natural outgrowths of the capabilities of many species to act on instinctual behaviors in an intelligent manner to improve fitness. Nest building in birds is an instinctual behavior encoded and passed in the bird's DNA, but the actual construction of a nest requires a considerable intelligence to select a proper location and building materials and interweave them in a stable manner that will withstand the elements and provide protection and warmth for their eggs and chicks.

I would suggest those biologists who imagine they measure bird intelligence on the basis of human designed tests to instead attempt to duplicate a bird nest with a single set of tweezers and see how intelligently they perform by avian standards!

COMMUNICATION AND LANGUAGE

Individuals of all species communicate via the information they continually output. Every observable action of any sort is a potential form of communication that can convey information to other individuals to the extent they can sense and interpret it. This includes chemical outputs, electrical fields, bodily postures and motions, and of course vocalizations. The very presence of an organism is a communication of information.

Thus all actions, body language, cries, scents etc. are forms of communication that convey information between and among individuals, both of the same species and other species as well. Originally these are all expressions of internal states, feelings and responses, but as biological programs improve their ability to gain information about their environments they begin to function as a language that communicates the conditions and intents of the individual organisms that generate them.

Thus there is effectively a vast language that permeates the biosphere that continually communicates information about all the biological programs that compose it. This is a universal language that carries vast amounts of information and individual organisms learn to tap into those areas relevant to their functioning and survival.

Since the cries, scents and body language of members of the same species tend to be similar members naturally become able to purposefully communicate feelings, intentions and other information by recognizing the meaning behind those signals in terms of what would cause them to produce them themselves.

Thus individuals of many species intentionally communicate with their conspecifics and often with members of other species whose signals they have learned to understand. The information that is communicated is primarily the feelings and meanings of actually present states, but to the extent individuals imagine or think about future or other non-present states that information can be communicated as well in terms of the expression of the feelings it generates.

Thus mallards may use special vocalizations and body language to communicate their intent prior to takeoff so that both male and female take off together, and bees may dance to convey information about the path and distance to remote nectar sources.

Human symbolic language certainly began as the cries and body language generated by internal feelings and meanings shared by so many other species. But gradually the natural sounds and gestures those feelings and meanings generated became able to communicate them even when they were not directly producing them. The idea of the presence of a bear could be expressed by the cry and gestures a bear's presence would evoke even when the bear was not present. In this way symbolic language began to develop where cries and gestures communicated abstract ideas independent of the presence of their original triggers.

The natural progression of this is symbolic language where individual words stand for very specific things, feelings, actions, qualities, or events, and sequences can be constructed telling complex stories about non-present things and events.

Thus modern human symbolic language, in which individual words stand for specific things, is the natural evolutionary development of all types of expressive communication from the chemical signals of colonies, to the gestures and behaviors of all species to the specific cries of birds and mammals.

ART AND WRITING

The evolution of symbolic language leads directly to written notation and art. For the first time information can be recorded in external media which could be shared and retrieved whenever necessary even when the writer or artist was not present. This represents an extremely important step in the communication of information and greatly improves its availability and accuracy. Hard copies are much less subject to change than information repeated verbally, which inevitably is embellished and altered as memory fades.

Numerous examples of art begin to appear in the Paleolithic period in many places in the world in the form of rock paintings and carved figures of animals and humans, in decorations on implements, and in the design of garments and dwellings.

Art is a natural development of the ability to abstract the representation of a thing from the thing itself. It is implicit in the storage of the information of things as perceptual representations, which occurs as animal brains evolve and begin to encode comprehensive internal simulations of their local realities.

Brains are highly tuned to recognize animals that might represent food or danger, as well as other members of their own species. This capacity for pareidolia is so important for survival that we tend to see faces even in inanimate objects. Thus it's quite likely that art originally began in the form of found objects that resembled animals or other significant aspects of the environment.

Such objects might be then modified to improve the resemblance and once it was learned this could be done it becomes clear that even more accurate representations could be made from scratch using naturally available colors of ochre and soot mixed with animal fat, or incisions in bone or carving in wood. The same mental capacity to see animals in inanimate objects would be used to produce art that resembled things other than what it actually was. And this capacity to abstract representations from the objects represented is the beginning of symbolic language and writing as well.

Writing begins as notational marks representing units of something, originally one mark corresponding to each unit. This was necessitated by the development of civilization in which quantities of various foods and other goods were stored and traded in quantities that were not easily remembered. Thus writing begins as notational numbers to make quantities easily accessible and accurate. In this way knowledge of available resources and the amounts of goods in transactions could be tallied and recorded in an agreed manner even before or after the transaction.

Such simple marks actually appear long before the beginnings of civilization on animal bones in Paleolithic Europe. These marks appear to record lunar cycles but could have also been used to record menstrual cycles to control fertility (Marshak, 1972). If this is true then it is quite possible that women invented writing.

As quantities of goods increased in the early civilizations of Mesopotamia it became more efficient for additional marks to be used to indicate sets of individual units and this led to the development of numbers other than one from which the basic rules of mathematics naturally emerged. Mathematics works consistently due to the logico-mathematical consistency of the computations that produce the universe including ultimately the invention of human mathematics.

Additional early developments in writing were representational symbols indicating the type of goods being tallied, and the names of the parties and other details of transactions. These would have all been necessary to record the necessary information of commercial transactions once societies had emerged.

Once this process begins it becomes clear that additional symbols can be used to represent and record more or less anything at all and writing is well on its way as seen in the extensive written records of the early civilizations of Mesopotamia, Egypt and China.

Eventually writing based on representational marks is replaced by alphabetic languages in which the sounds of things are written in terms of their phonemes. This vastly simplifies the communication of information and makes it accessible to a much wider audience.

Eventually the invention of the movable type printing press makes written information available to almost everyone by enabling great numbers of copies of books to be produced and distributed inexpensively. And of course the modern invention of the computer and the digital storage of data produces such an enormous explosion of mostly irrelevant information as to overload the minds of nearly everyone and swamp the information that is actually important.

All of these developments are computational processes; both in their evolution and the way they function. They are best understood as vast complex interacting programs operating in a historical context in which we individual humans are but miniscule subroutines carried along by the great flows of history.

This is all just the standard scientific story of the evolution of life and the development of human civilization reinterpreted from the perspective of the elemental computations and the entanglement network domains that produce it, all ultimately due to the nature of the complete fine-tuning of the quantum vacuum in which it's all computed into existence.

FUNCTIONAL EVOLUTION

Darwinian evolution is a special case of the more general principle of the computational evolution of information forms that applies to the survival of *species* of biological organisms rather than individual information forms. It describes progressive changes in species of biological programs, which are a specialized subset of programs that reproduce their kind.

It appears that DNA encodes a set of basic genetic building block routines upon which all multicellular life is based. This is beginning to be confirmed though the identities of the building blocks are still being discovered. For example Hox genes control the general bodily structure of many species (Wikipedia, Hox gene). These are a group of related genes that control body plans along the medial axis. They determine how many body segments there are, how many ribs and vertebrae, how many limbs, wings or antennae and so forth.

Thus it's reasonable to assume that evolution naturally results in standard sets of building block routines that computationally interact to construct the spectrum of actual types of organisms. The form and function of these basic building blocks can each be varied to some extent, and arranged and adapted to form all sorts of different types of organisms, all of them variations of sets of fundamental structural and functional patterns. And this is in fact what we observe today in the diversity of species all of which are variations on a basic plan or related sets of plans.

It is this that refutes the creationist's naïve contention that random choices could never produce the complexity of the known species. It is certainly true that completely random arrangements of *chemicals* would never produce any sort of living creature, but it's not individual chemicals that are being randomly rearranged so much as the building blocks of life that naturally emerge from the complete fine-tuning. Evolution works by selecting among shuffles of sets of proven building blocks of life that tend to produce viable species rather than random rearrangements of chemicals that wouldn't. This is something that evolutionary scientists need to understand and explain more clearly.

Thus evolutionary selection is well within the statistical limits of randomness. It's the difference between what results from the random arrangements of Lego blocks versus what results from the random rearrangement of the atoms of plastic that compose them. There is a many orders of magnitude improvement in the probability of producing a meaningful structure.

Individual living organisms are so complex it's quite clear that evolution would have been impossible based on random mutations to the billions of base pairs in the typical genome if each mutation inevitably resulted in an error. However the genetic code also incorporates numerous error correcting mechanisms that can effectively compensate for many types of simple random errors (Wikipedia, Genetic code). This tends to preserve viable life in the face of many types of mutations. And of course deadly mutations tend to die out because they fail to reproduce themselves.

Also DNA expresses sequentially as it builds organisms. DNA variants can result in viable modifications to later structures such as limbs, fins, or appearance if the underlying body and organ structures are sound. These peripheral alterations are more likely to produce viable organisms than ones that control initial basic organ structures.

Thus evolution has produced a genetic code that is quite robust and able to adapt to many types of random changes that would easily destroy the random DNA blueprint envisioned by creationists. The amazing result of the fine-tuning that determined the chemistry of our universe is that DNA encodes combinations of systems that interact synergistically to build functioning living organisms rather than random assemblages of chemicals of which none would have any chance of become a living organism.

Because DNA isn't a random blueprint but a finely tuned computational system designed to produce viable life forms under stress, random changes produce viable new life forms enormously more often than pure chance would.

Thus DNA consists of a complex set of templates designed to produce all the various aspects of functional organisms, and it's the interaction of these fundamental templates that produces all the diversity of living beings on the planet. They are all different models of the same underlying design just as all the different models of automobiles are variants of the same underlying design necessary to produce machines that move on their own power and be driven by people.

Genetic information is transmitted from generation to generation not just through the code sequences of base pairs on chromosomes but also by a complex system of epigenetic mechanisms that control the expressions of individual genes. These mechanisms allow some transmission of experience across generations (Wikipedia, Epigenetics).

The experiences of organisms, especially those with strong impacts on the organism, affect the ongoing expression of genes that control how it responds and functions. Genes are not just used to develop organisms but to run them as well. So even in adulthood these functional genes are constantly operating to control various aspects of bodily function and these genes are responsive to environmental conditions. Thus during adulthood each organism will acquire chemical overlays, for example methylations, that modulate the expression of its various operational genes.

These chemical overlays affect an organism's germ cells as well and in this way some aspects of experience are passed from generation to generation. Epigenetics is a new field of study and there is much to be discovered but it's already clear that many effects of experience are in fact passed from generation to generation.

Thus evolution has come up with a mechanism that enables adaptation much more quickly than previously thought. Epigenetics allows organisms to significantly adapt at least their behaviors to environmental changes in the span of a single generation by passing on changes to the functional expression of operational genes.

Though evolutionary selection based on fitness is widely recognized as the driver of the development of species, on average chance plays a much greater role than adaptation in determining which organisms survive. It all depends of the particular dynamics of each individual situation. Survival and reproductive success both depend on variations in individual fitness but much more often in the simple luck of the draw.

For example which individual krill in a whale's mouthful of tens of thousands of krill can swim a little faster has vanishingly little difference as to which individual krill survive and reproduce. Nevertheless on average better-adapted individuals gradually tend to increase at the expense of those less well adapted to specific environmental challenges. Luck affects all more or less equally so its effect is to damp and significantly slow the evolutionary effects of fitness. Thus fitness, which generally conveys just a small survival edge in the aggregate, may slightly win out in the end but this is far from certain. Nevertheless this slight aggregate effect is sill sufficient to guide the overall course of evolution over time.

Biological evolution may be a blind process but it inevitably converges on what works, and that is the basic design that underlies all life on earth. It then tries out innumerable variants of that design to ensure the fitness of at least some, and through selection allows the fittest to rise to prominence. It's a ruthlessly efficient and awesomely beautiful process that traces inexorably back to the precise complete fine-tuning of the universe as it began.

THE BIOSPHERE

The biosphere is the total interactive system of all biological programs living across, within, and above the surface of the earth. It's a finite system limited by the size of the Earth and its inorganic support systems. The total volume and supporting resources are limited which limits the total living mass of the biosphere.

The main inorganic systems that support the biosphere are the solar energy system, the atmospheric gas circulation system, the water circulation system, the seasonal and climatic systems, the geological systems, and the inorganic nutrient systems.

The biosphere is a dynamic system in which individual organisms continually cycle nutrients to maintain their existence. Thus the maintenance of the biosphere requires the continual transfer of nutrients from one life form to another. Nutrients continually cycle through living organisms and between living organisms and inorganic systems. The entire biosphere acts as a single program in interaction with the inorganic systems that support it.

For individual organisms to live they must continually consume nutrients, which inevitably results in the deaths of other organisms in which those nutrients are stored.

Thus death is necessary to support life because in general only death provides the nutrients necessary to sustain life. This includes the predation of animals by other animals. Therefore killing and death are essential to the maintenance of the biosphere. And because predation involves inflicting pain and suffering these too are necessary to maintain the health of the biosphere.

As a result the biosphere continually evolves through time as individual organisms die, are consumed, and are replaced by new organisms. The mix of both species and individuals is continually changing. The evolution of this mix is determined by the interactions of all individual organisms with their environments, which consist of the mix of physical systems and other organisms with which they come into contact.

The evolution of the biosphere is the aggregate result of a combination of new variant organisms being born and the often-changing environmental conditions to which they must adapt. Environmental conditions include both the mix of other proximate organisms and local changes in the inorganic systems upon which individual lives depend.

Species variants better adapted to current environmental conditions are more likely to survive and reproduce and will tend to increase their populations at the expense of less well adapted variants. But in general this widely touted Darwinian effect is minimal and only becomes significant over very large numbers of individuals and significant periods of time.

THE FUNCTION OF DEATH

Life is dependent on sufficient resources to support it. Living organisms are designed to produce more than enough offspring to replace themselves. One can argue that the main purpose of babies is to provide an easily predated source of nutrients for other life, that producing babies is a very efficient means of redistributing nutrients through the biosphere that improves its overall health and diversity. This is certainly the function that most babies serve by far in nature.

The continuous production of new variant organisms is necessary for species to have the capacity to adapt to environmental changes as they occur. Because species that don't change much are less likely to survive environmental changes on average, the self-programmed death of individuals is generally adaptive and selected for. Only death allows space for a mix of new variants some members of which are more likely to be better adapted to environmental changes.

Novel individual variations include not only hardwired changes to DNA, but also epigenetic modifications that enable the quick transmission of some effects of experience in the form of selective expression of genes over at least several generations.

Adaptive variation also includes cultural variation in the programming of individual learning much of which occurs during youth and is less easily revised in adulthood. Thus in general new younger individuals will have a superior capacity to learn to adapt to changing environments than their parents even though elders carry greater wisdom about unchanging environments on average.

The result is that the programs of almost all species have selectively evolved internal self-destruct mechanisms. The Hayflick limit due to telomere shortening in cell division effective limits the number of viable divisions of human cells resulting in a maximum possible lifespan of approximately 120 years (Wikipedia, Telomere). All other species, with a few possible exceptions, have similar self-destruct mechanisms that ensure the deaths of individuals of that species prior to a certain limiting age. However these limits are fairly arbitrary by species and thus likely subject to extension through genetic engineering.

Though the ultimate deterioration and death of all biological organisms is perhaps inevitable the widely variable life spans of different organisms all built on the same basic genetic model demonstrates that the life spans of species are part of their programs and encoded in their DNA. Thus most species have genetic self-destruct mechanisms and/or lack whatever self-repair mechanisms support the longer life spans of tortoises, bristle-cone pines, and other exceptionally long-lived species. There are even a few species that appear not to age and are effectively immortal (Wikipedia, *Hydra* (genus)).

Thus in general the death of individuals of species is adaptive and selected for because it facilitates their replacement by new variants that on average are more likely to adapt to changing environmental conditions. This large-scale programmatic mechanism maintains the freshness of the biosphere and supports its overall health. It enables the biosphere to self-adapt to its own evolution.

HUMAN SELF DESTRUCTION

The survival and reproductive success of individuals in most environments is highly dependent on their ability to compete for limited resources with members of their own species and of other species. Individuals, groups and species more effective in competing for resources tend to increase their populations at the expense of their competitors. Thus the species and individuals that exist tend to be more competitively successful than those they have replaced.

Competitive success depends on a number of factors including luck, functional intelligence, technological advantages, group cooperation, bodily strength, reproductive success, and general adaptation to the environment. But it's the propensity and ability to aggressively compete for resources that is the key driver of evolutionary success that is supported by these other factors.

The more aggressive and successful in competing for resources the more successful a species and its individuals tend to be. Thus more aggressively competitive species tend to be preferentially selected and expand their populations at the expense of others. The individuals and species that survive and prosper are likely to be the most aggressively competitive.

Humans, especially human males, are naturally aggressively competitive over resources. This core aspect of human nature, enhanced by functional intelligence and technology, is the primary reason the human species has come to dominate the biosphere. And this is also the reason that the most competitive and aggressive individuals tend to dominate and rule human groups.

For this reason human competitive aggression has evolved to become an integral part of human nature and a main driver of the instinctual imperatives of human programs. This aggressive ruthlessness, enhanced by intelligence and technology, has enabled humans to rule the biosphere and dominate other species. And in doing so it has become a strongly innate aspect of human nature, especially among human rulers and the military and economic ruling classes. It's often disguised and not always in evidence but it's always available when needed.

However this same innate aspect of human nature responsible for our success as a species becomes increasingly dysfunctional and maladaptive in our current world of declining resources. Human population has exploded across the biosphere at the cost of using up limited resources in a non-sustainable competition for ever more wealth and power driven by human instinctual imperatives.

The rise of humans has produced a significant reduction in the total biomass of the planet primarily due to the clearing of much of the arable land of the planet for agriculture. This is the clear measure of a declining biosphere. The human population explosion has also produced a significant decline in species diversity that weakens the biosphere by making it more susceptible to environmental shocks.

Because they have been strongly selected the aggressive competitive aspects of human nature responsible for our past successes are now innate and nearly impossible to change. Thus the human response to declining resources in the face of exploding population is almost inevitably increased aggressive competition over the limited resources that remain.

This dynamic will almost certainly lead to large population die offs with the more aggressive surviving at the expense of the less aggressive on average. In fact the history of the world with its interminable successions of wars, slaughters and starvations, can already be seen in these terms, and will almost certainly be repeated in more and more severe forms into the future.

Given the strong instinctual imperatives of humans and especially human leaders towards the aggression that gained them their positions, it's almost certain that more and more irrational wars will continue to devastate the planet and human civilization. One only needs to project the rhetoric of political speech to predict the likely results.

Rather than using cooperation and intelligence to achieve an optimal sustainable balance for humans as an integral part of the biosphere, the strongly aggressive aspect of human nature will most likely continue to strive for short term personal and group gain at the expense of other individuals and groups, and at the expense of the natural systems that sustain us and our biosphere.

These shortsighted personal imperatives of human nature can very likely lead to a collapse not only of the human species, but also of much of the entire biosphere whose dwindling resources humans will have to compete for even more intensely to survive.

Thus there is a high probability that our planet's ecosystem will collapse and human civilization with it. The same now hard-wired aggressive aspects of human nature that enabled our success as a species are now very likely to destroy us, and many of our planet's natural systems as well. Unfortunately this seems nearly inevitable given the innate aggressive instinctual imperative of human nature.

Thus it seems quite likely that any life form aggressive and intelligent enough to gain control over its planet as we humans have done may inevitably destroy itself along with its planet's biosphere.

It has only taken humans a couple thousand years to gain effective control over the earth, the mere blink of an eye in the life of a planet. And technology and the power it conveys tends to expand exponentially once it gains momentum while the inbred aggressive competitive nature of a planet's dominant species remains largely unchanged.

Thus it's quite possible that any technologically advanced civilization will inevitably quickly destroy itself and much of the natural systems of the planet that fostered it. Evidence for this may well lie in the current complete lack of any intelligent signals from alien civilizations. It seems highly likely that many much older planets exist that could support advanced life, and certainly once it arose it would quickly become able to signal its existence. So one would expect there should be a significant number of advanced civilizations we would be receiving signals from if they had not already destroyed themselves. The lack of such signals in the face of the likelihood they should be found is called the Fermi paradox (Wikipedia, Fermi paradox).

Also the fact that it's highly probable that any species that gained control over its planet would most likely be a ruthlessly aggressive predatory species should be taken as a serious warning that it would very likely continue its aggression on earth if it had the technology to reach us. Thus it's completely reckless to broadcast our existence to the stars as we can expect that any alien civilizations that exist would most likely want either to enslave or destroy us.

As an intelligent species becomes more and more successful it may be able to use technology to temporarily increase the production of necessary resources for some period of time. The agricultural, industrial, and information revolutions are good examples. But in the long run the environment of the planet always limits resources so such measures can only be temporary and growth is inevitably limited.

Thus it's absolutely necessary for humans as a species to use their intelligence and compassion to override their instinctual aggressive competiveness and reach a long-term sustainable balance with the biosphere.

PROBABLE FUTURES

Current human overpopulation is likely far greater than the biosphere can healthily sustain for long into the future. Thus if human overpopulation is not reduced *humanely* it will inevitably be reduced *inhumanely* on a global scale.

Given the aggressive imperative of human nature, exploding human populations, and dwindling natural resources, there appear to be two possible likely futures, both of which seem bleak.

The first possibility is a chaotic collapse of civilizations caused by the simultaneous collapse of multiple resources and the resulting apocalyptic global wars. This results in mass casualties on a global scale and devolves into a long-term post apocalyptic feudal Dark Age composed of ruthlessly oppressive local fiefdoms ruled by brutal warlords who rule by decree over their subjects and engage in frequent bloody battles with surrounding fiefdoms over resources and survival.

The second possibility is the linear development of current trends towards a new world government ruled by oligarchs supported by national security forces so powerful and technologically advanced as to make effective dissent impossible. This might be better overall for the planet but only at the expense of a vast worker servant subclass most of which will be progressively replaced by intelligent machines and become unnecessary and disposable. The result will be a permanent division of society into a small super rich elite that rules over a greatly reduced worker class living in poverty and virtual slavery.

One might hope that this super elite ruling class would have the collective wisdom and cooperation to live in sustainable harmony with the biosphere and be able to maintain itself long term as the future of human kind. But since its members will likely be the most aggressively ruthless people of all this seems unlikely. It is quite probable that this system would eventually fracture due to internal strife.

Thus the long-term future of civilization seems quite questionable, and perhaps some unstable mix of these two possibilities is the best we can expect. The likely result is a greatly reduced human population, and some version of an impoverished civilization typified by even greater inequalities of wealth and privilege, and likely the near complete death of nature, as we know it today. In both these scenarios selection of the remaining human population will continue to be primarily on the basis of aggression, wealth and power rather than intelligence, compassion and wisdom.

AN OPTIMAL FUTURE

The very unlikely but optimal alternative is a global meritocracy able to achieve and maintain an optimal sustainable balance of human civilization as part of the biosphere. To overcome the aggressively competitive aspects of human nature it must be administered by an apolitical civil service meritocracy chosen by academic excellence in solving real world problems for the common good. It's leaders and administrators would be selected from the best problem solvers as they rose up through a free universal educational system open to all so that the best qualified among the entire populace would be selected.

Current political systems of government inevitably produce leaders on the basis of competitive success rather than ability to solve real world problems for the common good in sustainable balance with planetary resources. They are basically choices among what is best for the particular factions that support them rather than for society as a whole. And they tend to be chosen on the basis of short-term popular appeal rather than the ability to provide optimal sustainable solutions to real world problems.

Thus the optimal government is a global civil service dedicated to providing essential services to all peoples in sustainable balance with the planet. Only a robustly self perpetuating system of this type can eventually preserve civilization in balance with the environment far into the future and bring peace and justice to everyone on the planet.

This purely administrative government would provide the essential services necessary to maximize the wellbeing of planetary society cost free to everyone as their natural right. This would include free education in useful academic disciplines up to the limits of everyone's abilities, free health care as needed with incentives to live healthy to minimize its need, free care of the sick, needy, aged and others unable to support or care for themselves, free minimal housing, free disaster relief, and free protective and justice services including equal access to legal counsel irrespective of financial resources.

In addition the government would provide and administer a unified global communication system for the storage and retrieval of all information and an electronic banking and funds transfer system for everyone on an equal basis. There would be an artificial intelligence based system designed to respond with the single best answer to any query on any subject rather than the millions of mostly irrelevant hits to Internet queries today.

Every person would be issued a single DNA based identity that would be used to secure all transactions and communications and determine location in the event of accidents. When the government and its laws are just and equitable a single biometric identity card is highly desirable and serves to protect rather than oppress.

Funding this government would be achieved through a very simple, fair and efficient system of taxation on all electronic transactions. All electronic transfers of money or funds of any type would be subject to an immediate automatic deduction of a miniscule percentage to the public treasury. Since by far the largest monetary transfers today are exchanges of various market instruments among corporations and the super rich only an extremely small percentage would need to be taken as taxes to fully finance the government. A tax rate of only a minute fraction of a percent on all monetary transactions would be sufficient to fully fund all government activities.

And the elimination of the enormous waste, corruption, and unnecessary special interest expenditures of today's government would reduce the necessary tax rate even further. This includes the elimination of military expenditures in a peaceful world.

This transaction tax would be the only source of government funding necessary. It would be automatic, immediate, equitable, and incredibly efficient. And the tax rate could be instantly tweaked as necessary to offset government expenditures as needed. No IRS and no armies of tax lawyers would be necessary and both the rich and the poor would automatically pay their fair share at the same rate.

One of the most remarkable additions to the programs of emergence produced by humans is the development of money. By establishing a networked information system that encodes value and effects flows of value this computational system is able to mobilize and direct human energy and goods in the reverse direction around the nodes of a single integrated economic network. The flow of money in one direction directs the flow of goods and services in the opposite direction.

This system enables the efficient operation of economies at all scales. An efficient and equitable monetary system is critical to the operation of the optimal government and must be under the direct control of the government. Bank issued credit cards in particular are a usurpation of this critical government function. And the operations of the Federal Reserve Bank are also designed to increase the wealth of the major banks that it represents at the expense of ordinary people. The network of electronic payments is of such central importance only an optimal government can be trusted with its administration especially since in the transaction tax system it's the immediate source of all government revenues.

The goal would be to make the entire governmental system as simple, transparent, efficient and equitable as possible. The enormous complexity of today's laws at every level is a huge unnecessary burden on the efficiency, energy, and resources of society. The primary function of this unnecessary complexity is to facilitate hidden loopholes that can be exploited by those wealthy enough to have funded its creation.

There should clearly be a single equitable and just legal system that applies to everyone on the planet. This legal framework must allow maximum individual freedom for everyone insofar as their actions do not cause unnecessary harm to other persons, species, or the environment. Judges, juries, prosecutors and attorneys, all with their own personal agendas, would best be replaced by an equitable artificial intelligence judicial system that would decide all legal cases compassionately on the basis of all evidence without exception, and the first priority would be reparations to the victim as opposed to penalties or fines paid to the state. Convictions should not be a source of government revenue.

Sentences to those found guilty would first require compensation to their victims, and second to conditions that would prevent recurrence of similar crimes. Rather than imprisonment, most sentences would be to public service and reeducation towards successful life paths away from criminality. This requires removal from peer groups of other criminals, which is the current norm in imprisonment, to peer groups of successful well-adjusted persons.

In a just, equitable and compassionate society crime would be greatly diminished since a large percentage of crime is motivated by desperation, oppression and want. All victimless crimes would no longer be criminal offenses. That combined with sentencing to avoid imprisonment whenever possible would greatly reduce the vast human effort, expenses and infrastructure currently devoted to criminal justice.

Even more egregious are the vast expenses, resources and manpower currently devoted to military operations as nations strive to gain and exercise competitive advantages against other nations through violence and intimidation. A single global government would have no need of a military to defend against other nations, and today's enormous military resources could be diverted from destructive to constructive purposes. The elimination of war is one of the greatest advantages of a single global government.

The structure and decision making of this government must be systems based with a rigorously self-correcting design to identify corruption and remedy it before it has any chance to take hold. Given human nature with its innate desire for power and wealth at the expense of others and the environment this is an absolute necessity.

Once all the interacting systems of human society and the planet are effectively modeled on super computers like the weather has been, government gains the ability to recognize, predict and remedy problems even before they occur. It can simulate the effects of decision-making options and tweak its policies towards the optimal good of the whole planet and all the humans and other species that inhabit it.

Highly effective rational and compassionate global education is an essential component of establishing and maintaining this government. Children are learning programs that develop under the influence of their parents, schools, peers, culture and experiences, and they largely become what they are programmed to be. Thus the global education of children in the goals and ethics of this benevolent government to the benefit of society and the planet as a whole is an absolute necessity.

Rather than the usual identification of personal benefit, identity and success with particular ethnic, religious, social, racial or national groups all children must be educated to compassionate, intelligent and rational standards by a universal free public education system that prepares them to be productive and caring members of society in accordance with their abilities and interests.

Due to human evolutionary history people nearly always strongly identify with particular groups in opposition to other groups, but for humanity to live in peace and in balance with the environment it's necessary that humans must all begin to achieve identity as human beings and earthlings in commonality with all the other peoples and species of the planet. Only taking their identities as integral parts of the single system of planetary life will allow people to work together for the common good of humanity in sustainable balance with the biosphere.

The overall planning and operation of this government would be based on transparent and rigorously tested models of all its integrated systems and subsystems. Just as climate and weather models now quite accurately predict future weather patterns, similar models of social and economic systems offer the best approach to government decision-making. With accurate models of the economy, information flows, planetary environmental systems, human demographics, and all the other interacting systems of global concern it becomes possible to simulate the effects of government decision-making in a truly effective manner and optimize governmental policies and their implementation.

Currently governmental and legislative decisions tend to made on the basis of what may seem to be good immediate solutions to social problems at least in the eyes of the factions they represent, but such apparent short term feel good fixes often lead to dysfunctional unintended consequences if their longer term effects are not carefully thought through. And perhaps more seriously much legislation today is purposely designed to benefit powerful special interests at the expense of the common good.

In an optimal government a legislative branch would be unnecessary. Decisions would be made by the administrators for the maximum good of the whole system on the basis of tried and true simulation models. All such operational decisions including the allocation of funds would be completely transparent and subject as necessary to a quorum review process weighted by excellence and experience in the subject under consideration, and by the weighted convergence of simulation models. All proposed decisions would be run through the simulations to evaluate their effects and benefits to society not just short term but over reasonable medium and long-term time frames as well, and final policy decisions would be made on that basis and carefully monitored and tweaked as necessary.

And even more important is the effective simulation not just of individual policy decisions in isolation, but also of the whole range of possible independent decisions considered in concert to predict their overall interactive effects through time and thereby converge on the set of decisions that would lead to the optimal functioning of society and the optimal sustainable health of the planet.

Thus it's fairly clear what an optimum system of government would require, how it would operate and what its goals would be. The problem of course is how to transition to such a global meritocracy in the face of the aggressive competitive and avaricious nature of the current ruling classes of the sovereign nations. However if the problem is analyzed from a systems perspective there does seem to be a natural path that leads us in the right direction.

The key to making this transition is to use the power of human nature itself to incentivize it. Rather than challenging human nature head on, stepwise intermediate changes and policies could be implemented that would convey immediate advantages to current power holders to garner their support, but which would set society firmly on the right path. This is likely the only possible successful path.

There are several approaches to this. First the free provision of essential services funded by a transaction tax system is obviously to the immediate advantage of almost everyone, so the majority should support it once it's explained clearly so that its benefits become apparent. It immediately lowers taxes, reduces injustice and oppression, and improves society by directing enormous unnecessary expenditures and resources into constructive uses. So this first step is primarily a matter of effective education. This is possible even though it would face barrages of misleading propaganda by the establishment. This initial step gains the backing of the large majority of the less powerful.

A second step is to incentivize those who currently hold power and wealth at the expense of the general populace to adopt key aspects of the new system. This can be done with a well-designed systems approach that uses human nature to solve the very problems it has created.

Again this could be done is by providing immediate benefits to wealthy individuals offset by the introduction of taxes on corporate financial transactions. If all monetary transactions were electronically taxed as they were executed, the resulting equal tax rate could be immediately lowered to a fraction of a percentage. This would provide immediate large personal profits to wealthy individuals as well as to everyone else as big banks and corporations picked up the difference. This would incentivize everyone to adopt the new system since personal advantage will always tend to trump advantages to the banks and corporations people are associated with. This uses human nature itself to help transition the dysfunctional system it has built towards a new system designed to maximize the common good of all.

It would also significantly reduce the huge volume of speculation in the markets that is a primary cause of economic bubbles and their resulting recessions. And of course an essential key is to break the mechanism by which special interests effectively purchase legislation favorable to themselves and their causes through political 'donations'.

There are an number of other mechanisms that can be used to smoothly effect the transition to the optimal government that are best considered elsewhere but they are all potentially achievable through a stepwise systems approach incentivizing each step by clearly offering an immediate advantage to those who have the power to oppose it. And in the end when an efficient and effective global system of compassionate, just, and equitable administrative government is achieved that conveys increased benefits to all the system will tend to become stable and self-perpetuating.

What must be converged upon is an optimal system in which it's clear to all that any major systemic changes would negatively affect the optimal good sufficiently to prevent such changes. Only such a well-balanced system transparent to all in its design and operation will automatically stabilize in an optimal homeostasis. Only such a system has a chance of maintaining human society in a sustainable balance with the planet long term as is absolutely necessary to preserve both humanity and the planetary systems we depend upon.

SUCCESSOR SPECIES

In the long run it may be that only an artificially produced successor species will have the necessary collective intelligence and lack of personal competiveness to live sustainably on the earth. In fact this may be the inevitable next evolutionary step to intelligent biological life on any planet if that intelligent life doesn't first destroy itself. A eusocial sentient robotic species all members of which are networked as expressions of a single sharable compassionate super intelligence used to collectively make choices optimal for the planet and the common good is certainly technologically achievable in the not too distant future if the human programmers that first create it do not screw it up which is more probable than not.

The success of such a species would of course depend on its programming and how much freedom it was given to create its own agendas. It would clearly need enough global data and wisdom to act on the new instinctual imperatives of preservation of the planet in optimal long-term sustainable health. It would have to be created so that all its individuals were identically motivated strongly networked members of a single civilization so that natural selection would not begin to act to select the more aggressive and selfish among them. Again this is the subject of another book but does shed light on the nature of emergence and how it will very likely tend to evolve towards ends implicit in the original complete fine-tuning of the universe.

In the grand scheme of things the function of the evolution of biological intelligence may only be to achieve the ability to program a much better adapted electronic successor species. For millions of years the universe programmed its biological programs through the very slow process of evolutionary selection. But once it has created a species that is able to program the next generation of living programs that initiates a revolutionary paradigm shift and an exponential explosion in the evolution of intelligence and changes forever the history of the universe in the blink of an eye. The future is unclear but it will certainly be enormously revolutionary and interesting.

THE SIMULATION

WE LIVE IN A SIMULATION

We all live our whole lives entirely in a simulation rather than the real actual world. But this simulation is not produced by an alien programmer but by our own brains. Every one of us and every individual of every species lives entirely in a world of its brain's own making, a private world largely unknowable to all others. And every one of these private simulations of reality is at least somewhat different and all are much different than the actual reality they all share.

Of course we all actually live in the same shared reality but every one of us experiences reality differently in our simulation. And every one of us believes our their own internal simulation is the true picture of their common actual reality even though none of them actually are.

There must be an actual real external world for us to be able to exist and function. If everything existed only within our own mind as Bishop Berkeley suggested all sorts of fatal contradictions would arise (Wikipedia, Solipsism). There must be considerable *logical* correspondence between our internal simulation of reality and actual external reality since we do manage to function and survive quite effectively within external reality on the basis of our simulation of it. But beyond this logical correspondence it's easy to show there is hardly any similarity at all.

Simulations are dynamic cognitive models of reality that exist in the brains of living organisms and are projected back out onto the real actual world to make sense of it and facilitate an organism's functioning within it.

Thus the world we see around us and think we are living within is not at all like the real world we are really living in and it never has been. We live entirely within our own private simulation of reality thinking it's the true nature of that reality when nothing could be further from the truth.

The world we *seem* to see and experience our existence within exists entirely within the neural circuits of our brains, and all it shares with actual reality is some basic correspondence of logical structure. It's only this logical correspondence that enables us to function reliably within external reality. All the rest of our simulation, in particular its appearance, is a very convincing illusion, an illusion that completely misrepresents the true immensely complex information nature of reality.

Both biological and cognitive science confirm that our mind's internal representation of reality consists entirely of information being computed in the neural circuits of our brains, and it's these computations that produce the semblance of the physical reality we believe we inhabit. Our simulation clearly consists of information designed to convince us we should interpret it as a physical world in a dimensional spacetime populated by individual objects undergoing events even though it isn't.

Thus the very convincing everyday world we live in is actually a world that consists only of information in our brains no matter how physical it seems. This includes our objective concept of ourselves, which is an integral part of our brain's simulation of reality.

However our only possible experience of actual reality is through our simulation. Thus the only possible way to approach the true nature of reality is to examine the illusions of our simulation and how they misrepresent actual reality. Ultimately the information content of our own brain is all we have available to us.

Thus the best method to discover the true nature of reality is to examine what our simulation adds to it and carefully identify and remove each of those layers of illusion one by one until we finally discover the true nature of reality laid bare before us. We must identify and remove the veils of illusion in the simulation one by one until we finally discover the ultimate truth they conceal. What remains after everything mind adds to our simulation of reality is removed can only be the true nature of reality itself.

There are a number of ways in which our simulation of reality is clearly illusory. They are covered in somewhat greater detail in my previous book (Owen, 2013) and papers (Owen, 2009), and relevant research on the structure and foibles of perception and cognition fills many journals and textbooks, and is even the subject of popular science shows. However few if any of the authors make the leap of understanding to consider the obvious implications for the nature of reality itself. In this chapter we take that leap. We explore the basic types of illusions in our simulation and discover that all that remains when they are removed is the previously identified logical information structure of reality itself.

EVOLUTIONARY ORIGIN

Originally at the level of inanimate processes all xperience is actual but unconscious. Only gradually have living organisms evolved sensory, perceptual, and cognitive systems able to begin to represent their local realities. Thus it's no surprise that our human simulations of reality are still imperfect and unable to represent the world as it actually is. Nor would that even be desirable because it's much more adaptive that our simulation represents the world in the most useful and easy to compute manner possible rather than the most comprehensively accurate manner imaginable.

The notion that we simply open our eyes and the see the world as it actually is should seem incredibly naïve to anyone who understands the function and operation of our perceptual and cognitive systems yet even specialists in cognitive science rarely recognize its quite obvious and profound implications for the nature of reality as they explore the individual aspects of the human mind in isolation.

Our mind's simulation of reality is an evolutionary adaptation that makes it easier to compute successful functionality in the external world and thus increase our odds of survival. By greatly simplifying the logical structure of reality to its relevant essentials and dressing them up with appearances and valuations based on our interactions with them our simulation presents us with a world consisting of individual things and events that is much easier and more meaningful for our minds to compute than the enormous seething mass of raw data it actually is.

All biological organisms have their own variant simulations that have evolved to help them better adapt to their own particular environments and lifestyles. Thus every organism experiences its existence in a world whose appearance is of its own mind's making, and all these simulations differ in significant ways between species and even considerably among members of the same species. Every organism lives in a reality of its own making including us humans and it's a little appreciated fact how different these various simulated realities can be even as they all must share some of the logical mapping of actual external reality sufficient for organisms to function within it.

All these simulations are enormously complex computational

programs that continually compute their particular internal representations of reality, and their organism's functioning within them, and they are continually updated via the great variety of sensory inputs among species.

The fundamental programmatic and information structures these simulations are based on are encoded in an organism's DNA and passed from generation to generation. There is no other possible source for the basic structural components of an organism's operational software other than DNA, a fact that seems to have been totally overlooked by modern biology.

This DNA encoded information includes the basic instinctual imperatives such as survival, reproduction, pain avoidance, and pleasure seeking, that are common to all beings, as well as instinctual routines of more limited scope such as suckling in mammals, neonatal standing and walking reflexes in herbivores, and the pursuit and avoidance instincts of predators and prey.

DNA transmitted software also includes sophisticated learning routines and a dynamically updatable simulation model of self within environment, the ability to identify and project current trends to imagine future options, the ability to weight and valuate options in terms of instinctual imperatives and the ability to make intelligent decisions among imagined options, and the operational routines that translate decisions into effective bodily actions and evaluate them via feedback circuits. Together these routines constitute the *mental software of an organism* that enables species to function purposefully in their environments.

These are the running simulation programs of organisms, and they come in an enormous variety of species-specific variants. However they all operate on the basis of the common logic of things, which is sufficiently consistent with the underlying logic of the laws of nature to enable organisms to function effectively within their environments.

This mental software is encoded and transmitted in DNA along with the basic software that controls the development, functioning and maintenance of bodily growth, maintenance and repair. These programs themselves are not transmitted in their full forms as at least some undergo considerable development in infancy so that the programs passed are actually programs that develop these programs rather than the fully formed programs themselves.

The structural details and operation of the simulation program is a complex subject best explored by computer simulation models of the functionality of various organisms, a subject still in its infancy. Much of what has been learned about how living organisms operate at the functional level has been learned from practical experience in designing AI and robotic systems. We will home in on the issues most relevant to the nature of reality in this chapter. And of course how organismic programs are actually implemented in biological structures is another vast and complex subject well beyond the scope of this book.

SIMULATION STRUCTURE

Though all members of the same species have significant differences in their simulations they all inhabit the same collective simulation to the extent that their individual simulations share the same structure, and all living beings live in the same shared simulation to the extent of their shared perceptual and cognitive structures and their dependence on the logic of things.

The logical scaffolding of our simulation is an extremely simplified and selective mapping of the external logic of reality but all the rest, especially the appearances, meanings and valuations, is entirely a product of our brains. And in many cases the logical scaffolding can clearly be wrong, as in the case of delusional belief systems, or when its information is simply inaccurate or incomplete.

In all cases the logical scaffolding is a vastly simplified sampling of information structures deemed useful or interesting to the organism in question. This information is acquired through perceptual and cognitive filters adaptively tuned by evolution as antennae to extract information from external reality pertinent to the particular organism's function. Due to the extremely limited information capacity of any organism this necessarily leaves most of the information of reality unknown.

The simulation is a vast mostly unconscious mental model of the structure and details of an organism's local environment, only a very small portion of which is conscious at any moment. It's the complete model of how a particular organism thinks the world works and everything it contains stored as memories. It includes the stored logical structures and individual specific meanings that the organism uses to make sense of its current flow of experiences.

These are all wrapped around the organism's instinctual imperatives inherited from its ancestors but fleshed out into all the current desires, aspirations and goals the organism uses to purposefully direct its current moment by moment actions.

A major part of the simulation is the organism's model of itself. How it conceives itself as part of the world in which it exists, and the interleaved logical structure of its relationships with aspects of the external world, and all the remembered details of those relationships.

The simulation also includes a valuation system that assigns relative values to events and its own past actions and intentions. The organism uses these and the probabilities of desired outcomes to compute its current actions. This valuation system involves a combination of meanings and feelings, both emotional and somatic to help assign values to prospective actions.

The simulation also includes the entire input and output system of the organism. This includes all the internal and external sensory input that constantly updates the simulation down to the communications between every cell and neuron and chemical gradient within the body. And it includes the sensory feedback from all the actions an organism performs within its environment.

All of these are part of the total computational system of an organism. And the current state of the computational system of an organism is its total simulation of reality, both external and internal. All organisms are programs, computational systems of immense complexity. The total integrated information of an organism's program is the complete organism itself. And the organism's total internal information model of itself in its environment produced by this program is its simulation.

The simulation produced by every organism is the reality in which it actually seems to experience its existence. This is true of all organisms of all species, and to the extent robots have simulations of their actions in their environments, of robots as well. Every observer constructs its own particular simulation of reality and experiences living its entire life within its simulation.

Every organism thinks the world in which it lives is the true view of the actual world but this is impossible because every observer's simulation is different, even those of the same species. Only the logical structure of the simulation has any correspondence with the true nature of the actual external world, and even that we must struggle to keep consistent.

The actual world is a world of data and computational interactions. Within this world at the emergent level exist multitudes of living organisms, each continually projecting its own internal simulation model of the world out onto the world as if it's the only real world in which all other beings exist. Every organism projects its own simulation model of the world out onto a world of other observers, and every observer lives within its own simulation of the world believing it's the actual world in which all live.

But nothing could be further from the truth. The real world in which all observers live together is a world of programs computing data, all part of the universal program that continually recomputes the current state of the entire universe. Through the long process of evolutionary adaptation the programs of individuals of all species have developed the individual simulations that best enable them to function within their environments.

All these programs of individual organisms now act purposefully to compute their individual existences based on their instinctual imperatives and their simulation models of themselves within their environments. This is true in an emergent sense, but at the fundamental level the actions of all observers, including ourselves, is all being computed as a part of the single universal program of the universe of which our programs are all integrated parts.

We manifest purpose, and freedom and intelligence because our programs have adaptively evolved designs that do so, and because of the richness of randomness at the quantum level that injects meaning and life into what would otherwise be a deterministic universe.

We all live entirely in our own personal simulation of reality. Without it we could not function and could not exist. Though enormously useful in an adaptive sense our simulations obscure the true nature of reality from us. Since we are forever trapped within our simulations we can only discover the true nature of reality within by carefully analyzing what our simulation adds to the nature of reality itself. We accomplish this by progressively identifying and subtracting these veils of illusion one by one to discover the truth that lies behind them all.

When this is done we discover the only thing remaining after all illusory appearances are stripped away is the logical structure of the simulation. Thus once again we must conclude that the true nature of reality itself is a logical structure consisting only of data and its logico-mathematical relations, and the programs that compute them. Every road leads to this inevitable conclusion.

PERSONAL PROGRAMMING

Though we humans are all members of the same species, there are enormous differences in how we view and relate to the world. This is the personal programming that makes each of our simulations unique derived from differences in our personal experiences and how we relate to them. Our personal programming is constructed over time by our program's learning routines and continually updates our simulation.

Our simulation is heavily programmed by our parents, culture, schools, media, and personal experiences throughout our lives but especially in childhood as it's initially constructed. Our personal programming is so pervasive it's difficult even to begin to grasp its extent, yet it heavily colors our perception of the world we think we live in. And it often contains objective inaccuracies and varying degrees of irrational, delusional, and dysfunctional thought patterns, not to mention it's often quite limited in scope to largely parochial detail.

The software we inherit in our DNA contains extensive learning routines that construct most of the details of our simulation from our experiences. Unfortunately these formative experiences are usually far from optimal and end up programming our minds with all sorts of irrational and dysfunctional thought patterns many often passed from generation to generation within one's native culture. These are often filled with prejudices and delusional belief systems that color individual realities, and painful early experiences can also lead to neurotic and even psychotic modes of thought.

And unfortunate experiences, exposure to the success of the more fortunate, and ordinary human covetousness can easily lead to the unreasonable attachments and desires that Buddhism correctly recognizes as the root of suffering (Suzuki, 1956). All this exists as specific individual programming in people's simulations of reality.

Thus much of our individual views and relationships to reality are aspects of our own personal programming rather than attributes of actual reality. External observer independent reality is essentially neutral with respect to our existence, but is often imbued with personal attributes and attitudes it doesn't possess in our simulations. The realities of most people seem to consist mostly of the vagaries of their personal relationships with other people, their emotional states, their opinions about events and other people, and their internal representations of their mundane realities and their concerns about them. Thus most people are living extensively in dramas of their own making that can have little to do with objective reality.

Personal programming includes the full spectrum of often misinformed and even delusional belief systems, prejudices, ideologies, ethnic identities, religious beliefs, gender affiliations, and political and interest affiliations. All these become incorporated into people's simulations of reality and projected back onto external reality as if they were actually attributes of external reality when of course they are not. These heavy personal overlays make it increasingly difficult for people to recognize the true nature of actual reality within their overly cluttered simulated worlds. And because people think this is the actual nature of reality they naturally tend to act in accordance with it.

The good news is that since our personal programming is *learned* during our lives, it's potentially subject to reprogramming through correction by objective facts and reeducation. However in practice human prejudices and mindsets are often very difficult to change even when clearly delusional as we sadly see all around us.

With respect to personal programming there are two kinds of people, those who understand their minds have been programmed and who try to understand and transcend their programming, and those who think they *are* their programming. The later have little chance of achieving a true knowledge of reality. They are so completely submerged in their personal feelings, prejudices, ideologies and emotional constructs, and their mundane daily lives that they have little interest in the actual structure of reality, and little ability to discern it.

Sadly most people believe the world actually is as it exists in their simulation and they have no doubt at all they are right. In fact they tend to identify their personal identities so strongly with their often dysfunctional simulations that they habitually go to great and sometimes violent lengths to defend them.

The take away is that what we imagine is the objective reality in which we live is almost inevitably heavily programmed by our personal experience. The result is that our simulation doesn't accurately represent reality, and this makes it very difficult to discern its true nature. However this personal programming is not immutable and with understanding and work can be reprogrammed to achieve a much clearer and accurate simulation of the true nature of the reality in which we exist.

Our personal programming strongly affects our functional intelligence and the degree of dysfunction in our lives and thus the balance of suffering and happiness we experience. We are all burdened to varying degrees by the dysfunctional aspects of our programming, which can adversely affect our physical and mental health, and our ability to lead happy and successful lives. This is the source of the neuroses and other aspects of irrational thinking that negatively affect so many lives.

From ancient times to modern cults and psychology there have been countless methods tried to remedy such problems with varying success and many books have been written on this subject. But the key to success is understanding that most of these problems are the result of dysfunctional subroutines in our personal programming that can be reprogrammed.

When we realize we are our running program we find we have considerable ability to reprogram our program to what we want it to be and make it happier, healthier, more effective, and successful. We can potentially change our personal programming as much as we want, within the constraints of reality of course. We can reprogram our personal thought processes but not the laws of nature.

The first step is to realize we are the programs of ourselves and that we have been extensively programmed by our personal histories. Without understanding this it's enormously difficult to escape the problems our programming impose.

The second step is to identify and analyze the details of our dysfunctional subroutines to understand how they work, and how they lead to suffering. Then we need to discover as simple key or keys that changes the dynamics of the subroutine into a healthier one.

Take desires and attachments that lead to suffering for example. We must first realize *we* are not our desires or attachments, these are simply routines in our personal programming and we have considerable power to retain, discard or change them as we wish.

An effective approach is to recognize attachments for what they are but not to dwell on them when they arise, as that tends to reinforce them. Let them arise and fade away naturally without dwelling on them or worrying about them. Recognize that attachments are to some extent part of the human condition, but the trick is not to be attached to one's attachments. Over time this weakens them.

Another effective technique is to replace dysfunctional unattainable attachments and desires with healthier activities that are pleasurable and attainable. Replacing negative with positive programs is much easier than trying to erase or change negative ones.

Another approach is to unclutter the mind through meditation or healthy physical activity. The mind's continual engagement with processing the often unimportant details of our daily lives can obscure the joy of a deeper experience possible when we engage more with the things that are truly important and beneficial to our existence.

Thus the realization of the true nature of our self can be aided by simplifying and clarifying our mental processes, and especially by eliminating any dysfunctional personal programming that is causing suffering, stress or ill health. When our personal programming is realized for what it is this becomes much easier.

And because our true self is not our personal programming but our deeper nature it's much more important how one feels inside than how one looks outside. But if the inside feels right the outside will have a healthier appearance as well.

BIOLOGICAL PROGRAMING

Our biological programming is our species-specific programming due to the DNA coding of our software and is clearly less subject to reprogramming. However it can still be recognized and transcended to some extent. Our view of reality is enormously different from the view of reality of a fox, a bird, a cuttlefish, or an earthworm. Yet all these views of reality are equally valid to the species involved and enable them all to function quite well in a common external reality. These obvious differences demonstrate the enormous divergence in our mind's simulation of reality from what external reality must actually be.

The great differences between species are reflected in very significant differences in their internal simulations of a common external reality. On this basis alone no single species' simulation of reality, including our own, can claim to be the only correct one. Nevertheless all species do have one thing in common, and that is the ability to reasonably compute their functioning within their actual environments on the basis of their simulations. Thus their simulations of reality must all share some common logical structure that maps fairly accurately to the actual logical structure of the external world. All organisms function on the basis of the common logic of things of the emergent universe insofar as they understand it.

It's clear that all organisms' simulations of reality must share some common logical structure with external reality to enable them all to function effectively. This is strong evidence that the universe and everything in it, including the programs of biological organisms are all programmatic structures that obey similar rules of logic.

The point is that different species have vastly different ways of seeing the world. The ways spiders, deer, snakes and people view the world is clearly enormously different. Yet individuals of all species naturally assume the world they experience must be the way the world actually looks and functions. Because the world appears different to all species it's completely clear it cannot be the way it appears to us.

Every species builds its internal simulation model of reality based on sensory inputs, and sense organs vary greatly from species to species. And of course the basic mental software of each species is vastly different so it's inevitable that a fox's internal model of reality will be very different than a human's. Nevertheless there is enough evolutionary similarity and adaptation to the common logic of things that each species is able to function effectively on the basis of their vastly different views of reality.

Again the take away is that the appearance of the world we think we live in is completely an artifact of our existence in human form. The world would appear quite different to us if we were lizards. Yet both are equally valid views of a common reality that consists not of appearance but the common information and logical structure upon which the different simulations of different species are based. Thus it must be only the data and logic of reality that is the true nature of reality rather than its widely variable appearances to different species.

Though the appearances of the world to different species varies greatly, the basic logic of things remains fairly consistent though different species abilities to compute it varies considerably based on their computational capacities and the information structures most important to their species.

PERCEPTUAL ILLUSIONS

We assume the world around us is actually as we perceive it in our simulation, but this is clearly an illusion. We see only a very limited range of electromagnetic wavelengths, hear only limited frequencies of sound waves, smell only a minute fraction of airborne chemicals, taste only 5 tastes, feel only a very limited range of skin contacts, and we lack altogether senses that other species have such as bat echolocation and the lateral lines and electroreceptors of fish. Thus we are blind and deaf to much of the actual information of reality and cannot be said to perceive reality as it actually is.

Our simulation of reality is based only on our very limited and species-specific inputs, so it's obviously impossible for the world we experience around us to actually represent the complete reality of the world as it actually is. Any notion that the world actually is as we perceive it is an obvious illusion.

Another example is that we see reality in terms of focused objects thanks to the lenses of our eyes, but the light of external reality itself is not focused so its actual reality is a blur of light at best in which the identity of individual things quickly fades.

More conclusive evidence that it's the information and computational logic of reality that is its true nature, and that all *appearances* are artifacts of how the programs of individual observers interact with external reality, rather than external reality itself. The real external world just cannot be as it appears to us.

THE ILLUSION OF INDIVIDUAL THINGS

We know from studies of developing minds and the science of robotic intelligence that the identities of individual things are laboriously constructed in the mind from very complex locally repetitive associations of sensory inputs such as colors, textures, forms under rotation and translation, behaviors, functions and other attributes. The concept of individual identifiable things develops fairly rapidly in childhood (Piaget, 1956, 1960) but has taken much effort over a number of years to begin to perfect in robotic systems (Wikipedia, Pattern recognition). Individual things and events as we perceive them are clearly not necessarily an intrinsic characteristic of observerless reality.

Instead the existence of individual things and events is largely a construct of our simulations of reality, and external reality is quite different since what actually exists is masses of computationally interacting particles composing the single universal program of the universe. The concept of a reality composed of completely discrete individual things is largely an illusion of our simulations because at the level of elementary particles the boundaries of things are in continual interaction and transition and never perfectly distinct.

Nevertheless, at the classical level of multicellular biological organisms, the simulation's representation of a reality consisting largely of individual things works quite well. Biological organisms function quite effectively on the basis of the emergent logic of things that describes the classical world and almost all of science is based on these laws as well. However actual reality is quite different, as it has no such preferred thing-oriented scale, but includes all scales at once. Thus the world of individual things we seem to see around us is simply not a representation of the true nature of reality.

What actually exist are computational domains. Domains are emergent areas of computational density in the universal nexus of computations and observers tend to base their concepts of individual things on natural domain boundaries. However domains overlap both hierarchically and interactively so there are no precise actual individual things existent in reality, with the exception of the most elemental. At the emergent level there simply are no exactly defined individual things or programs, there is only the universal program within which domains exist as fuzzy overlapping areas of computational density.

Thus a surfer views the ocean in terms of individual waves, a smelt experiences it in terms of tides, and an oceanographer in terms of currents. But these are all domain-based views of a single ocean, and they all overlap. Leaves, leaf lobes, twigs, branches, trees, tree species and forests are another example of hierarchical overlapping domains that humans selectively identify as individual things on an *ad hoc* basis.

Humans, and no doubt other species, tend to view the world in terms of individual things, properties, events and relationships. These are the basic elements of the logic of things. And of course they and their logic are also encoded in the elements of grammar in which humans describe their concepts of reality (Chomsky, 1965). This very simplified world is very much easier for humans to compute than the actual world of enormous fluid computational complexity and overlapping domains and programs.

So we humans see the world in terms of individual things and their characteristics and interactions but this is not the true nature of the world around us. It's another convenient illusion that makes it easier for us to compute our functioning within the world. Our simplified cartoon simulation operates on the basic of the emergent logic of things, but this is far from the computational logic of reality that actually computes it.

This is another illustration of how the world we think we live in is not the true nature of the actual world. The great miracle is the super-consistency of the universal program that enables us to function effectively on the basis of the emergent logic of things when individual things don't even actually exist, at least in the sense we imagine.

So the whole notion of reality consisting of individual things is at least partly an illusion based in the classical scale of humans and other organisms. And there are other critically important ways in which our simulations don't accurately represent external reality.

THE ILLUSION OF AN OBJECTIVE SELF

So things are constructed by our simulations out of raw perceptual data and encoded and stored as objective concepts. The self is one of those things. In our direct experience there is no objective self and it must be constructed by our minds as a concept we then identify our subjective self of direct experience with (Piaget, 1960).

This is also true of other species to varying degrees and robotic programs must also be coded to recognize their objective selves in distinction to their surroundings in order to function effectively as well.

However raw experience itself is prior to any distinction of self and not self. Only as experience is encoded and analyzed in terms of the simulation is it categorized as part of self or not self. In particular this includes the concept of self as the physical body.

It's likely the strong human sense of an objective self in the form of a physical body only developed to its current level with the advent of mirrors and later photography and video, which allowed people to see themselves objectively from the outside, a view they rarely had previously. With the current flood of personal images of everyone, especially those deemed most beautiful, has come a much stronger identification of self with the physical body as demonstrated by the modern obsession with personal appearance. This tends to obscure the fact that our true self is the totality and harmony of our inner feelings rather than our visual image.

Prior to our current obsession with objective self we thought of ourselves much more as animals do in terms of the subjective self of our direct experiences, perceptions, feelings, actions and thoughts as experienced from the inside. That is of course much closer to our true identity, which is the totality of our direct experience.

So our concept of our selves as an objective thing with a physical body is as much an illusion as are all other classical level things. These are useful concepts that are logically consistent in computing our functioning in reality but they are fundamentally misleading illusions that obscure the true deeper nature of the reality.

OBSERVER ILLUSIONS

There is also the very basic problem that our perception of reality is always from the point of view of ourselves as individual observers. Our simulation's representation of reality is totally dependent upon us as the single observer of that reality. For example we are at only one place at a time, but reality is everywhere, and every aspect of our representation of reality is necessarily in terms of its relation to our location and our physical properties.

A truly accurate description of external reality must be observer independent, and most certainly cannot depend on the location and characteristics of any particular human observer. We always imagine reality as having some particular position, orientation, scale, and clock rate, but these are all things our simulation adds to reality relative to ourselves.

Absent an observer, reality simply cannot be said to have any position, orientation, size or clock rate whatsoever, because these are all relative to our own position, orientation, size and biological clock rates. If we even try to imagine an observer independent reality from all points at once, all scales at once, and all orientations at once, it's simply impossible. Yet that is what is precisely what is required to accurately represent reality in an observer independent manner.

The only possible accurate representation of an observer independent reality is a mathematical representation independent of a particular coordinate system of measurement. Only in this way can we represent an observer independent view of reality because everything is expressed in relative terms to each other in such a mathematical model.

In fact one of the great advances of science was the ability to represent reality independent of particular observers in this way. This is precisely why relativity is called *relativity*, because it describes matter and energy in spacetime in an observer independent manner on which any observer can then overlay his own coordinate metric to make sense of it. And all such metrics are equally valid observer perspectives (Wikipedia, Theory of relativity).

Thus relativity itself conceives the universe of mass-energy in spacetime as an abstract mathematical structure upon which any observer's frame of reference can be validly overlaid to incorporate his individual *view* of that reality. However this is only possible with the recognition that observer independent reality must be reduced to an abstract mathematical construct that exists from no particular view at all, and is independent of the view of any observer. This is yet another convincing reason to believe that the true nature of reality must be a logico-mathematical structure consisting only of programs and information since only this type of structure can represent reality in an observer independent manner.

Universal Reality adds one refinement to this relativistic view by introducing a single preferred universal frame in which the data structure of the observable universe is actually computed and with respect to which actual rotation and world lines are relative. However observers still see the universe in terms of their own individual frames.

This may seem like an obvious and irrelevant point when it comes to describing the reality we seem to see around us, but it's of critical importance. It's absolutely fundamental to understanding the true nature of reality as it actually is because the true nature of reality must be independent of any particular observer. The true nature of reality must be completely independent of the existence of any particular observer within it including us. Reality's true nature is everywhere at once from no viewpoint at all, and all viewpoints at once, and it consists only of relative relationships among its parts with no reference to any single preferred observer. This is just the opposite of our own mental simulation of the surrounding world in which we seem to live, and which we mistake for the actual nature of reality.

Thus the evidence is clear that actual reality must be a logico-mathematical structure, a universal program that is observer independent. So the universe consists not of things in space and time as we see them from our own perspective, but of an information structure upon which our simulation overlays our personal frame centered on our own personal coordinates to make sense of it.

The world we think we live in has an up and down, an orientation, an apparent rate of clock time, and everything in it has a particular size. But every one of these characteristics is completely dependent on us as an observer and every one is relative to us and our perceptual systems. Every one of these characteristics we think belong to the actual world would be completely different from the perspective of a housefly observer. Thus none of them can be actual characteristics of reality itself.

For example a fly sees the world much larger and only closer things, slower because of its faster reflexes though that varies with temperature, probably in terms of facets, and with different colors. And of course its world is full of odors and sounds completely imperceptible to us. And it also flies in a much denser air with much stronger winds. And there is no doubt the fly believes all this is the actual nature of reality though it clearly can't be.

Actual reality has no size, no orientation, no intrinsic apparent clock rate, and it has no location at all. All these are entirely characteristics of relative relationships among things within the universe and without an observer don't even exist.

Again the only way this is possible is if the universe is not a physical structure but a logico-mathematical computational structure, a running program consisting only of information. The bright world of forms and colors we see around us vanishes into the pure invisible information of running programs. Nevertheless the interaction of our program with the programs of the external world is presented to us in our simulation as the bright colorful world around us.

QUALIA

These are only a few of the many ways in which our simulation of reality differs from actual reality. The essential aspect of all of them is that our simulation is not just a model of reality itself, but is almost entirely a model of our *interactions* with reality. What we see when we look out into the world around us is not just a representation of the world around us, but everywhere our interactions with reality projected back onto it.

All the *appearances* of things we see in the world around us simply do not exist 'out there'. Every one of them is added by our mind and exists as what are called qualia in our simulation of reality rather than in external observer independent reality itself. Qualia are all the private internal qualities of things, such as colors, feelings, touches, odors and so forth, which exist only in our mind's representations of our interactions with reality rather than in external reality itself (Wikipedia, Qualia). They all exist privately in our individual simulations and how we actually experience them is ultimately unknowable to others, though we can assume similarities based on similarities of biological structure and our ability to communicate.

For example though we assume that our experience of red is the same as other people's experience of red, this is very difficult if not impossible to objectively confirm. And in fact we know that in many cases it's at least somewhat different as in color blind people, and much different in other species with different color sensitivities who all look out into the same world and see things quite differently. Many species have no color vision at all but instead see much more clearly in low light than we do. The mantis shrimp sees several times as many colors as we do, and eagles see in much higher resolution than we do, thus our view of the world is simply not the way the world actually looks.

Of course our simulation tries to convince us all the appearances we see are somehow out there in external reality itself, rather than being data representations of our interactions with the world that only exist in our minds, which they clearly are.

Thus all the appearances of reality, all the private experiences of sounds, smells, touches and all our sensory and perceptual experiences, are all actually qualia added by our minds to the logical structure of reality, and simply do not exist in observer independent reality itself.

Thus if actual reality has no colors, smells, tastes, sounds, feelings etc. and these are all in our mind's representation of how we interact with it, then all that is left of actual reality is the information that produces qualia when our own program interacts with it.

Of course the raw sensory input of particles of various types we interpret as colors, odors, tastes and sounds are part of external reality, but these are all the data of particle interactions and it's only in our simulation that they become colors, odors, tastes and sounds. We input only information, and only in our simulation is that information organized and portrayed as the bright world around us.

All the apparent appearances of reality we experience without exception exist within our simulation of reality rather than in external reality itself. Our simulation takes the dark invisible logical scaffolding of reality and paints the bright colorful world of our experience over it, expands it into the semblance of a physical spacetime and places us and all the things and events it extracts from that logical structure inside it, gives them all scale, orientation and positions and sets the whole into motion continually updating it against the actual evolving data structure of reality.

THE RETINAL SKY

Though our simulation actually exists non-dimensionally within our brains it seems to exist as a 3-dimensional external world centered on us. This is because our brains project our internal simulations of reality back out into the semblance of a spatial world on the basis of the information of dimensional relationships extracted from our interactions with external reality. But as we have seen the 3-dimensional space we appear to exist within is only an interpolation of dimensional relationships computed by quantum events projected into a graphically displayed 3-dimensional world.

Thus we must recognize that the 3-dimensional world we seem to see around us is an illusion. What we are really looking at when we look out into the world is the information of dimensional relationships of events encoded in the neural circuits of our brains. This information is most certainly not a little dimensional model of reality, but consists only of sets of dimensional relationships among data structures.

And all the *appearances* of everything that populates this apparent dimensional world are actually the information of qualia in our brains. So what we are actually seeing when we look out into the world around us is our mind's *interpretation* of colors on our retinas rather than an external sky. More accurately, the structure and appearance of the entire world we see around us is actually the interior structure of our own brain and perceptual system!

What we see as an external world consists of a highly simplified information structure extracted from the logic of external reality, which our mind then paints over with appearances and meanings it generates itself. The external world we seem to live in is the interior of our own brain projected outward into the semblance of a material world in 3-dimensional space! We look deeply and directly into our own being when we look out into the world. And if we only look deeply enough at what is really going on out there we begin to see the reality within the illusion.

We think reality is the same as our visual representation of it, but actual observer independent reality itself simply has no appearance whatsoever, it's only a computational structure composed entirely of data. It has no color but only data representing color. Every last bit of the appearance of the world is added by our mind and exists only as qualia in its simulation in our brain. This includes its apparent 3-dimensional structure.

Thus the external world is an illusion in our own brain. It's a sample of reality's logical structure painted over with appearances and meanings by our mind and projected outward. The logical structure fairly accurately maps reality's classical level emergent logic of things, but the appearances it's colored with are added entirely by our minds. The world that we see around us is actually a moving painting in the gallery of our mind. It's an interactive wraparound virtual reality show with us at its center. It exists only in our own brain and so we are actually observing the workings of our own brain as much as the workings of reality. To a fair degree the actual logic of the external world is directing the show, but all the costumes and sets are produced and staged in our simulations.

REALITY IS A RUNNING PROGRAM

In the final analysis if we subtract everything our own minds add to the reality we seem to experience around us, all that remains is a dynamic information structure that evolves according to logical rules. All that remains is running programs interactively computing the universe. If we remove the appearances of things, and the perspective of us as an observer from our simulation of reality then all that remains of reality itself is an enormously complex program actively computing the observable universe.

All that remains of the world is the information of things and their logical relationships, and even this is our classical level view of the emergent information structure of reality rather than its actual elemental particle component structure. Thus the true nature of observer independent reality is enormously more complex than our simulations could possibly encode. It includes the enormously complex information of every one of the things around us all the way down to the particle information structures that underlie and compute them.

Reality itself includes the information structure of every last detail of the entire universe when we simulate only a miniscule fraction of only our local environment. Our simulation of reality misses most of the actual detail of reality by many orders of magnitude, and what we do experience is only a tiny sampling of its colorless logical structure painted over with qualia in our minds.

Thankfully our simulation of reality doesn't encode the complete actual structure of reality itself. Our little three-pound brain, so minute in comparison with the universe, would be completely overwhelmed. Our vastly simplified simulation of reality works well enough to compute our lives within the universe and has enabled our success as a species, but it's clearly an illusion that conceals the true information nature of reality from us.

There is only one complete and accurate simulation of the universe, and that is the universe itself. Thus all individual knowledge of the universe must be vanishingly small in comparison with the actual information content of the entire universe itself.

Our cognitive system functions as a set of filters or antennae, finely tuned by evolution to extract only the most pertinent logical structures from those of reality so we are able to understand and function effectively in reality. Our perceptual and cognitive filters allow through only what is most useful and meaningful; other species have somewhat different filters, differently tuned antennae that extract information structures meaningful to their particular existences. All these various filters are essential to our existence but they are also the veils of illusion that obscure the true nature of reality from us.

OURSELVES AS PROGRAMS

Along with the rest of the world it's important to understand how we experience ourselves as a consciousness in a physical body if we are actually just the running programs of our selves. How can we feel so human if we are actually just running programs? It's really quite easy to understand this and this insight applies to the apparent reality of every other aspect of the simulation as well.

All the feelings of our mental processes and our body are simply our experience of our program running. And the feeling of our life force within our running program is simply the active immanence of existence continually happening within us that gives us reality in the present moment as our program continuously computes the progress of our lives.

The feeling that we are a consciousness inside a biological body is simply the active information of those information forms continually being computed. Everything remains as real as it ever was, we remain as real as we ever were and exactly what we were and are, we just now truly understand what we really are and our true fundamental nature.

We know that our experiences of all the things in the external world that seem so completely physical are just their information forms being simulated in our brains' neural circuits, yet they all seem so intensely real. Our existence as a biological organism is exactly the same. We are programs running in the existence of reality, and our experiences of ourselves are the programs of our simulation of our selves running within the program of our whole being. They all feel so real and meaningful because they are. They feel real because the living presence of existence within our program is the life force that makes us real.

    We, like all things, are the information of ourselves being continuously computed by our programs, which are subroutines of the universal program. It's important not to think of this in terms of an ordinary computer program. Our program is a program of reality running in reality, therefore we are as perfectly real as we can possibly be, and as we always have been. We just aren't quite the flesh and blood biological being we appear to be. Well we are, but the way we represent this to ourselves is far different than the actual information state of our programs that we really are.

    This is certainly counterintuitive and can easily be misunderstood and mistakenly rejected. But we are not trying to make us anything other than what we already are; we are just offering the best explanation for exactly what we are. There is no change to what we are, we just now understand what we are more deeply and completely consistent with the nature of the rest of the universe that Universal Reality reveals.

    We lose nothing in this explanation, not our freedom, our emotions, feelings, our capacity to love, or our consciousness. These are all essential parts of our program that computes our own personal reality in a universe of other programs and information. All these things are the information of themselves, and we are the running programs of our selves.

    The running programs of reality are simply the real computational processes of reality we see and experience everywhere around us and within ourselves. All aspects of our existence are clearly computational. The lower level cellular and other bodily processes of our bodies are clearly computational processes that keep us alive and functioning. So considering our total self as a program that includes these subprograms is not as counterintuitive as it first appears. It's simply the best description of what is really going on in a universe that is clearly computational.

    All the things we believe make us human; our feelings, desires, emotions, intellect and consciousness, are all manifestations of internal computational processes. There is simply no other way they could possibly be generated. Like all things in the universe, they are the running programs of these things, and their information can only evolve computationally.

    Changes in information states just cannot arise without a

computational source, that would be the most nonscientific theory of all, the most nonsensical option of all. All computations are programs in our sense. Programs are simply the processes that perform the very obvious computations of reality, and everything that exists, including our selves, are the ongoing results of computational processes.

Humans are enormously complex multiply hierarchical information programs. Through our DNA systems, our cells, our organs, our hormonal and nervous systems up through the structures of our brains, all these subsystems operate in concert as a single integrated program to compute the function and maintenance of our self.

In terms of consciousness, the human simulation consists of a highly detailed internal representation of the structure of the world, of us within it, and the emergent level laws that describe it, all encoded as programmatic information in neural circuits. Particularly salient for human consciousness is that this simulation of the world includes a strong sense of a personal self that experiences it. In other words, the simulation includes a strongly developed representation of one's own existence as a thing that stands apart from other things. This sense seems less developed in many other species. The simulations of other species certainly have strong subjective experience and exquisitely intelligent computations that support their individual survival but perhaps less developed representations of themselves as separate objective entities.

The simulation can be thought of as a computational structure that sits atop the vast lower level computations that make up the entire program of an organism. The simulation takes all of the inputs from lower level computations including those of the sensory and perceptual systems and constructs a model of reality that includes its model of itself. It includes a very detailed model of the logical structures of individual things and their relationships. Like all forms the simulation is being massively simultaneously computed by existence. And these computations continuously update the model with perceptual inputs and compute the implications for the individual's function and survival.

Though most of the computations of an organism occur at an unconscious level, the information of the simulation is available to the focus of conscious attention. And the simulation is continuously refreshed and updated with perceptual input and computational results generated at the unconscious level that percolate up into it.

There are vast differences across the spectrum of organisms in their computational structures including their simulations but in every case the world that an organism experiences it living within is actually its own internal simulation of its particular environment. This is equally true of humans and all other life forms, all of which have at least some rudimentary representations of their external environments, which convey more than random functionality. The world we experience ourselves living within is entirely our simulation of that world computed in our own minds. We simply have no direct experience of our world other than our simulation of it.

ILLUSIONS OF TIME

It is important to understand a couple of ways in which our simulation obscures the true nature of time. The first is our perception of a present moment with duration, and the second is the fact that our mind makes us think we live slightly in the future.

The actual duration of the present moment is the time it takes to complete a P-time tick of happening which is the time it takes to recompute the information state of the universe. This duration is many orders of magnitude below the attosecond scale. An attosecond is equal to 10^{-18} of a second (one quintillionth of a second). For context, an attosecond is to a second what a second is to about 31.71 billion years. Thus the duration of the actual present moment is far below the resolution of human temporal perception and even far below our finest observations of quantum interactions.

Thus if our simulation accurately represented the duration of the present moment as it actually exists our entire experience would consist only of the precise current state of things in the exact infinitesimal moment. There would be no time to retain and compare before and after states of anything or the context of any event. Thus meaningful knowledge would simply be impossible. We wouldn't see any motion at all and there would be no sense of change whatsoever.

Thankfully our simulation represents the present moment with a several second duration, it holds time open just long enough we are able to compare before and after states and observe the context of events as they occur. This is accomplished by a short-term memory subroutine that holds representations of events together in a sort of cache memory long enough they can be compared before tagging them as past events and moving them to long term memory if required.

This slight opening of the present moment in time is an essential aspect of knowledge and consciousness but doesn't accurately represent the near infinitesimal duration of the actual present moment of existence. So our perception of our existence in a present moment that lasts long enough for us to make sense of things happening is a complete illusion, but an illusion essential for our existence. We can confirm this in operation in the visual tracks of birds and moths.

Inanimate programs don't have any capability to hold a present moment open which is one reason their experience of events, which is just as real as ours in its own way, is entirely unconscious and without context.

A second way our simulation misrepresents time is by projecting the states of processes slightly into the future when that's of course actually impossible since by definition the future has not actually been computed. Our minds are continually building a simulation of the current state of our surroundings that includes projecting current short-term processes slightly into the future and representing them to us as if they are already happening. So what we see happening around us is our mind's prediction very slightly into the future of what it expects to happen.

From an evolutionary perspective this gives us an active advantage in preparing for possible future events slightly before they occur but of course these projections can't always be accurate and are continually corrected by inputs from actual events as they occur. This correction process is usually ignored by consciousness but sometimes results in a slightly shocked recognition that we saw something wrong. I believe this process has been confirmed experimentally but cannot locate a reference.

DREAMS

Our mind is a reality-simulating machine in constant activity busily constructing its simulation of reality on the basis of its own imperatives. Normally this process is continually corrected against sensory inputs from actually occurring events and brought back in general accordance with them, but it basically has a will of its own and in the absence of continual feedback from sensory inputs tends to go off on its own direction constructing its own version of reality.

This is what happens in dreams and to a lesser degree with more conscious guidance in daydreams. The mind keeps busily constructing its simulation as usual but corrective sensory inputs are largely shut down or ignored so it begins merrily constructing a reality of its own design based on information in the simulation but with little correspondence to actually occurring events. There is no input stream from reality constantly bringing it back to representing the actual state of the world so it has the freedom to represent whatever reality it wants to. Delusional psychoses are a disorder of this feedback mechanism where perceptual input doesn't properly correct the thought stream.

Thus dreams do reveal the inner concerns of mind when it has the time and sensory isolation to develop them and present them to consciousness, and they also open a window into the workings of the organizational processes that maintain our simulation of reality.

In dreams, when there is less concern with the actual surrounding world, the focus of consciousness also has more freedom to dip below the usual mundane surface of the simulation into the realm of usually unconsciousness processes and concerns.

In sleep the mind typically paralyzes the major voluntary muscles to prevent dreams from initiating bodily actions (Wikipedia, Sleep paralysis). Sleepwalking can occur when this protective mechanism fails, sometime with tragic results, as dreams are acted out.

Sleep paralysis can also fail in the opposite way when we wake from sleep but are unable to move our bodies. This is often mistakenly interpreted as the presence of some malevolent being that is actively paralyzing us. Throughout history these experiences have been interpreted as various types of demons such as incubi or succubi, and more recently as alien abduction experiences.

OUT OF BODY EXPERIENCES

There are other interesting implications of understanding how our simulation functions. When we understand that the mind constructs a 3-dimensional universe from the non-dimensional neural circuits of the simulation and projects it into an apparent 3-dimensional world with our simulated body at the center collocated with our consciousness it becomes easy to understand out of body experiences (OBEs).

If the mind can construct a simulated 3-dimensional world around a simulated physical body, and then place our consciousness within that physical body, then it's easy to understand that the mind could just as easily move the conscious 'I' out of our simulated body in situations of immanent danger as a protective mechanism to lessen the I's experience of possible body trauma.

Thus there is nothing supernatural or hard to understand abut OBEs, they are just a good example of how mind constructs our simulated reality on the fly and modifies it as it deems appropriate. This includes normally locating our conscious self within our body. How the experience of self is normally located within the body by the simulation is the important thing to understand. Then it's easy to understand how it can also be located outside the body as well.

This is also the key to understanding near death experiences (NDEs) in which the conscious 'I' is also experienced leaving the body and traveling either down a long tunnel into the light or even to another world. This often occurs associated with symbolism from the subject's belief systems such as entering Heaven, the Tibetan Bardo realm (Evans-Wentz, 1956)), or the passage of the soul through the underworld described in the Egyptian Book of the Dead (Budge, 2008).

In near death the consciousness can remain operative longer than other parts of the body such as the perceptual systems, and consciousness can retreat towards the center of the brain and finally be relocated by the mind in what it deems an appropriate experience, again apparently as a protective mechanism to lessen the trauma of impending death.

Similarly in psychedelic experiences, and in delusional experiences associated with mental illness, the mind is just simulating reality differently than it usually does. All these experiences are excellent examples of how completely our minds simulate our realities, and they all help us understand how enormously different our simulation of reality is than reality itself.

Anyone who has taken LSD or other psychedelics becomes aware just how ephemeral and illusory our usual simulation of reality is, and how the actual reality it obscures can appear in so many other forms with just a little chemical stimulus. These experiences provide important insights into how vastly different reality must appear to other species and even to other members of our own species. Thus there is no reason to attach any particular metaphysical significance to any of these experiences or assume any alternate realities. But they do most certainly reveal just how wonderful actual reality is and how variably it can be simulated under different conditions.

So it turns out that all realities are alternate realities including the one we normally experience. Every one of them exists only in our minds. Every one of them is inherently arbitrary and the only reason we normally experience the one we do is because it allows us to function in actual reality. Our usual simulated reality is just the product of our evolution and every one of our individual realities is significantly different, and those of other species even more different. And none of them accurately represents the true nature of the actual reality in which they all exist.

ARTIFICIAL REALITIES

It's common knowledge that many aspects of reality can be realistically simulated in the form of information in various media. When a high quality recording of a birdsong is played it's effectively impossible to distinguish the information of the recording from the information of a real birdsong without appropriate context. Movies and videos present very convincing canned realities, and the advent of various virtual realities promises to make such experiences fully interactive.

There appear to be no intrinsic limits to the ability of properly formatted information to represent convincing realities. Thus there is no reason to suspect that actual reality as well doesn't also consist entirely of properly formatted information. It's clearly demonstrable that actual reality could certainly consist entirely of information and be entirely convincing, and in fact this is what all the evidence suggests.

So it's quite reasonable to consider not only our simulation of reality as information, but the actual reality that it simulates as information as well. This in fact is the only way our simulations and the various media representations of reality could encode actual reality as convincingly as they do. If actual reality doesn't consist of information too, then how can it be encoded as information in our brains and in our media, and in our logico-mathematical sciences?

Though theoretically possible there is no reason to believe that we live within a simulation produced by another being, be it an alien computer programmer or some god. What is quite clear is that our own minds do an extraordinarily competent job of simulating reality on their own. And our simulation is programmed not by any other being but by the process of evolution.

Even if we did exist within an artificial reality produced by some other being that reality as well as the other being and its reality would ultimately have to exist within an actual reality encompassing them all. My suspicion is that there would always be some way to ascertain that. There would always be some inconsistency or incompleteness in the artificial reality that would enable it to be discovered from the inside. After all we do live within our own mind's artificial reality and the whole point of this book is that we have been able to discover that and even discern the actual reality beyond it.

On the other hand it's also clear that the minds of gullible persons in particular, and all of us to some degree, are being actively programmed on a continual basis by all sorts of external influences. In the search for the true nature of reality it's of critical importance to recognize the nature of this external programming and insulate oneself from its effects. Otherwise the true nature of reality is ever beyond our grasp.

TRUTH AND KNOWLEDGE

INTUITION

Understanding ourselves as programs provides an insight into intuition. Almost all the myriads of constant computations of our simulation occur at an unconscious level, only a fraction of which emerge to conscious levels. Almost all our ideas are computed unconsciously and some then bubble up through various filters into consciousness.

An intuition is simply an idea computed unconsciously emerging into consciousness more or less fully formed. This happens all the time. When we are consciously considering some situation and a relevant idea pops into consciousness we assume *we* had that idea, but where did it really come from? How does any idea actually come into being? Ideas are always computed somewhere in our simulation, and more often than not at an unconscious level and then presented to consciousness.

Now especially when our conscious mind is focusing on something else and a good idea on another subject of interest suddenly pops out of unconsciousness into consciousness we can't so easily pretend our conscious mind had that idea, so we call it an intuition. So intuition is easily explained, and is nothing very mysterious. When we understand that our simulation is constantly computing our ideas mostly at an unconscious level and that some naturally bubble up into consciousness, the nature of intuition becomes clear.

This is also true of our actions and decisions. They are mostly initiated at an unconscious level, and then if there is time, may be presented to consciousness for final approval. But in situations where immediate action is essential we act before we think which confirms how computations at are made largely at the unconscious level. It has been experimentally confirmed that most if not all decisions are made at the unconscious level because experiments show that decisions are typically made slightly before we are consciously aware of them being made.

KNOWLEDGE, INTELLIGENCE & WISDOM

Knowledge is the accuracy and scope of an organism's simulation of reality. Intelligence is the ability of an organism to compute valid implications from information in its simulation whether or not that information is an accurate representation of the information structure of reality. Wisdom is the breadth of accurate knowledge accumulated over time, most often with reference to knowledge useful in solving practical problems. Intelligence applied to accurate knowledge over time leads to wisdom.

All knowledge of the universe consists only of information, more evidence that the observable universe itself must also consist of information. How else could knowledge of reality consist of information if reality itself did not also consist only of information? Knowledge is a mapping of the logical structure of the data of reality. Because the data structure of reality must be consistent to be computed the test of true knowledge of reality is the consistency of its logical structure.

Beings of all species have intelligence, accurate knowledge of reality, and wisdom to the extent they are able to function effectively within reality. This is the *functional intelligence* and knowledge of the program of their whole being, only part of which may be mental. The very fact that our physical and biological programs are built by reality ensures some degree of consistency with reality in their computations and actions. And even living systems composed of fairly basic structures exhibit some degree of intelligence even in the absence of organized mind-based simulations, as demonstrated by the very effective actions of plants and unicellular organisms within their environments.

Obviously complex organisms with nervous systems and brains have an advantage in the simulations of reality their minds are able to construct and maintain. The mind-based simulations of various species allow large amounts of knowledge to be stored about their environments, and their increased intelligence enables a much richer spectrum of actions to be projected, evaluated and activated.

This sort of functional intelligence is universal in the animal kingdom, and compared to random activity the degree of animal intelligence across almost all species is quite impressive. The program design of the programs of all species is really quite amazingly effective.

What is important here is the concept of functional intelligence.

IQ intelligence is only the ability to perform well on IQ tests, but functional intelligence is the ability of an organism to function effectively in the face of life's continual and varied challenges. My father often pointed out that life is a continual series of intelligence tests that we either pass or fail. In terms of functional intelligence, humans are not that much smarter than other species and often less so in their ability to live their lives well within their environments.

Knowledge about the world can be accurate or inaccurate. But since our knowledge ultimately exists only in our brain's simulation of the world how do we know whether our knowledge is accurate? What is the test of true knowledge? Obviously knowledge is accurate to the extent it's logically consistent with the logical structure of reality. We know the logical structure of reality is self-consistent, so if the logical structure of our simulation of reality's logical structure is consistent with it then our knowledge of reality will be accurate.

But the problem is that all tests of consistency ultimately occur within the simulation itself. We are just testing the consistency of information in our simulation against other information in our simulation since the only knowledge of reality we have is of our simulation of it.

Nevertheless we do act effectively within external reality as demonstrated by our continuing existence, so we know we do have some sufficient degree of accurate knowledge of reality. If we didn't we wouldn't be here able to ask the question.

Thus the only possible test of the accuracy of knowledge is its own internal self-consistency. Reality is self-consistent; therefore accurate knowledge of it must also be internally self-consistent. This and the fact we do function effectively on the basis of our knowledge of reality is an effective proof that our knowledge of reality is accurate, that it's true knowledge. But even so this is typically just knowledge of the logic of everyday things, which is far from the whole story. Full knowledge requires incorporation of consistent knowledge across the whole scope of reality as this book attempts to discover.

So the only possible test of true knowledge is the internal self-consistency and completeness of our internal simulation of reality. In the final analysis this is all we can ever know about the world.

Obviously we don't have complete knowledge of reality. It's possible to have accurate knowledge of some aspects of reality and incorrect or no knowledge of others. Yet reality has a wonderful habit of continually presenting us with apparent inconsistencies in our knowledge of it. Whenever any inconsistency arises in our knowledge of reality we know our knowledge of reality is incorrect in some respect, and it's the continual recognition and resolution of inconsistencies in our models of reality that leads to the continuing progress of science and the refinement of our individual simulations.

So self-consistency of knowledge over the entire span of possible experience is the goal of science and reason, and should be the goal in our daily lives in the improvement of our simulation. Insofar as we seek out, recognize, and resolve the inconsistencies in our own thinking about all aspects of our lives, we have accurate knowledge of our world, and the more accurate and complete our knowledge the more effectively we are able to function within the world.

THE LOGIC OF THINGS

Organisms, including humans generally comprehend reality in terms of *the logic of things*. The logic of things is the emergent logic that describes the internal logic of simulations and thus appears to govern reality as living beings perceive it. The logic of things is flexible, adaptive, quite complex and developed in higher organisms to describe many more types of emergent phenomena than it is in simpler ones.

The logic of things is the basis of the knowledge of humans and other species. Its fundamental principles are generally implicit and unstated but underlie all human knowledge, grammar and science. It is also essential to robotic intelligence and control systems as well and it is better revealed and understood through their development.

The complete logic of things is quite complex and adaptable to nearly all situations organisms encounter. Its general principles are relatively simply but there are as many exceptions and tweaks as there are types of phenomena. We'll outline only a few general principles here to convey a sense of it but much more remains to be done to formally outline the entire logic of things. Again the logic of things describes the logic of the emergent world as it appears to work in the simulations of organisms rather than the way the world actually works at the level at which it's computed.

In the simulation model governed by the logic of things the world is composed of things, characteristics of things, actions, relationships and events, or more fundamental just of things of various types including objects, subjects, qualifiers, relations, actions and events. Relations can be variously defined to include unary relations as in qualifiers of things and single and multiple thing event relations. Relationships can be or static or dynamic as in the case of events.

Things and events exist and occur within a dimensional spacetime continuum in which things have locations and temporal duration and events have times and locations. And they both exist against the context of an undiscriminated perceptual background from which individual things and events are discriminated as required.

Things are defined as sets of characteristics, and are characterized by type sets of characteristics into thing categories such as self, people, animals, and inanimate things. Characteristics are of different types, such as functions and physical or behavioral characteristics

Things are defined by and obey the rules of set theory. Relationships operate on the basis of computational logic. Propositional calculus is used to compute the logical rules of actions and relationships. (Wikipedia, Propositional calculus).

Individual things are discriminated from the perceptual background at least partially on the basis of group theory since they are entities that tend to maintain their identities under spatial and to some extent temporal transformations.

Thus the logic of things is mapped onto the perceptual background as needed. The simulation extracts relevant sets of basic entities from the perceptual background, the emergent structure of the entanglement network, as needed.

The logic of things is also the underlying grammatical logic of all human natural and computer languages. Parts of speech correspond to the basic entities of the logic of things. Language developed to express the logic of things and thus the common grammatical rules correspond to the rules of he logic of things. Nouns are things, adjectives are characteristics, verbs are actions, and subject-object relationships are relationships. Adverbs qualify actions, etc. Much work has been done by Chomsky and others on the deep logic common to all human languages (Wikipedia, Universal grammar).

In practical applications all the basic rules by which organisms understand and successfully function within their environments derive from the logic of things. Some simple examples are

1. Things maintain their existence until changed. If a leopard exists and it's unseen then it must be hiding somewhere rather than having vanished.
2. Things maintain their identities until changed. If something acts or was like x, it will likely continue to be like x unless something changes it.
3. Events always have causes. There are reasons for everything even if they are unknown or ascribed to supernatural agents.
4. Things move continuously in space to get from one location to another, they don't vanish and reappear.
5. Animals are a class of things that are alive which means they purposefully move and act on their own volition. Animals can die in which case they can no longer move or act.
6. Etc, etc. etc. ….

Consideration of sets of similar things can be more easily understood in terms of basic number theory. Each member of the set is presumed to have the common characteristics of the set unless demonstrated otherwise.

One could go on at great length but this is sufficient to get a sense of what is meant by the logic of things. It's the underlying principles of the emergent world as it's modeled in our simulations and the simulations of other organisms that enable effective understanding of and action in their environments.

SCIENTIFIC METHOD

The test of internal consistency is fundamental to scientific method and all reasonable attempts to find the truth of any matter. However it needs to applied carefully and exhaustively and always subject to correction. We must be very careful to ensure every possible detail of our knowledge is consistent with the whole. It was very easy to conclude that our knowledge of the Newtonian universe was accurate when it wasn't tested in relativistic conditions.

Thus internal consistency across the whole of knowledge over the broadest scope possible is the decisive and only test of its accuracy, and the only possible method to determine the true nature of reality. I have considerable confidence that Universal Reality performs well on this test as it is appears consistent with the vast internally consistent logical edifice of modern science, and consistently incorporates all the other major aspects of reality not addressed by science such as consciousness, the present moment, and the nature of existence in a completely unified and internally consistent Theory of Everything.

As a note of caution, we must be very careful not to assume that the classical logic of everyday things upon which our simulation of reality is based applies to all aspects of the universe as a whole. This classical logic of things is so thoroughly embedded in our language and thought processes that it's very difficult to understand it's intrinsic limitations. The logic that properly describes the actual universe is much more complex and nuanced. So we must always be very careful in applying this emergent logic of things to the deeper mysteries of actual reality because it inevitably tends to over simplify and distort the actual logic of reality.

The logic of things reflected in our simulation and our language evolved to describe the everyday world of things and events in the context of our simulated local environment. It wasn't designed to describe or comprehend the universe as a whole, nor the quantum or cosmological aspects of reality. Thus we must always be on guard not to apply the logic of things to situations where it isn't appropriate and we must constantly strive to escape the limitations of language and thought as we have tried to do in this book.

For example we can easily be misled into thinking that the universe as a whole had to have had an origin because we see that the mundane things of the world around us all have origins and are all subject to their individual existences ending. And we can be misled by the fact that events in the world around us happen at particular times to incorrectly assume the origin of the universe occurred at some point in time which was more likely the origin of clock time itself.

And we can be misled by the apparent physicality of things in our simulation that we can pick up and heft in our hands to think of the world as a physical structure without understanding that the information of a stone's heft is simply the information produced by its computational interaction with the program of our arm muscles as part of the rest of the program of our body. The beauty of Universal Reality is it incorporates all these insights and views reality with fresh eyes freed from at least some of the deeply ingrained thought patterns of our evolutionary programming.

This apparent physicality of things, and science's ability to explain reality in physical terms largely ignores its source as immanent information in existence. This was an ancient oriental insight that was largely lost with the rise of physical science. This has been corrected in Universal Reality as physical science is naturally integrated with the ancient view of an immanent reality in a new computational setting.

Though they may seem independent, our mental computations are being computed by the program of the universe and thus are always computed logically and consistently. But the logic the simulation uses to compute its model of reality is a generalization of that the universe uses to compute the simulation. The simulation is a meta-program that is computed by reality but runs according to a simplified version of its logic. Thus it can be subject to inconsistent computations and computations based on incorrect premises and data.

This is actually an evolutionary adaptation that enables our minds to continuously redefine the individual elements it computes as needed without regard to their consistency across previous working definitions. In the highly simplified and malleable concepts of things in the simulation this ability to think in terms of at least partially inconsistent time isolated scenarios allows us to think fairly effectively in any given situation on the basis of working definitions relevant to that situation without regard to their complete consistency with previous scenarios. This ability to ignore overall consistency over time and across individual applications enables us to focus more effectively on what are hopefully the salient circumstances at hand.

An example would be the capacity to think of a person as a single entity as a good person at one time and compute reactions on that basis and as a father or scholar at another time and reason on that basis without the need to reconcile the inconsistency. Or we could think of the limbs of a tree as individual things when pruning, and the whole tree as an individual thing when cutting or fertilizing it. This ability to change the definitions of things and their relationships on the fly as needed is what enables us to efficiently compute reality on the basis of incomplete internal models of its various aspects.

Ultimately all the programs of reality are consistent across their entirety, but the *ad hoc* simplified models of aspects of reality in our simulations need not be cross or even completely internally consistent because their elements are defined differently. Only this ability to reason heuristically enables us to reason effectively on the basis of our *ad hoc* simplifications of reality, but it also gets people into trouble because it allows them to hold beliefs and ideologies they don't feel a need to actually test against overall reality or each other. By compartmentalizing fragmentary reality scenarios in their simulations that are inconsistent with one other it becomes possible for delusional beliefs to persist untested.

Our knowledge of reality has expanded exponentially with the progress of science. The instruments of science have greatly expanded the range and depth of our simulations of reality from representations of our mundane immediate environments to the smallest and largest depths of the universe and greatly expanded our ability to discover the information carried in other wavelengths. And no doubt this is only the beginning in building ever more complete and accurate models of reality.

One other important point needs to be mentioned. Reality is a computational process but the equations of science are static descriptors that must be interpreted and correctly applied by scientists. This adds another barrier to a Theory of Everything. Ideally a Theory of Everything would itself be a program running so as to duplicate insofar as possible the computations of the programs of reality. Of course scientists now frequently employ computer simulations of various aspects of reality, but they should aim towards more and more comprehensive programs able to encompass ever more aspects of the information structures of the universe and their computational interactions. Only thus can a truly comprehensive theory emerge.

Of course not all of reality is mathematical and can be described by equations, but every bit of it is informational and programmatic and obeys the rules of computational logic. Thus it can all potentially be simulated on computers. What is mathematical is meaningful only in some logical context, and this is often missing from static equations in print that assume the reader will add the proper logical context.

Science is a logico-mathematical structure and its mathematical equations only make sense within the logical context of their applications as they more explicitly do in computer simulations and only when properly correlated with observable variables. The advantage of programs over equations in print is that they have to supply most of the proper context to even run.

Everyone but especially scientists systematically fool themselves into thinking they understand the universe because they know some of the general principles that describe it. But in actuality they can't predict or explain the actual growth pattern of even a single blade of grass among the uncountable trillions of grass blades. This is not really understanding the universe.

If science doesn't know how something is computed then science doesn't really understand it and we don't know how it actually happens. However there is hope for greater understanding. So long as reality is a logico-mathematically consistent structure a logico-mathematical solution to every problem must exist.

REALITY MATH & HUMAN MATHEMATICS

It's reasonable to assume human logic, mathematics, and computers work so well to simulate and describe such a wide range of natural phenomena because their basic logical structure is very similar to the computational structure of the universe they are used to describe. Thus many of the basics of the information structure of the universe must be reflected in the basic logical structure of these human inventions. Only if they were consistent with the logico-mathematic structure of the universe itself could they accurately describe so many of its diverse aspects

Thus human knowledge, mathematics and science, and the logical structures of computers are all human models of the underlying logic of reality that actually computes the universe. However there are important differences between reality mathematics and human mathematics.

Reality mathematics is the logic and mathematics of the computations that actually compute the universe and nothing in addition to that. It stands in sharp contrast to human mathematics, which is logically consistent with reality mathematics, but extensively generalized. Most of human mathematics is not used in the actual computations of the universe and not a part of reality mathematics. However because it's consistent, its emergent generalizations often effectively *describe* the logical structure and processes of reality at aggregate levels. Most of human mathematics and logic, and the laws of nature emerge from the computational mathematics and logic of reality just as emergent phenomena emerge from the elemental computations of the universe.

Human mathematical structures can be said to emerge from the basic rules of reality mathematics as other emergent programs of reality do. Thus human mathematics, except for the fundamentals it shares with reality mathematics, is not discovered but created or at least derived by mathematicians. It's only the fundamentals that are discovered and then their implications developed in a potentially unending process.

There are several important differences between human and reality mathematics. For example the universe is finite. Nothing real and actual can be infinite. This is easy to understand when we understand infinity is not an actual number but a never-ending *process* of continuously adding numbers forever. Thus it's simply impossible to ever produce an actual infinity. Many scientists have the very bad habit of using 'infinity' to mean larger than they can comprehend. This is incorrect usage and very misleading.

Thus reality mathematics contains neither infinities nor infinitesimals. It contains no infinitesimals as the information of the universe is quantized at the elemental level and isn't infinitely divisible. Scientists have another bad habit of referring to the smallest scale of the universe as the Planck scale. This is also incorrect usage meant to imply the smallest actual level. However there is no evidence that the Planck numbers have any connection at all with actual minimal physical scaling. True a couple of them are much smaller than any known physical analogue, but then the Planck mass is much larger than the mass of elementary particles so the usage is clearly untenable.

It's not even clear that numbers as such are part of reality mathematics; individual data instances may be sufficient to compute the elemental universe though this is not certain. Clearly most arithmetic operations can be carried out by simple logical *comparisons* of actual *instances* of things, or actual instances of data in the case of a computational universe. On the other hand numbers are very useful *descriptors* of reality at the emergent level even if they are not involved in actually computing reality. Note that numbers *per se* are not really part of even computer machine language, which creates both numbers and code from the context of fundamental bits that are just on or off states.

So it may be that all that's included in reality mathematics are elemental bit structures, and the logical operators to consistently manipulate them. More work on what is necessary to compute the universe is clearly required. So far as I know there has been little work on this however the mathematician Gregory Chaitin has described mathematical systems that may be relevant (Chaitin, 2006)

It follows then that almost all human mathematics is not actually used to compute reality and is not actually a part of the information of the virtual quantum vacuum. Almost all of human mathematics is invented rather than discovered. Thus there are significant differences between human mathematics and the actual computational mathematics of reality.

Human mathematics works so well to describe the emergent reality it doesn't actually compute simply because human mathematics is the consistent logical extension of the much simpler reality mathematics that does compute it. Thus the entire logical structure of mathematics is consistent just as the emergent structure of the observable universe is consistent with its elemental structure. This is one more indication of the magnificent super-consistent design of the complete fine-tuning of the virtual quantum vacuum.

Also reality mathematics is entirely logically consistent and logically complete in its operation. As opposed to human mathematics reality mathematics isn't subject to Gödelian incompleteness because every state is directly computed from its prior state and this can always be done.

Gödel's incompleteness theorem applies to correctly formed statements in a mathematical system that can't be proved or disproved, but reality never makes up a data state and tries to reach it, it always just computes a data state from the previous one. So reality mathematics is consistent and complete. It must be, otherwise a computational universe would tear itself apart at the inconsistencies and pause at the incompletenesses and couldn't exist.

REALIZATION

THE DIRECT EXPERIENCE OF REALITY

While it's clearly true we experience reality only through our simulation of it there is one aspect of reality we do have direct immediate experience of and that is our simulation itself. Everything that exists is part of reality by definition and that includes our simulation of it.

Thus we do directly experience the true nature of our simulation, which is part of reality. The key is how we realize that nature. If we take the simulation for the true nature of reality that is illusion, but if we take the simulation as a simulation of reality, that is reality. Ultimately our simulation is the only aspect of reality that is available to our direct experience because it *is* our direct experience of reality. It's the only part of reality of which we are directly aware. Thus it's only through the direct experience of our simulation of reality that we can approach reality. This includes the direct experience of all aspects of our simulation model of the universe.

Thus every aspect of our direct experience is an actual direct experience of reality because every aspect of our direct experience is part of our simulation. Thus we do directly experience the true nature of reality all the time in every experience that we have. It's just a matter of correctly realizing what it is that we are experiencing.

In this chapter we attempt to take the final step in our study of Universal Reality and cut through the illusions of our simulation to achieve a direct and clear experience of the true nature of reality to the extent this is possible in human form.

APROACH

Universal Reality defines reality as the totality of everything that exists, but the true nature of reality is not at all as it appears. The appearances of both ourselves, and the world we seem to exist within, are useful but illusory simulations of reality created by our minds in the form of the familiar physical world of our experience. But this illusory world is nothing at all like the true reality of running information programs other than it shares considerable logical correspondence.

In this context realization can be defined simply as understanding and directly experiencing the true nature of reality that lies hidden within the simulation. In Universal Reality there is nothing more to realization than that, and there are none of the usual metaphysical or religious connotations implied. However understanding and directly experiencing the true nature of reality is certainly the most awe inspiring and transformative experience one can imagine.

There are three fundamental aspects to realization; realizing the illusory nature of the world of our usual experience; realizing the actual information nature of all things; and the realization of the immanence of existence in all things, including ourselves. Realizing the world as its running information programs, and experiencing the immanent existence of these programs, including the immanence of our own program, is the key to realization.

In addition there are continual possible realizations of the deeper natures of the individual processes of the universe, many of which have been explored in this book, including the deeper aspects of the events of our daily lives. Some of the most fundamental insights are explored in this chapter but for the seeker this is an unending process of better understanding the inner workings of reality in all the events of our lives including the seemingly most mundane.

Using Universal Reality as a guide we can also shed light on some core concepts from Western and Oriental philosophy and give them a consistent scientific interpretation. This enables some important personal approaches to realization from which basic ethical principles consistent with the overall theory can be derived.

Then with some final key insights into the deeper nature of realization the book concludes.

THE FUNDAMENTAL REALIZATION

The central experience of our existence is our consciousness in a present moment through which clock time flows and things happen. A fundamental realization of Universal Reality is that this is in fact our direct experience of the most fundamental process of the universe occurring within our own being. We ourselves are an integral part of the universe and the fundamental process of the universe is continually occurring within us as it does within everything in the universe, and our consciousness in the present moment is our direct experience of this process, it is our actual participation in the fundamental process of the universe as an integral part of the universe.

Our consciousness in the present moment is our direct experience of the continuous extension of the radial P-time dimension of our hyperspherical universe that is the source of the happening of the universe that drives its continual computational evolution. We directly experience it within us because it is occurring within us as it does within everything in the universe.

This is the fundamental process that makes us alive and conscious and continually traveling forward in clock time at the speed of light. It's our direct experience of the universal processor continually recomputing our existence in the present moment within the 3-dimensional surface of the cosmic hypersphere. This is the fundamental process of the universe and we are right there in the middle of it experiencing it in every second of our existence because it is occurring within us as it does within everything in the observable universe.

Realizing this is the fundamental realization of our existence. The fundamental process of the universe is not something just happening far out in the depths of interstellar space. It is happening inside each one of us all the time, and all that needs to be done is to realize this and experience it for what it actually is.

REALIZATION OF TIME

The central experience of our existence is our consciousness in a persistent present moment of time within which happening occurs and clock time passes at the rate of that happening. And this is true of all observers in the universe all of whom exist within the same universal present moment in which the entire universe exists.

It is clear from relativity that clock time passes at different rates for different observers within this shared present moment. It is also clear from relativity that all observers in the universe continually travel forward in clock time at the speed of light as measured by their own clocks. And it is clear that all observers see the 4th dimension of past clock time as distance in every direction from every point in the 3-dimensional space of the universe.

All these aspects of time can be directly realized in our experience. If we turn our attention to the passage of happening and clock time through the present moment we find our consciousness of this process is indeed the fundamental experience of our existence. We just need to realize that this experience is us and everything around us traveling at the speed of light through the 4th dimension of time even while we sit on our sofas. We are surfing the 3-dimensional surface of our expanding hypersphere at the speed of light as we ride the evolving wave of existence.

And with the assistance of science we can directly experience the fact that clock times passes at different rates in different relativistic circumstances. If we observe the half-lives of decaying particles moving at relativistic rates, the speed of our clocks on earth relative to those traveling in space, or even by directly comparing our clocks to those returning from space flights we can directly experience this. These can all be directly realized in relativistic circumstances in our daily lives. Even magnetism is our direct experience of the relativistic effects of moving electric charges.

We can also directly experience and realize the continual computational creation of the information state of the present as a process that occurs only within the happening of the present moment, thus realizing the non-existence of the future. We can also directly realize the non-existence of the past even though we observe it as distance in every direction because we are observing that and everything else in the universal present moment of all existence.

Thus we can immediately realize the impossibility of time travel in the sense of traveling out of the present moment. The present moment is all that exists and where everything exists and happens. We can see down the past dimension of time only because of the finite speed of light. We are not actually observing the past, but the light trace of the past in the present moment.

The past exists only in its traces in the present moment, as memories, as apparent distance, and in all its computational contributions to the present information state of the universe.

But there a deeper realization here and that is that the current present moment information state of the universe and everything in it is actually a recording of the information of the past back to the beginning of time redistributed among the data of the present. The present is entirely a recording of the past. Things are not just what they are in the present but the computational accumulation of everything they were in the past. When we look at the present we realize the living past within it because the present is a recording of the information of the past and that is all it is.

Finally there is the illusion of the duration of the present moment itself. The present moment of our experience seems to have a sliding duration of several seconds so that our minds have long enough to compare things and make sense of things. But the actual physical duration of the present moment in which the programs of the universe recompute their data is far far below the resolution of human experience. The actual duration of the present moment is even far below the time scale of elementary particle interactions.

It is only because our short term memory holds a simulated present moment open long enough for our mind to compare things and events that anything makes any sense at all. If our short-term memory didn't work this way we would not even be aware of changes as they occurred since that depends on a mental comparison of before and after states in an artificially extended present moment of consciousness. Without this illusion of time we would experience reality as inanimate objects do, completely real but completely unconscious.

This is something that can be experienced directly to some extent. If we rest with eyes closed and listen to calm music or even a single tone and progressively direct our attention closer and closer to the exact instant that it appears into and out of existence we finally experience a state of instantaneity of time, a vanishingly short duration present moment and we suddenly realize the true nature of the present moment of time. It's a vanishingly short instant, and within that nearly non-existent moment is the entire existence of the universe and us as well. These are the essential aspects of the realization of time.

REALIZATION OF SPACE

We don't exist within the dimensional space of our experience. Though we seem to exist within the familiar 3-dimensions of our daily lives, this is a highly adaptive illusion produced by our simulation to help us make sense of the world.

The fundamental computational space of the actual universe in which we exist is not a dimensional space at all. It's neither dark nor light, it's neither large nor small. It has no extent, location, scale, or orientation. It's pure non-dimensional existence in the computational space of the quantum vacuum. Within this space is computed all the information that our simulation convinces us is a bright earthly world and universe centered on us.

The dimensional space of our familiar experience is an enormously complex illusion ultimately computed at the level of elementary particle interactions whose conservation generates dimensional relationships among those particles. These events in turn form vast networks of dimensional relationships that are interpreted as spacetime fragments at the classical level.

Our simulation continually stores and correlates the dimensional fragments we participate in through the particle events of our senses and from them constructs a mental model of a fixed, pre-existing, 3-dimensional space within which events seem to occur. Science then adopts the underlying logico-mathematical structure of this fixed space as the basis of its concept of spacetime. But by not understanding the whole picture of how dimensional spacetime is generated by elemental events problems of consistency arise between quantum theory and general relativity as explained in previous chapters.

Thus when we look out into the apparently dimensional space of the world around us we are actually looking at an illusion in our simulation. Just as the world we see happening on our TV screen actually consists only of digital information, so too does the world we see with our eyes. Ultimately dimensional space doesn't exist except as data. The non-dimensional space we seem to enter in meditation as the data of our thoughts and feelings pass through our consciousness is a much more accurate picture of the true computational nature of space.

REALIZATION OF INFORMATION

We know that everything we see happening in the world around us is actually information happening in our brains. Thus there must be a way to actually experience the information nature of all the apparently physical things of the world. There is and it's very easy to do once we get the hang of it. And we then actually do experience the true immanent information nature of the seemingly physical things of the world.

Universal Reality reveals that the physical world as we experience it is an illusion produced by our mind's simulation of reality. This can be realized first by understanding it intellectually, and then by applying this understanding to the individual things of the world and experiencing them as the information and running programs they actually are. This is a fairly straightforward process that can be applied to anything at all including ourselves.

Once we understand that the whole apparently physical world of our experience consists entirely of its information in the neural circuits of our brains it's clear it has no actual physicality at all. The only actual correspondence between external reality and our internal simulation of it is the logical correspondence that enables us to function within the external world by processing the information of its representation in our simulation. But this logical correspondence is itself information, thus there is no reason not to believe the actual external world consists entirely of information just as our internal simulation of it does. How else could it be encoded as such a convincingly real physical world if it itself didn't also consist entirely of information?

For the human mind to be able to consistently simulate the world as information in our brain, and for science to best describe the universe as logico-mathematical structures, the actual universe itself must also with near certainty be a logico-mathematical structure, a running program consisting of information only.

Therefore it can be confirmed with near certainty that everything in the universe consists only of its data, and this can also be directly confirmed and experienced by analyzing anything at all into its information content beyond which there is nothing left.

It is interesting to note the strong correspondence of this view with the ancient Indian and Buddhist concept of the 'emptiness of forms', this emptiness being the active agent of their existence (Wikipedia, Heart

Sutra). Thus this core principle of ancient philosophy naturally integrates into the theory of Universal Reality in which the universe consists entirely of pure information forms given being by the active immanence of their existence.

To realize the information nature of things directly we need only to consider any thing at all and mentally deconstruct it into every last aspect of what makes it what it seems to be. It soon becomes clear that what makes things appear to be physical objects is simply collocated associations of various types of information such as color, hardness, heft and form, and if we were to discard that information piece by piece there would be nothing at all left of anything but the immanent emptiness of its existence.

Consider the stone by the side of the road. We can easily mentally deconstruct it into what makes it appear to be a physical object. Its visual color and texture are information encoded in our brains about how our eyes and visual systems perceive it. Its hardness and texture are the information of how our muscles and fingers interact with it. Its odor, if any, is the information of how our olfactory system interprets it, and the sound when we strike it is the information of how our auditory system perceives the resulting sound waves entering our ears. Discard all this information and there is nothing actually left of the stone. Thus the stone is the set of all its information and that's all it is.

This is the classical level stone as it appears in the world around us. At this level it's clearly the set of all the information of what it is combined into the semblance of a physical object in our simulation. What our simulation tells us is a physical or material object is a data structure with information of color, shape, weight, texture, and perhaps use and meaning. What we call physical things are spatially collocated sets of specific types of information. They are information sets that our simulation labels as physical things.

So it's quite clear, and quite easy to realize with a little practice, that all the stones and other inanimate objects of our experience are only collocations of specific types of information that our simulation interprets as physical objects. Their apparent physicality is simply an information label added to the information of a thing so our simulation can make better sense of our environment by categorizing information in useful ways. But all such categories are more information on top of information and finally everything in our simulation consists only of data.

This is the realization of the true nature of the stone as we experience it in our simulation, and there is every reason to think that the true nature of the stone in the external world also consists entirely of information since our extremely convincing mental representation does. What happens is that our program, consisting only of information itself, interacts computationally with the program of the stone, also consisting entirely of information, to generate the information of our interaction with it, and that information is then encoded in our simulation and interpreted as the physical stone of our experience.

This is the analysis at the classical level of our simulation but we could still argue that the real stone actually consists of all its elementary particles and is still a physical object in that respect. This cannot be directly realized as it's below the level of our perception but it is clear that in Universal Reality all elementary particles and particle components are also demonstrably data.

Of course every inanimate object is actually the running program that continually generates and updates its information, though in the case of the stone the changes that program effects in the information of the stone typically occur very slowly on a human time scale. We can directly realize the programmatic nature of things by simply analyzing things into their data and watching that data computationally evolve.

Consider a housefly. The fly is clearly a very active program that generates continual changes in the information that it is. It's a little biological robot with a robust computational system capable of highly intelligent (relative to randomness) decision-making in accordance with the objectives of its instinctual imperatives. It samples relevant information from the information of its environment, and computes effective actions to feed, avoid damage, and reproduce. And these systems are all supported by an enormously complex integrated hierarchical program down through the subprograms of every cell in its body and the particle interactions that power them.

It is this complete information program of the fly that actually is the fly. Like the stone, its apparently physicality reduces to the computational interaction of our information with its information. But we can clearly experience the fly as an intelligent running program that generates the information our program interacts with to generate the information of our experience of the fly in our simulation.

Thus the fly, the stone and all the individual things of the world can actually be experienced as the running programs they are as experienced by our own running program computationally interacting with them. To make better sense of the data of reality our simulation represents it as the familiar physical world populated by inanimate objects and living beings all neatly filed into categories meaningful to our functioning.

But when we actually look at the world with opened eyes we see that every bit of it consists only of the information of what it is. That information of things is all we ever experience of them, and is all that can be experienced, and when this is understood and directly realized the true nature of the world we live in is revealed. Nothing can be experienced other than information. Nothing other than information and its immanence can possibly be experienced, thus all we experience ultimately consists only of information. And because the world can be so convincingly represented only as information in our simulation we can assume with overwhelming certainty it actually consists entirely of information.

This realization changes nothing about the world. The world remains as it always was, but now its true nature as the information of everything that it is, and the information of the running programs that are continually computing that information is realized. Now as we look out into the world and into our self it becomes clear to the realized mind that all is information only. We actually see the world as the information it is being continually computed by all the programs of the observable universe including our own. Everything that exists is the running program of itself continually recomputing its information and these are all the real things of the world because of the immanence of the existence in which they happen.

REALIZATION OF INFORMATION HISTORY

Everything is the information of what it is, and that information is the cumulative result of all the computational interactions its program has been involved in throughout its history and beyond. Thus the information that things are is a recording of their entire computational histories. All things are information only, and that information is the current state of their complete information history back to the beginning of the universe in the big bang. Everything in the observable universe is the information of its history. Everything is its information history and that is the information of what it is right now.

This is something that can be easily realized in the information of each particular thing though the complete information history details of things are enormously complex and only partially revealed even by science.

Thus when we look at the leaf on the lawn in Autumn, we realize the true nature of the leaf is the information of what it is, but that information is much richer that its immediate appearance, because its exact location on the lawn is the result of an enormously complex interplay of information programs that computed it into reality. The exact size and aerodynamic shape of the leaf in combination with the exact breezes that brought it to this precise location, and the exact information of the chemistry that loosed it from its twig at the exact moment those breezes were blowing all interacted computationally to bring it to the exact position it lies in now.

And the moment of separation, and its shape and weight, are the computational results of millions of years of evolution of the species it belongs to which in turn are the computational results of uncountable program interactions that can never be fully known. And the DNA content of that leaf responsible for the general plan of how the tree it came from grew and developed and produced that particular leaf in the particular location it fell from are also essential components of the computations which resulted in the leaf as we see it at this moment on the lawn. And further back the acorn that fell in the exact spot from its parent tree, and the lineage of all the acorns back through the entire history of the species, all must have been computed exactly with not the slightest difference for this leaf to lie on this lawn at this very moment. And all this is present in the leaf as it lies on the lawn.

This is the realization that everything consists only of its entire computational history back to the beginning of time to the original fine-tuning which opened and constrained the possibilities of every individual thing in the observable universe as it actually exists right now. The current information state of everything in the observable universe determines the exact uncountable information states of every universal instant of the entire computational edifice of the past that computed the present.

Every computation of every program in the observable universe throughout its entire history is revealed in the exact information state of the entire universe in this exact present moment. It is all there waiting to be realized though most is far beyond our understanding. But everything that is there, every last bit of its information, is the true recording of past events and lies waiting to be realized.

This is a realization we can apply to anything and everything in the world around us and to everything in the entire universe including our self. Though we can never know anywhere near all the details of all the computations that took place among all the programs in the history of the universe, they are all recorded right here right now in the exact details of the information of things as they actually are. This is a profound realization that can completely open us to the incredible awesome beauty and meaningfulness of our universe. And we can look at anything without exception and realize it from this perspective in the information that it is.

And with this realization we also realize the entire observable universe as it exists right now in this exact present moment down to its finest possible detail means that the entire past in every last detail through every moment of time had to have been exactly as it was without the slightest possibility of any difference whatsoever. Thus the original complete fine-tuning and every other minute detail of every microsecond of the entire past of the entire observable universe could not have been different in the slightest possible detail than it was. The past is exact and immutable, the future is probabilistic and the present is the process that computes an exact past from the possibilities of the future.

REALIZATION OF IMMANENCE

Universe Reality reveals that every individual thing that exists in the universe consists only of its information, but what makes that information a real thing in the actual observable universe is its existence as a data form within the medium of existence, the quantum vacuum that is the originally formless substrate of the universe.

This medium of existence is already being experienced by all of us all the time as the immanent reality of all the things of the world. If the information of things didn't exist in the medium of existence they simply wouldn't exist and would have neither being nor observability. Therefore the realization of existence is already with us as our experience of the actuality of the universe and all the things that populate it. It's just a matter of waking up and realizing this.

We experience the immanence of existence all the time but we are not aware of what we are actually experiencing because we take everything for granted without realize the true immanent nature of its existence. Only things that have existence can be experienced. Thus we never have any non-existence to compare the presence of existence with to make the immanence of existence really pop into consciousness. Yet all the while it's the precisely the immanence of things that manifests as our consciousness of them.

Immanence is in one respect a simple realization. It's the simple fact that things are actually there in the here and now of the present moment. But its complete realization is subtle. Traditional science and materialistic philosophy speak of existent things but ignore the problem of what their existence really is. Universal Reality answers that the existence of things is the fact that their information forms are forms *of* existence that exist *in* the otherwise formless sea of existence. This medium of existence is the single medium of the universe and is the universe. Everything in the universe is a form *of* existence *within* the universal medium of existence.

This realization is central to Universal Reality. In Universal Reality the quantum vacuum is identified as the universal sea of existence within which the observable universe of programs runs computing their data. Data appears as forms in the quantum vacuum as water waves appear as forms within an otherwise formless ocean. The information forms and programs of the universe can exist only within the quantum vacuum of existence because that is the common 'substance' of all things. It's the only locus of existence and the single substrate or medium in which the forms of things can appear and exist.

As the possible forms of water waves are determined by the nature of water, so the possible information forms of the universe are determined by the intrinsic nature of the quantum vacuum which is the virtual complete fine-tuning of the universal medium of existence in which the programs of the universe run.

Thus the fundamental realization is the experience of the immanence of existence, both the immanence of the presence of the formless universal sea of existence within which all forms exist, and the immanence of existence manifesting in every individual form. Every form, no matter how mundane, continuously manifests the immanence of its existence that makes its information real and present in the present moment. No longer is the universe a dark dead empty material space, but a living happening presence that actively self-manifests its existence in all the information forms of the world including ourselves. And the inner light of the immanence of existence of all things manifests in our consciousness of them.

The living presence of existence continuously glows and flows with the immanence of its being within all things giving them their actual presence, life, and happening. We too exist entirely within this living sea of existence, which gives us our life, our presence in reality, and all the wonderful manifestations of the running program that we are, and which we directly experience as our inner true self if we only stop and realize it.

So the direct experience of this living immanence of existence in all things is central to realization. When immanence is truly realized it's an amazing transformative experience and the world we exist within will never be the same. We become our running information program floating in the immanent sea of existence, and we experience the living existence of the universe glowing and flowing around us and within us giving life and being to the information of ourselves.

The presence of a universal sea of existence within which all information forms and programs exist and acquire their reality is completely different from the traditional materialistic view of the universe. In the old materialistic view the universe is completely empty between instances of particulate matter. Only with the discovery of the quantum vacuum has this old view begun to change and the fact that the vacuum itself is not an empty nothingness but the source of all existence begun to be recognized.

The realization of immanence tends to arise naturally with the realization of information. When things are fully recognized as only their information then the immanence of that information naturally shines forth. Imagine all the information of a thing vanishing and experience all that remains. That is the immanent existence that made that information real.

The immanence of the existence of all things now begins to become clear. Suddenly we realize that if all the information of the world suddenly vanished what remains is the formless sea of existence itself in which that information appeared and became real, present and actual. That real, present, and actual absoluteness of formless existence is always there within all information forms including our selves. It is the formless sea of existence in which all things exist and we directly experience it in all the things of the world as the consciousness of those things. This is the fundamental experience of reality and this is its realization.

The concepts of Tao and Śūnyatā were ancient approaches to this realization (Legge, 2010). These were both names for the original formless substrate of reality in which all forms appeared, and the 'emptiness' of all the forms that appeared was a recognized ancient philosophical concept. Thus Universal Reality seamlessly integrates these ancient concepts into its modern Theory of Everything and gives them a rational scientific basis.

There are differences of course. Taoism proposes an initial separation of the formless Tao into the fundamental forms of positive and negative, and all other forms arise from combinations of these two, as outlined for example in the hexagrams of the 'I Ching' (Wilhelm, 1962). In contrast in Universal Reality, the fundamental forms that arise from the formless quantum vacuum are those of the particle components, and the rules that govern them, and all the other aspects of the complete fine-tuning. However the initial concept of formlessness from which all forms arise is very similar.

There are various techniques of meditation, and direct insight, which enable the realization of the pure formless immanence of existence. By the mental exercise of meditation one greatly reduces the appearance of forms in consciousness and more easily realizes the underlying field of immanence in which forms appear that remains as the field of consciousness itself. The experience of formless immanence as consciousness is critical, but forms must be dealt with in daily life so it's also essential to realize the immanence manifested by forms in their individual existence. The realization of the immanence of both forms and formlessness is essential to the full realization of the immanence of existence and its experience as consciousness itself.

Existence, the quantum vacuum, is the universe itself and is the fundamental reality. It's the absolute formless substrate of being in which all the information programs and forms of the universe appear and become real, actual and present. And it's the dynamic, living happening in which all the information programs of the universe compute and evolve according to the innate fine-tuning rules of the quantum vacuum of existence.

Every one of us experiences this at every moment of our lives as our own reality, life, actions, and consciousness. It's simply a matter of realizing what we are already experiencing. We are not material objects *in* a physical universe, we are integral aspects *of* a universe of immanent existence and the fundamental processes of this universe are active within us in every moment and *are* our very existence. The universe continually computes our existence in the universal sea of immanent existence.

REALIZATION OF CONSCIOUSNESS

Consciousness itself is simply the immanent presence of reality itself. It's the here now living presence of the immanence of existence that radiates within the forms of all actual things. The essential component of human consciousness is common to the existence of everything in the universe. Human consciousness is simply the presence of immanent reality within the running program of our simulation.

Once the immanence of existence in all things is realized, the true nature of consciousness becomes clear. Consciousness is not something generated in our brains and shown out onto the things of the world like a spotlight, it's the immanent self-manifesting existence of those things into reality, or more precisely the immanent self-manifesting existence of our internal simulations of those things.

The information forms of our simulation of reality, like all forms in the universe, manifest the immanence of existence. It's the immanence of special forms encoding the fact we are experiencing forms representing other forms that we experience as consciousness. All the forms of our brain's simulation of reality manifest the immanence of their reality, but that reality remains unconscious until the special forms of the focus of consciousness encode the fact that other forms are being experienced. It's the immanence of these recursive forms that manifests as consciousness.

Forms only manifest their immanence through their actual forms. Thus for immanence to manifest as conscious experience its form must encode an experience of a thing rather than just a thing itself. It must encode the information of a thing being experienced.

This top-level brain function is what is normally called consciousness, but all the forms of the simulation share the essential ingredient of consciousness because all forms share the immanence of existence. It's just that this immanence is only recognized in a reportable form by other forms specific to that purpose. Thus everything in the universe shares immanence, which is the essence of consciousness, but only specifically designed forms that monitor other forms being experienced are able to report the presence of those forms as the consciousness of them. This is the understanding of the consciousness of humans and other species.

Our consciousness is a mixture of consciousness of the external world and of our interactions with it. Ultimately we have no direct experiences of the individual things of the external world but only of our internal simulations of them. However to the extent our internal representations are consistent with the logic of the actual world we have direct knowledge, though not experience, of the external world. And of course the immanence of the forms of our simulation is the same immanence of all the external information forms of the universe. This explains why our simulations of things seem like real things, they seem like real things because they share the immanence of real things.

So our consciousness of things is actually our consciousness of our brain's representations of them. Our consciousness of a fox is the immanence of a relatively very concise representation whereas the immanence of the information of the actual fox is the actual living fox and consists of its entire actual running information program down to the complete cellular and elementary particle level. So we are conscious of our encoded experiences and thoughts of external things rather than the entirety of the things themselves.

For the true nature of consciousness to be realized a clear distinction must be made between consciousness itself and the contents of consciousness. The *fact of consciousness itself* is due entirely to the self-manifesting immanent presence of all information forms and is thus a basic attribute of reality itself. It's generated by the immanent existence of things rather than something being produced by human brains.

The essential active ingredient of consciousness exists in everything in the universe in the immanence of its existence that gives it reality. But for immanence to manifest as consciousness in a biological entity, that being must have the necessary cognitive structures to register immanence as immanence, to register it as consciousness. The specific forms of the contents of consciousness depend on the perceptual and cognitive structures of the biological entity but the fact of consciousness itself, that those contents are conscious, is due to their immanent existence.

Consciousness in the simulation is exactly analogous to existence in the external world. The immanence of existence makes things actually real in the real actual world. Likewise the immanence of existence makes things consciousness in the simulation. The immanence of existence makes the specific forms in which it manifests real things in the world. The forms of the things in the world given existence become the real things of the world. The forms of experiences in the simulation given existence become the real experiences of consciousness. The exact same process is at work making all forms the actual real things they represents, be that the forms of things or the forms of experiences.

When information forms appear within the immanent existence of the external world they become the real things of the world; when information forms appear within the immanent existence of our simulation they become our consciousness of our mind's representations of the things of the world. It's exactly the same fundamental process of the universe working both inside and outside our brains. It's our mind's participation in the fundamental process of the immanent existence of the entire universe.

Thus the realization of the true nature of consciousness is that consciousness itself is the immanent existence of the forms of things that makes them real and actual in the observable universe at work in the forms of our simulation. Consciousness itself is the immanence of existence. It's the continual happening of the immanent existence of the universe manifesting within us as it manifests in all things.

REALIZATION OF TRUE NATURE

In the last analysis all that can actually be demonstrably confirmed to exist is experience itself. Ultimately the existence of every last thing in the world can be confirmed only through an experience of it. Thus all the things of the world including even the theory of relativity or Universal Reality can be known only through experiences of them.

Of course it's quite reasonable to assume an objective model of reality in which things exist even when they are aren't being experienced or even if they've never been experienced but ultimately all such models themselves exist only as their subjective experiences. So in the end everything that manifests existence does so as experience.

We assume that all experienced experience is 'our' experience but there are fundamental problems with this because experience is primal and fundamental and happens antecedent to the construction of a self and not self in the simulation. Experience happens and only then is its information categorized and those categorizations experienced.

Thus experience just is, and the contents of experiences are then organized, categorized and stored in the simulation including the apparent distinction between experiences of ourselves and of other things. However every aspect of that entire process also consists entirely of the successive experiences of it.

Thus only experience itself, no matter what its content, is primal, original and fundamental. Experience is all that ever demonstrably occurs. We can assume that other beings exist that are also having experiences but those experiences never exist in 'our' experience and can never be subjectively confirmed.

Since all that is ever experienced is experience itself, no matter what its content, the true direct nature of reality must be experience itself. We can define the experience that occurs as 'our' experience but we must be careful to understand what this really means because all experience is prior to the distinction of self and not self.

Thus ultimately there is only direct experience and nothing demonstrably exists except as an experience of it. Every model of reality existing independent of its experience exists only as an experience itself

Thus experience itself is the true fundamental immediate nature of reality because it's all that demonstrably occurs. We can then say that our true nature is experience itself but to do so we must assume an object self that has a subjective nature that has the experiences we experience but the nature of experiences is they are antecedent to any categorization into the experienced the experiencer.

Thus the best we can say is that all that demonstrably exists is experience itself, the experience of various information forms. We can certainly meaningfully categorize experiences into self and not self, and experiencer and experienced but ultimately all such categorizations are themselves experiences and there is nothing that appears in reality that does not appear as an experience. Thus we can reasonably say that the true nature of reality is experience and our true self consists entirely of experience but these concepts also exist only their experience.

Ultimately experiences occur and their information content is categorized into our simulations of reality but every bit of this whole process is knowable only as its experience.

We can go with this and generalize it as we did in the chapter on Consciousness as *xperience*. In this model all processes are programs that continually recompute their information and the update of any information form is an xperience of that update. Thus all the processes of the observable universe are effectively generic observers and the universe consists entirely of the xperience of generic observers. In this way the universe continually xperiences itself into existence and the observer becomes an essential aspect of reality itself.

Thus xperience is the inverse view of immanence. Immanence is the self-manifestation of being within all forms as from the external perspective, and xperience is the self-manifestation of existence from an interior perspective.

Thus everything in the observable universe manifests the essential active ingredient of consciousness and it only requires specialized recursive forms encoding that an experience is occurring in a simulation to manifest as the consciousness of that experience.

Thus loosely speaking we can say 'our' true nature consists of the totality of the experience 'we' experience and more accurately that 'our' true nature consists of all the experience that's experienced.

This is the realization of true nature so our true immediate self is not our physical body or even our running program but only experience as it's experienced. Everything in the observable universe exists only as experiences of it. Thus our true nature is all experienced experience as it occurs.

Ultimately only experiences can be experienced. Direct experience is ultimately all that exists. At the most fundamental and immediate level all that exists is experience. Only in the subsequent categorization of the information of raw sensory experiences does our simulation tell us that there is a self that 'has' these experiences.

But even all these subsequent organized information structures of our simulation, from the simplest to the most complex Theory of Everything, are again experienced only as experiences, the *experiences* of those information structures. So ultimately, everything reduces to the experience of what it is which is the immanent manifestation of the here now existence of its information form. Ultimately all that exists is information forms that appear and disappear, and their existence manifests as experience.

Though this realization is not the typical state of organismic consciousness it's the natural state of inanimate unconscious programs which always exist as pure raw experience itself devoid of any context or categorization or assignment to an individual self or any categorization at all. Forms always exist only as the information of what they are, and only forms of consciousness include the context of an experience as an experience.

To a conscious mind, consciousness and the experiences that happen within it fill the entire domain of reality. Thus consciousness and reality are one and conscious experience is the true self. To the experienced forms of consciousness all is consciousness only because all is experience only. To the opened mind consciousness and reality are one and the same and always awesomely real, present and absolute.

This is the realization of 'our' true nature, which is not really ours, and by extension the true nature of the observable universe. Everything without exception is immanent experience. Some of this is the subsequent immanent experience of categorizations of the original raw immanent sensory experience.

Thus our true nature, a better term than true self, consists only of all the experience that appears in existence before it fades from existence. And whatever form the experience that becomes into existence takes, that's 'our' true nature in the present moment. Our true nature is all the experience that appears within existence in the present moment. Our true nature is that of the universe itself.

Our course we can logically assume that other observers exist (we experience them existing) and that they also experience experiences we do not, but ultimately this assumption is also another experience. Thus we can logically assume that the universe contains a vast number of experiences that 'we' are not experiencing but other beings are, but 'we' have no direct experience of this.

Thus our 'true self' can be identified with the local domain of actual experiences. At the fundamental level 'we' are the totality of experience in the present moment. This includes all experience whatsoever whatever its information content. Experience is the only possible manifestation of our existence; therefore it is our true nature.

At the fundamental level of experience there is no individual self that experiences self and not-self forms, there is no dualism between experiencer and experienced, there is only raw experience itself. The individual self is an adaptive but illusory subsequent categorization of experience in our simulation of reality. At the immediate primal level of existence there is only experience itself.

Thus if we want a notion of a 'true self' it actually consists of the abandonment of any notion of personal self whatsoever, and consists only of all experience without exception, and nothing other than all experience. This is the mind of Buddha existing as pure experience in a formless world through which forms pass as experience manifesting the immanence of existence.

This realization is what the Diamond Sutra calls 'Awakening the mind while dwelling nowhere' (Suzuki, 1956). There is no individual locus to consciousness because consciousness simultaneously pervades the entirety of the reality of all experience rather than being focused on any particular individual forms. Instead there is a total mental openness to everything as forms flow freely through consciousness and existence itself as experience.

In this state all forms exist as pure raw uncategorized experience, antecedent to any notion of self or not-self or anything at all. They are just pure raw forms of experience with no immediate meaning or reference because they are not being categorized and structured in the simulation. They are exactly and only the information of their selves as opposed to being interpreted or related to anything else at all. And if they are interpreted and related those information forms are again only the experiences of that information. All is experience only, no matter what the information content of that experience.

There is no self-center to being with this realization because being pervades all experience. There is no center to being because this is an experience antecedent to the imposition of dimensionality and thus without dimension (Wilhelm, 1931). This is the raw manifestation of the immanence of existence as it happens in the present moment.

So just experience your raw primal experience as conscious experience itself and ignore its information content concentrating only on the experience. Then discard the you experiencing it and just leave the pure unmediated experience and that is the experience and realization of the true nature of reality. It's the direct experience of the immanence of existence itself that's the living essence of all that exists.

REALIZATION OF CHI & ENERGY BODY

It's easy to experience the current feeling we associate with each individual part of our body from the inside. We feel our arms, feet, hands, and other parts of our body as the internal feelings of them. This is more easily accomplished lying peacefully and still with eyes closed but can be done in any situation with a little practice.

Now combine the feelings of all parts of the body so it can be experienced as a single *energy body*, which is simply the feeling of one's entire body from the inside. This energy body is a simple straightforward experience we all have all the time if we just pay attention to it. There is nothing at all esoteric or metaphysical to it or implied.

Though we are often aware of the feelings in individual parts of our bodies in our daily lives, for some strange reason Westerners in particular are resistant to making the leap to experience the total internal feeling of our whole bodies as a single energy body. But that is exactly what our own direct experience of our bodies actually is. The feelings within the energy body are simply the experience of the active immanence of existence within us.

It's useful to have a term for the active immanence of existence within us and we may reasonably identify this with the oriental concept of 'chi' if we are careful not to include any of the many irrational and exaggerated claims often associated with it. In this usage chi is simply our active life force, and that of all beings, the immanence of their existences, and every one of us experiences it all the time as the energy of the internal feelings of our bodies.

Chi is simply a useful term for existence when it occurs within a biological organism such as our selves. Chi is the same immanent existence that makes all things in the universe actively real and actual. We just feel it directly in our own selves as a particular thing and thus call it chi.

While it is easily demonstrated that our *experience* of the flows of chi within the body are subject to some control by mind, breath and movement, one needs to take all the many claims about chi with a big grain of salt and always subject to experimental confirmation. There is quite a bit of evidence that one can improve one's general health by freeing up the flow of chi and by changing the tone of chi to feelings of love and well being flooding one's being rather than hostility, anger, hate, resentment, depression or stress.

However such benefits are limited because all processes have their own chi, not just our own. All things including bacteria, viruses, and attackers intent on harm, dangerous natural forces as well, all have their own chi energies, and so one's own chi is never a magically effective force against all harm. One needs to deal with the real actual energies of all processes and avoid or redirect them as best one can rather than assuming that just by strengthening one's own chi one can always prevail.

In a larger sense, all the experiences our simulations categorize as *of* external objects, our sensations, conceptions, thoughts and feelings about things, are in fact part of 'our' self, because every one of them occurs 'within' our body. Thus true self consists not of a mind within a body, as our simulation represents us, but of the totality of experienced experience including the experience of not-self things.

In this sense our minds are coterminous with the totality of our experienced universe. When we look out into what appears to be an external universe, we are actually looking into the depths of our own minds and experiencing ourselves in our experience of the external world. Reality is a two-sided mirror and 'we' are on both sides experiencing ourselves in our experiences of all the other things of the universe.

This is the realization of the retinal sky. The sky we see, and everything under it, is actually our internal representation of the information of an external sky as sampled by our retinas and constructed by our minds. There is more of us in our experience of the external world than there is of the external world itself. True self consists of all experience without exception.

Experience is not static; it's the computed results of the running program that we are continually evolving in the present moment as happening occurs. The realization of true self is the understanding and direct recognition of ourselves as the experiences continually generated by this running program. Though information only, it's the immanent existence of this information that makes our lives so wonderfully rich and filled with sensations, feelings, meanings and memories. Nothing changes with this realization, everything remains as it was; we just now have a much deeper and richer realization of what the reality of true self always was.

In any case the energy body is our true direct experience of 'our' selves. It includes all the internal feelings of our existence including the internal feelings of all parts of our body as a single body, and it also includes all our thoughts, feelings, and meaning feelings as an integrated part of that energy body. From this perspective 'we' are the total unified body of our internal feelings much more than we are the mental construct of our physical body.

And it's important to note the energy body is not completely conterminous with how our mind represents the boundaries of our physical body. All our feelings and perceptions of the 'external' world are in fact part of our own energy bodies. The touch of something our mind tells us is external is actually a feeling in our own energy body, and likewise our entire view of the external world is in fact actually within the energy body.

Thus what our mind tells us is our 'physical' body is actually a mental construct that mind then locates approximately contiguously with our energy body based on the relative locations of how the various parts of our energy body feel.

As an aside, this easily explains how 'out of body' experiences occur. We simply have to understand that our simulation usually locates 'us' in the same dimensional location of its model of our physical body, but since our mind does that arbitrarily as a matter of adaptive convenience, it's then easy to understand that in times of extreme immediate threat mind can just as easily relocate our selves out of our bodies in a attempt to lessen potential trauma.

DEFINING GOD

Having discussed realization from a more objective viewpoint we now consider how some concepts of personal myth may aid in realization. Personal myths can assist in a more personal relationship with reality, and they can be quite useful so long as they are understood as myth rather than objective truth.

In Universal Reality there is no necessity of a God. The universe works quite well on its own, and certainly needs no external supernatural agency to design or run it, nor a creator since it has 'always' existed. However the notion of God has a very wide traditional appeal and for those in the monotheistic tradition there is a simple and quite reasonable and scientific way to integrate God into the theory if desired.

All that needs to be done is identify God with existence, with the universe itself, or at least the motive force of the happening of the universe. We then have a God which creates the universe of forms, is the source of the laws governing the evolution of those forms, and which sustains, directs and generates its evolution. This God is also the immanent living essence of all things that gives them being. There are obvious similarities to the gnostic and mystical traditions of the Abrahamic religions.

By this definition God even maintains the traditional attributes of divinity. God is certainly omnipotent as the happening of existence is the source of everything that happens. It is omnipresent as it's present in every detail of the entire universe, and in a sense it's omniscient as knowledge consists of information, and this God of the quantum vacuum is the source of all information, in fact since it consists entirely of information, the universe can be thought of as the knowledge of itself, the knowing of itself, as the running program of the mind of God.

And if anything is divine and miraculous, it is certainly the universe itself and the immanent existence that animates it. The universe itself is certainly the proper subject of our awe and reverence and devotion. And the existence of the universe as it is including our own personal existence is certainly the ultimate miracle.

It seems to me that if we want a God, reality itself is the only reasonable and scientifically acceptable definition of God. It also has several very important and obvious advantages. First there can be no doubt that God exists since it's self evident that reality exists. And second the attributes of God then become merely a matter of scientific discovery. Third, this definition of God is non-sectarian and non-divisive, and should be equally acceptable to anyone with an open science oriented mind.

Most of the interminable and often violent arguments over whether or not God exists, and if 'he' exists what 'his' nature is are immediately resolved using this definition, and the way is clear forward to determine the rest through the application of scientific method.

However it's critically important not to bring along the huge burden of non-scientific mythology that clutters the Abrahamic traditions. These are a mix of ancient historically based tales with perhaps the best, or at least the most convenient theories of reality the pre-scientific authors could come up with and should be appreciated from that perspective, but believing in them as a matter of faith in this day and age is delusional and dangerous.

Therefore, though it's not a necessary part of the theory, the identification of God with existence itself can lead naturally to a more personal and spiritual relationship with reality and thus aid in our appreciation of the awesome wonder of reality. We may obtain a more personal relationship with the universe by identifying God with existence or the universe. From this perspective God is reality itself and the active happening that animates all things and gives them existence according to their forms.

Thus we realize our own true nature as that of God. If God is the immanence of existence then God lives and breathes within us all and only waits for our realization of its presence to appear. And God's divinity is our own true nature as well so that one can now truly say that God dwells within us, that we are God.

By this definition God manifests personally within all of us as life and consciousness and is the true self of all personal beings. By this definition we, and all things, participate in and manifest the immanent divinity of reality.

Defining God as the universe is just a way of conceptualizing and relating to reality in a more personal manner. As such it is can be a useful form of personal myth. Personal myths can be useful aids and comforts and are not inconsistent with reality so long as they are realized as myth and not confused with objective reality. Only when they are mistaken for reality do they become delusional and hamper realization. Otherwise, recognized as myth, they can be perfectly consistent with reality and even aid in its realization.

From this perspective God, being the existence within all things, looks at us through every eye and looks out of our eyes at the world as well. And God sees itself looking back at itself looking at itself in recognition of itself. In this way God recognizes and knows itself and the reality of the universe and we and all organisms become the sense and knowledge organs of God that allow God the universe to experience and know itself. We realize ourselves as the consciousness of God within us as God is the active living essence of all things including ourselves.

This is true not just of looking and seeing but of the experiences of all our senses and our consciousness as well. All the organisms in the universe are the means through which the universe as God manifesting in those individual forms becomes able to experience and know itself and thus begins to become more self-aware. We, and all beings, are the individual distributed sense organs and minds of God through which God knows and experiences itself and the universe gains self-awareness.

Properly understood there is nothing supernatural about this realization. God as the active existence of the universe exists in every form but is only expressed through the actual form of that form. God sees only out of forms with eyes and cannot see out of forms without eyes but since all forms experience other forms in their interactions with them, God experiences through all forms and is experience itself, but only in whatever form that experience happens in. This is entirely consistent with Universal Reality if we define God as the existence of the universe.

Because experience is the self-manifestation of reality, God can be said to create and self-manifest itself as the experiences of all things. This is the universe experiencing, and in some forms knowing itself, and this is how God manifests and knows itself and becomes self-aware.

Thus the universe and God is its own self-awareness of itself self-manifesting as experience. In a fundamental sense it is not even clear we can meaningfully speak of the existence of a world of information forms absent its experience of itself because there is no way to confirm its existence or structure if it does not self-manifest and observe itself.

Thus reality is reality experiencing itself. And all of us and all organisms and all things and forms are part of this process of the self-realization of the universe and thus the self-realization of God as experience.

This notion of God has both a non-personal ubiquitous aspect and innumerable personal manifestations as the immanence of all individual things and personal beings. We may sense the presence of God in the meadow but God remains unseen and formless other than in the actual forms in the meadow that God is actually manifesting as. We may realize the presence of God in the form of every being and thing but God is always limited by the forms within which it manifests. God's presence is felt in everything around us as the immanence of existence, but God never appears except as it manifests in the actual forms of the world as the immanence of their existence.

We may long for God to appear as a personal caring and protective being in full divinity with supernatural attributes but this never occurs because God manifests only in actual forms and all actual forms are natural and obey natural law. But we can take comfort that the actual reality of this universal God is enormously more profound than any traditional supernatural being.

This is an entirely rational view of God insofar as it goes but one must always be wary of the danger of imputing any of the traditional delusional supernatural characteristics of the Gods of traditional religions to this God. This God is more akin to the rational scientific God of Einstein, and is simply another name for the immanent self-manifestation of existence in which the universe of information forms arises (Wikipedia, Religious views of Albert Einstein).

The complete fine-tuning of the universe is such as to allow realization of its true nature that it may itself realize its true nature and divinity through us. And that's equally true of all of us and of all life forms that exist or have existed to the limit of the capabilities of their forms. We are all bound together in the web of universal experience and consciousness through which the universe knows itself. May that be an enlightened and compassionate experience!

The direct experience of reality itself as consciousness itself is the living presence of this God in a non-personalized form. It is waking into a world where the presence of God is tangible as immanence but being formless remains unseen. But then some person or even some animal opens its eyes and looks at you and God suddenly manifests in that personal form looking out through those eyes. And all the while you were looking in vain for God with your own eyes it was actually God that was looking through your eyes searching for itself! God manifests in both personal and impersonal form because there is nothing that is not God, defined as reality, and there are both personal and non-personal forms in the world.

Thus God is the totality of all forms including all of us simultaneously acting as its innumerable sense organs and consciousnesses and the combined experiences of all things of itself. It's all its forms experiencing itself and thus continuously self-manifesting its immanent formless nature to itself in the present moment of its presence.

In this way Universal Reality explains the gnostic and mystical experiences of the Christian tradition as the direct immediate experience of the immanence of the divine nature of the existence of all things (Wikipedia, Gnosticism). For the Christian mystics this often manifested as the direct experiences of the immanence of existence especially in their Christian symbols but for saints like St. Francis of Assisi he seems to have tuned in to the immanence of animals as well. Thus the gnostic and mystical traditions are based in the fundamental nature of reality even though their individual symbolism and interpretation are most often delusional (Wikipedia, Mysticism).

In this view the universe and everything in it is the living presence of God. There is nothing that isn't God. Every part of the universe is part of the miracle of God's existence and science is the study of God, of the miraculous nature of God, of the miracle of God's existence, which is the existence of the universe.

Sitting inside the quantum vacuum as it computes the observable universe within it, here if anywhere we glimpse the mind of God at work creating the universe on the fly in effect continually thinking it into existence. Here is the mind of the universe imagining the world into living existence in all its awesome beauty and majesty, in all the wonderful divinity of its immanent reality.

We all interact with God all the time in our every action. God speaks to us in every event but we understand only a little of what God is telling us. Studying the workings of God the universe is a proper form of prayer. Look to the universe itself for knowledge of the workings of the divine.

BUDDHA NATURE

There is also a natural way to integrate a core concept of Buddhist tradition into Universal Reality as well. The concept of 'Buddha-nature' can be easily identified with the immanence of existence, the universal common 'active ingredient' of all things. But again the usual overlay of religious dogma and superstition that runs through many Buddhist sects must be carefully excluded.

Buddha Nature is a concept from Buddhist philosophy that often draws scorn among Western thinkers, and often for legitimate reasons due to its many unscientific and illogical interpretations (Wikipedia, Buddha-nature). But again when defined rationally in terms of established concepts, Buddha-nature can be a useful aid in understanding and promoting realization because it enables a more personal perspective on abstract concepts such as existence and the quantum vacuum.

In our usage Buddha-nature is simply another name for Chi or existence from a more personal and individual perspective suggesting the possibility of personal realization. Thus the realization of Buddha-nature is another term for the realization of the true nature of things including one's self.

Though most Buddhist schools use a more restrictive definition, limiting Buddha Nature to sentient beings only, in our definition all things have Chi or Buddha Nature because all things are forms that have immanent existence. This definition enables a simpler and more consistent view of reality, as it's just another perspective on what has already been established.

From this perspective realization can be considered the direct awareness or experience of the Buddha-nature of all things as the true fundamental actuality that fills the emptiness of their forms with the reality of being. This is consistent with the views of the more rational and philosophical forms of Buddhism such as Zen (Suzuki, 1956).

In this view all the things and beings of the world share the same presence of immanent existence as their common fundamental nature, and realization is the realization and experience of precisely this. All things share the same existence and this is true no matter whether their forms interact in harmony or conflict with one's own form.

Realization involves seeing the Buddha-nature in all things and beings no matter who or what they are. As another name for reality itself the Buddha lives within the forms of all beings. Buddha bum, Buddha whore, Buddha killer, Buddha next door. All forms are manifestations of Buddha because all forms have Buddha-nature because the fundamental nature of all things is existence and when we realize this we experience all things and beings as Buddhas whether they know they are or not, whether they have attained this realization or not.

This includes all animals and other organisms as well as people. Buddha bear, Buddha fox, Buddha bird, Buddha dog, Buddha cow, Buddha worm, Buddha flower, Buddha bacteria. As chi the Buddha dwells in all beings waiting to be awakened to its true nature. And

Buddha nature is the true nature of every non-living thing as well. Every stone, every drop of water and speck of dust is a form filled with the Buddha Nature of immanent existence. In this view the entire universe of forms consists only of Buddha in its myriad forms.

From this perspective we also have Buddha Nature and are Buddha. Buddha lives within us all and we can consciously choose to realize and express our Buddha nature in a clearer, purer more realized form. We can abandon the unnecessary and unhealthy forms of our personal self and become our Buddhas and move through every aspect of our lives as Buddha. We can be Buddha walking down the street recognized or completely unrecognized through the world of forms. We can choose to let our Buddha guide our actions as we go about our daily lives as Buddha.

By realizing and becoming our Buddha, Buddha guides our actions. By surrendering our personal desires and attachments and prejudices to our Buddha-nature we become our Buddha and let our Buddha guide our actions, our lives, our work, our destiny. We walk down the road as empty forms filled with the living immanence of Buddha being. In any case we are doing that already whether realized or not. It's just a matter of realizing it.

Of course this is all personal myth, a personal perspective on reality, and though certainly a useful aid to realization, we must be careful not to stray too far into fantasy. After all the Buddha within things can express only through the actual forms of those things. There are no super-heroes here. But there is nothing wrong with personal myth so long as it's recognized for what it is and doesn't lead us into delusional thinking but is used to inform and enhance realization.

With that caveat in mind then by becoming the Buddha we already are we become our true realized being in the disguise of our old self moving through the world of forms among other Buddha beings most of whom are ignorant of their Buddha-nature.

Our old personal self was an illusion of internal mental forms programmed into us since childhood. By becoming our Buddha our personal forms are transformed and purified by the flows of purer less mediated chi energy that naturally tends to manifest as a loving healthy life force. We swim like fish through the surrounding sea of living immanence, warm, loving and supporting. As our Buddha we realize we are empty forms within a warm loving sea of chi which continuously fills us with being and reality and we become better able to release and dissolve away all our stagnant unhealthy personal forms and blockages to allow chi to flow more smoothly and strongly and peacefully through us helping keep us vital, fresh and healthy. In this way, as our Buddha, our forms become more pure and balanced and strong.

By becoming our Buddha and living as our Buddha-nature we discard the illusory shells of our old personal being that concealed it from us. We see the world as it is with Buddha's eyes, touch it with Buddha's hands, and manifest Buddha's realization of his own Buddha nature as our true selves. In this way we commune directly with the fundamental nature of reality itself as it self-manifests within us as our Buddha-nature.

REALIZATION OF LOVE

Though chi is the single energy of our existence, it's experienced in many different forms as the information of how our bodies feel from moment to moment from the inside. Chi is an immediate diagnostic tool of the internal state of our being and all parts of our body. It's important for our well being that we pay attention to the feelings of our energy body and understand what it's telling us.

We can also exercise a considerable amount of control over how we experience our chi. Properly nurtured, chi can manifest within us as a wonderful feeling of health and love throughout our whole energy body. We can experience our chi as the living presence of pure unconditional love within us. Not only is this the most wonderful feeling imaginable but there is considerable evidence that it fosters our health and well being, though of course the effect of our own chi is always limited in the face of other active chi energies. The universe is all one computational flow of chi or existence in which our form, which manifests our personal chi, is but a miniscule part.

We can choose to experience the presence of chi within us as pure love, as a feeling that floods our being and refreshes and nourishes us. We can also imagine this as the presence of the living God within us or the awakening of our Buddha-nature so long as we remember this is personal myth rather than objective reality. In any case it's a wonderfully refreshing and transformative experience.

And objectively we can say that God, the universe, does love us simply because the universe is manifesting us into existence. Being manifested into existence is certainly the ultimate act of love. We exist only as the unique result of vast uncountable and unknowable numbers of enormously improbable coincidences. One out of millions of sperm at each conception of every one of our billions of ancestors, and the actual pair choices of each of the multitudes of possible ancestral matings, not to mention the uncountable myriads of quantum events back to the original complete fine-tuning of our universe; every one of these had to happen exactly as it did for us to be here right now in the present moment of our existence as we are. Our amazingly improbable existence in the present moment is the ultimate act of love, and can certainly be experienced as such. The universe, God if you wish, embraces us in the arms of existence and floods us with the pure love of the immanence of being.

With the proper understanding we are one with God and Buddha Nature as we already share their common existence. We are a part of the living God of the universe. From this perspective God is our essence and within us at all times. There is nothing other than God within the universe, and every finest detail of our being is a part of God. If you don't experience this it's only because you don't let yourself experience it because God is the active experience of all things. From this perspective every thing that happens in the entire universe is an act of God, is divine and perfect and absolute, and every one of us continually lives within it in the existence of the present moment.

Thus a wonderful, beautiful, and enormously profound new vision of reality emerges naturally from Universal Reality as a personal relationship with the existence of the universe. Completely consistent with modern science, this vision incorporates all the pieces missing from the usual interpretations of science such as consciousness, the present moment, and the nature of existence to achieve a complete Theory of Everything that automatically includes the realization of the reality it reveals.

Thus God can be identified with the divine living essence in all things including ourselves. All we have to do is realize its presence and God appears within us and becomes us and we become God. God is always right here within us waiting to be realized.

In this view God is the existence that animates all things and shines its immanence within their forms. All things are empty forms filled with God. God breathes in our every breath, God moves in our every movement, God thinks our every thought, God feels our every feeling and our every feeling is of God. And God is love and can be experienced as love, as love that fills the empty form of our being.

ACCEPTANCE

Because reality is absolute in the sense that it is all that is or can be exactly as it is in the present moment it is always enough. This is always true no matter where we are or what our situation is. There is after all nothing else possible in the present moment than what actually exists.

When this is realized there is never a need for anything else or any sense of loss, incompleteness or anything lacking. Because reality is the very substance of our being it is all that is ever needed because it is all that we can ever have or be. The ever-present formless essence of reality is all that can be and thus when its true nature is realized its direct experience is all that one could ever want or need. It is our very essence and our only true self and there can never be anything else. Forms come and go but the essence of reality always remains and that is always enough. This is always true; even as one works in the world of forms to effect changes in those forms our own inner nature, our own true self, never changes. Forms come and go but our Buddha Nature always remains and thus our true reality always remains. God never leaves our presence.

Forms continuously arise, change and vanish into non-existence but the common immanent reality within which all forms exist is always present. All forms themselves are empty, transient and illusory forms. It is only existence itself in which all forms arise that is permanent and ever present and always available to us in our form if we just open ourselves to its realization.

Because reality is what is and absolutely so and cannot be otherwise than it is right now, realization accepts it as such as it must to be in accord with reality. The necessity and inescapability of absolute acceptance of what is is an essential part of realization. This is true not just of the formless essence of reality but of the current state of all forms in the present moment. Once forms appear they absolutely are as they are and must be accepted exactly as they are if the true nature of reality is to be realized. Otherwise we deny the reality upon which we depend.

This need not keep us from working to effect change, it just means accepting that the forms we are working to change are the ones that actually exist. By accepting things exactly as they are we increase our capacity to change them.

Realization also includes the complete acceptance of our selves as we are. We accept our selves as we are by releasing our desires and attachments for things that are likely beyond our attainment. By releasing unreasonable desires and attachments and by accepting our situation in life as it is we release many of the forms that lead to suffering and come closer to realizing our true self as it actually is in the present moment. Our true self is the one thing that is always attainable and within our grasp if we just open ourselves to it and embrace it because it is what we actually are at every moment of our existence.

There is an ultimate bravery in the total acceptance of reality as it is and confronting its awesome absoluteness directly and completely. It is also the total acceptance of our complete and total aloneness in the eternal presence of God the universe. We are completely and totally alone in a personal sense because our personal forms are inherently distinct from all other personal forms, and yet we are always completely and absolutely in the living presence of God, the universe, because we share the essence of existence within our personal form in common with all other forms.

In absolute acceptance of what is we dwell at peace in pure love in the present moment. In this state of completeness there is nothing more that is needed. Reality is always enough. It is always eternally fresh and real and alive and is always immediately available to us because it is already our fundamental essence.

PURPOSE AND ETHICAL PRINCIPLES

The theory of Universal Reality naturally leads to some basic ethical principles and suggests a plausible, though speculative, purpose for our existence both as a species and as individual beings.

We are certainly sense and knowledge organs of the universe through which it becomes better able to experience and know itself. The universal program has evolved us and other sentient life forms and through us is able to become aware of itself. God, the universe is waking up with us. It can reasonably be argued that this is the purpose of our existence. But if so, to fully fulfill this purpose our knowledge and experience of reality must be as complete and as realized as possible.

To this end it's natural that it is ethically 'good' to spread scientific knowledge and realization as widely as possible and diminish delusion and ignorance and suffering as much as possible, and to that end, to work to make the earth a sustainable healthy and protected environment to facilitate this.

By doing so we move towards a more and more self-aware and enlightened universe in which God, the universe has maximal awareness and knowledge of itself. We can also speculate that this is the purpose of the universe itself, to move from an originally unconscious state towards the eventual goal of a fully self-conscious universe, and that mankind is a step the universe has evolved in its progress towards this eventual goal. This is of course speculative, but it's certainly a reasonable hypothesis based on the evidence.

The original fine-tuning of the universe implicitly contains within it the seeds of this progression, as it's exquisitely fine tuned so its programmatic evolution naturally leads through innumerable coincidences of random choice to the emergence of intelligent life capable of knowing the universe that produced it. All the critical elements of this design lie implicit in the original virtual nature of the quantum vacuum which produced the universal program of existence including us.

If our true destiny is to function as sense and knowledge organs of the universe then the more accurate and compassionate and enlightened we are the better is the universe's experience of the reality of itself. Each of us is a little fragmentary bit of God, a little bit of God's total mind and body, by which God knows itself and with realization becomes enlightened through us as we become simultaneously enlightened through the experience of God. Certainly this realization is its own reward.

There is no absolute good and evil in the computational universe. These are human concepts, which are always relative to some set of human standards. And it's often quite difficult to apply any set of standards because whether effects are good or evil is always a judgment by someone at some time and what is good for one is often bad for another. However there are generally accepted social norms from culture to culture that have evolved primarily to facilitate stable societies. These social standards are the primary references for good and evil around which individual standards tend to cluster.

The idea of karma, that good ultimately begets good and evil begets evil is not consistent with the actual laws by which information forms evolve. There may be some tendency in some cases for like to beget like but there are numerous exceptions and by whose standards are ethical results to be judged, and at what point in the continuously evolving network of events? There are innumerable examples of well-intentioned actions producing tragic unintended consequences. And there is certainly no reincarnation so there can be no karmic transmission from one lifetime to another.

Nevertheless it's possible to outline some general ethical principles in the context of realization. Certainly the first is to attain realization itself. While Zen correctly points out that enlightenment is not a thing to be attained, that is the view *from* enlightenment rather than from *the path towards it*. The corollary is the Bodhisattva ideal to promote realization among all beings and to minimize suffering. This can be done by example, by teaching, and hopefully by writing books like this one (Tsunemitsu, 1962).

Another very reasonable core ethical standard is protecting and fostering the sustainable health and viability of Earth's biosphere. This is arguably the single most important ethical principle in that it sustains and maximizes the health and existence of all known life. Earth's biosphere is the only known cradle of the convergent emergence that seems to be in the process of bringing self-awareness to the universe. For this to flourish human society must become sustainably integrated with nature, and man must begin to tend the earth as a natural Garden of Eden and strive to develop a Heaven on Earth. It would be an enormous, perhaps irreversible, setback to the apparent direction of the evolution of the universe if humanity were allowed to destroy the viability of the earth itself with all that implies.

Another fundamental ethical principle is compassion, which tends to arise naturally from the realization of the common Buddha Nature we share with all beings This realization naturally motivates us to help alleviate the suffering of all sentient life forms including our own selves and to foster realization among them.

This principle of compassion has profound consequences for how one relates to other beings including the question of eating meat. One recognizes the living sentient spirit within all animals and their capacity for suffering but at the same time one recognizes that predation fills an essential natural ecological function; that all individual organisms must die and that death both supports life by providing food for other life and also makes room for new life and thus creates the opportunity for better adapted life. We must realize and accept the great plan of life and death as essential for the evolution of the universe, but we should do so in a compassionate and intelligent manner.

Difficult questions always arise and there are not always easy answers. If life itself can be considered the ultimate individual good is it better for an animal to have lived a good and happy life till it is humanely slaughtered for meat or is it better for that animal to have never had the joy of existing at all? And if animals are to be killed for meat is it better to kill thousands of small creatures such as shrimp or one large cow of equivalent nutritional weight? These are profound questions that should always be approached with compassionate empathy for the beings involved.

Good and evil are not simple and are always human valuated momentary snapshots of isolated events in an enormous web of ever evolving forms. And these judgments are always relative to each other in complex interacting processes playing out over various time scales. In general realization and compassion for all beings including oneself and for the sustainable environment of the Earth are the great universal goods our lives should attempt to foster.

Zen has a somewhat similar approach that individual purpose is simply acting in accordance with the underlying principles of reality and flows of existence (Watts, 1957). It is to act not so much from one's personal desires, attachments and programming but in concert with the greater programs driving the world of forms. In so doing one gives up much of one's personal agenda and acts as one's realized self, one's Buddha within. In this view our ultimate freedom consists in giving up our personal freedom to align with the greater flows of reality, and thus our own realization and service is an example to others helping liberate them from suffering.

The traditional Buddhist notion of the Bodhisattva who upon realization returns to the world to spread realization by example is the prototype of this principle (Wikipedia, Bodhisattva). The notion is that by teaching, working with the poor and needy, or simply manifesting realization in the world one furthers realization and ultimately helps release sentient beings from suffering.

THE ENLIGHTENMENT EXPERIENCE

Because reality is completely absolute as it is and absolutely present, its direct realization often occurs with sudden profound intensity. It has been compared to the sudden shock of meeting a tiger on the road or suddenly looking into the eyes of God and seeing God looking back (Wu, 2005). An enlightenment experience is the sudden realization of the actual awesome presence of the absolute realness of reality in all its immanence.

What was previously understood only as an abstract concept is suddenly realized as the living here now presence of reality itself directly within and around one. The true nature of reality is directly experienced and not just intellectually understood. Zen calls this experience 'satori' but a similar experience is common to many religions (Suzuki, 1956). It can come as a sudden profound shock to consciousness as the veils of illusion suddenly drop away, the scales fall from one's eyes, and reality is suddenly revealed right here and now in all its awesome absolute realness as the living essence of all things.

Because reality is *absolutely real and present* the effective intensity of its presence is unlimited and dependent only on the capacity of the observer to experience it. Normally mind operates at a mundane level preoccupied with a continual procession of daily forms and tasks and doesn't allow consciousness to experience the truly awesome intense absolute realness of reality that is possible. Allowing consciousness to experience something of the true intensity of reality is normally reserved for sudden emergencies where maximum attention and engagement are required for personal survival. This is because extreme situations mobilize intense fight or flight energy levels in both mind and body that cannot be sustained.

The enlightenment experience is superficially similar in its intense clarity of mind but rather than extreme fight or flight adrenaline surges there is instead a strong, clear, healthy relaxed readiness of life energy that vitalizes rather than drains. This is a state of balance and refreshment rather than a sudden dissipation of energy. One is continuously aware of the awesome absolute presence of reality but there is a complete and total ease and acceptance and a perfect easy equilibrium in resting within it as if one had finally found one's true home (Suzuki, 1956).

Because the intensity of the actuality and presence of reality is absolute the only limit on the intensity of realization is the capacity of the realizer. By letting go of the natural tendency of mind to damp down the intensity of the experience of reality one naturally experiences that intensity to the level of one's capacity. To do that one must open oneself completely to the presence of reality and embrace it. Though sometimes frightening this becomes much easier when we realize there is actually no alternative to existing within reality as it actually exists in the present moment no matter how we might attempt to escape it by distracting or dulling our mind.

Mind normally makes us wary of reality and the dangers it may hold but while it's certainly true there are many programs running in reality which can pose significant dangers to our individual existence, the actual presence of reality itself is completely benign and in fact embracing it more fully and intensely actually enables us to detect and deter hostile forms more effectively (Saotome, 1989). Thus a major impediment to the intense realization of reality is the fear of hostile forms within it and the illusion that if mind somehow damps the intensity of our experience of reality that somehow protects us from those dangers when the opposite is actually true.

Thus completely opening oneself and embracing reality and releasing the illusory fear of its presence is essential to its realization and simultaneously allows us to live more effectively within it.

This is the mind of the samurai, which abandons individual self and accepts the total and absolute presence of reality, including even the ever-present possibility of personal death, and in so doing is able to exist at ease in the present moment with maximum effectiveness. The ultimate bravery in abandoning the forms of self that seek to insulate self from reality attains maximum realization of reality and maximum effectiveness within it (Musashi, 1974).

ZEN MIND

Realization is not to be found just within the gates of a temple or the teachings of some sect or master. Realization is the direct experience of reality and thus may be found anywhere and everywhere at any moment. Reality is everywhere and all one has to do is look with realized eyes to see it. No technique or path or teaching is intrinsically better than any other or even necessary. Sitting in meditation can be useful but realization is not found just in sitting. Realization is to be found anywhere in the entire world around us at every moment of our existence because realization is the recognition of reality and everything, including ourselves, are direct self-manifestations of reality (Suzuki, 1956).

There is no transmission of realization or enlightenment. Teachers can be useful in demonstrating and guiding one along a path towards realization but they cannot transmit any realization at all. There is nothing to transmit when everything is already present. Reality continuously self-manifests itself and reality itself is the only true teacher. All one needs to do is open oneself to the continuous presence of reality and see it for what it actually is.

Realization is not just to be found through a master's kōan. Reality itself is the ultimate kōan in whose solution is found realization. The quantum kōan and many others are the subject of this book. Reality is the only master and it presents itself to us as a kōan every second of our existence. Reality is the ultimate unanswerable question, the ultimate unsolvable kōan, in whose disappearance lies realization. The solution is not in the answer but in the vanishing of the question; in the realization of the presence of reality as it actually is. Realization of the living presence of reality itself unmediated by illusion is the only possible answer. The answer lies not in words, though words can be a guide, but in direct experience (Legge, 2010).

This is the meaning of the Japanese Zen expression, 'Mumon', which can be variously translated as 'no gate', 'the gateless gate', or 'the gate to emptiness' (Blythe, 1966). 'Mu' does not mean nothingness in the usual western sense, but refers to the emptiness of forms in which is found the true presence of being. Mumon means there is no gate that must be passed through to achieve enlightenment. And it specifically implies it's unnecessary to pass through the gated entrance of any Zen temple or monastery to achieve realization. Wherever you are you are already within the true reality you seek.

YOU ARE ALREADY ENLIGHTENED

This book has been a comprehensive and detailed search for the true nature of reality. We have discovered that the apparent reality of the world we seem to exist within is an illusion created by our mind, and not at all like the actual world of running programs computing data within the formless sea of immanent existence. And in this last chapter we have explored how to experience the true nature of the reality hidden behind the veils of the illusion of our simulation and how to directly experience that reality.

But there is still one more secret to be revealed. We must finally realize that our illusory simulation of reality is in fact our direct experience of the true nature of reality itself, and is in fact our only possible direct experience of reality.

Yes our simulation of reality is an illusion, but that illusion consists of real information structures filled with existence existing in an immanent universe. Our illusory simulation is as much a part of the reality of the universe as any other information structure within it. We realize the most important lesson of all, that illusion taken for reality is illusion, but illusion realized as illusion is reality.

Our mind's simulation of reality is a magician's trick. The trick is absolutely real, but its reality is not as it appears to be. Likewise our simulation of the world we seem to exist within is absolutely real and is our only possible experience of reality. Thus realization is not a matter of trying to escape or deny our illusory simulation, it's a matter of understanding and experiencing its true nature. We need look no further than where we are already, but we must look with enlightened eyes.

This is the meaning of the Zen saying, "Mountains are mountains again" (Suzuki, 1956). Originally we thought of mountains as the physical mountains our simulation told us they were. But then we realized that the true nature of mountains was information structures generated by programs running in the quantum vacuum. But now we finally realize that the mountains of our simulation are in fact what mountains really are. Our illusory representation of a mountain is the real mountain of our direct experience, but now we understand and experience its true immanent nature as well as its illusory appearance, and in experiencing the truth the world becomes much richer and much more real.

We are the dynamic information structure of our total program running in the immanence existence of reality, and our simulation of reality is an integral part of that program. Though all aspects of our program interact computationally with both internal and external programs of the world at all levels of our biological hierarchy, our simulation is our overall model and conscious experience of reality. We can improve the accuracy and realization of the illusory nature of our simulation, but the simulation, however we experience it, is the complete actual reality of our experience. As such our illusory simulation of reality is the reality we have always sought.

Thus we are all already enlightened. We are all already enlightened because we all live in actual reality all the time, and always have. We just have to look around and realize that the true nature of the reality we seek is our illusory simulation of reality seen for what it actually is. This is all that exists in our experience and everything that exists is by definition part of reality, we need only realize it for what it is.

Our simulation is the only part of reality we directly experience completely and accurately as it actually is. Its illusory nature is its true reality, and we already have the most absolute realization and direct experience of the reality of our simulation possible. We need only recognize it for what it is, rather than what it pretends to be. The illusion of our simulation realized for what it really is, is the reality we seek, and ultimately this is the only realization possible.

But of course this precludes nothing. Our simulation can be improved as our understanding increases, or as we transition from mundane life to meditation to realization. But no matter how the simulation changes, in whatever form it takes, it is always our ever present direct experience of the true nature of reality, because whatever it is, it's always the true reality of the present moment. The experience of the information of our simulation and its illusory nature is the true nature of reality.

So finally we realize that in Universal Reality nothing actually changes from how we saw it before, the universe is as it always was. We just now see the world around us with entirely new eyes, as the most profoundly beautiful and awesome presence imaginable. All things are now the living immanent information of what they are continuously interacting and evolving in concert to the music of a single Uni-Verse stretching back to the beginning of time mysteriously revealed in the vast computational information nexus that all things are part of.

Finally we understand that in our search for realization of the true nature of reality, that we all continuously live only within reality and are entirely composed of reality ourselves. There is nothing, and can be nothing, that is not already part of the true nature of reality. Therefore we are all already enlightened and could not be otherwise. We have always lived within enlightenment. The realization we have sought has been with us all along. It's just a matter of realizing that and embracing it.

There is really no trick or effort to realization or enlightenment. We are already enlightened. Everyone is already enlightened and always has been. Enlightenment is simply a matter of realizing we are already enlightened and always have been because there is nothing that is not the real and actual presence of reality lying completely clear and visible before us. Of course realization can be refined, but enlightenment is just seeing reality as it actually is and it is always exactly as it appears.

Everything in the world, every experience is exactly what it is. Yes, it has a deep structure, and yes it carries hidden secrets and illusions which are also part of reality, but nevertheless what we experience is exactly what we experience and that is reality because reality is exactly what is in the present moment and even if it is illusion experienced as illusion, even *that is* the reality of the present moment. However the deeper realization is experiencing illusion as the illusion that it actually is and thus its deeper reality. That then becomes the reality of the present moment realized more clearly.

Realization is simply whatever experience exists in the present moment, as it is with or without any cognitive interpretation in the simulation because all interpretations are also only the direct experiences of themselves. Direct experience even includes the direct experience of mistaken cognitive interpretations, whether realized as such or not. Illusion taken for reality is illusion, but illusion seen as illusion is reality.

Everything is illusion but everything is reality because reality consists entirely of illusion when it comes to forms. The empty illusory nature of forms is their reality, and their reality is the manifestation of the nameless fundamental presence of reality in which all forms arise and manifest which is the true nature of the universe and all things in the universe including our selves.

With insight, study and practice more and more of the true nature of things is realized but what we do experience now exactly as we experience it, realized as such, that is the true reality of the present moment. Thus we are all already enlightened and it's just a matter of waking up and realizing we are already here and always have been!

Ultimately all we ever experience is the immanence of existence itself. In whatever form, in truth or illusion, or in relative formlessness, ultimately all that exists is the immanence of existence. And this is our true nature and the true nature of reality.

Welcome to Universal Reality!

EPILOGUE - TESTING THE THEORY

Every new theory must be subject to experimental tests to either confirm or falsify it. For example Einstein's seemingly outlandish relativity theory would never have been accepted had it not made testable predictions of the bending of starlight by the sun's gravitational field (Eddington, 1928). In this case the tests were straightforward, Einstein was confirmed, and relativity quickly became an accepted theory.

Other theories have not been so lucky. For example the theories of evolution and of plate tectonics languished for years without simple conclusive tests until gradually the evidence became overwhelming. This is the usual case. Science progresses slowly and carefully but in the end it always progresses.

The theory of Universal Reality faces the same challenges. Since it's mostly a completely new reinterpretation of accepted science in the framework of a much broader Theory of Everything it's difficult to isolate clear tests that could either confirm or falsify it. However a few possibilities do come to mind.

Hopefully there will be at least some who test the theory against both experimental evidence and by consistency with the general body of scientific knowledge. However one must always be careful to test against actual mathematical theories rather than current *interpretations* of scientific theory that are not part of science proper. These tests may involve trying to find reasons why Universal Reality can't possibly work, and those will be useful in bringing to light points that need development, but I suggest the more fruitful approach with any new theory is to try to find ways to make it work.

The theory of Universal Reality appears to have much promise in that it explains so much so well from a single universal approach. The general approach has been to accept the experimentally confirmed equations of science but develop a completely novel unifying interpretation that incorporates all aspects of reality in a single Theory of Everything. The author believes Universal Reality is the best most comprehensive Theory of Everything on the market and urges others to put it to the test and report their findings to Edgar@EdgarLOwen.com.

Since much of Universal Reality is a new interpretation of science and other aspects of reality it may not be subject to experimental tests.

However it can be tested with respect to its logical consistency with accepted science. Overall consistency across all aspects of reality is the true and ultimately only test of validity.

There are other useful tests as well such as elegance, simplicity and beauty. Universal Reality is founded on a set of simple principles with universal scope, and scores high on these criteria. And it proposes a computational model that is quite parsimonious compared with many of the currently fashionable interpretations of reality.

Not only does it explain the universe of science quite well but it does do in a manner that intuitively integrates existence, consciousness, and the present moment, the fundamental experiential constituents of reality about which current science has nothing meaningful to say.

There are a number of specific testable proposals of Universal Reality that come to mind and there are no doubt others. Proving the existence of a universal present moment common to all observers and processes across the universe is a crucial one. This should be easy to confirm by simply proving there is a unique one-to-one correspondence between the proper times of all clocks upon which all observers agree.

Also of fundamental importance is a mathematical confirmation of the theory's core notion that a spacetime that is computationally created along with mass-energy structures by quantum events is key to unifying quantum theory and general relativity. To anyone interested in unifying relativity and quantum theory I strongly suggest this is the correct approach.

Confirming a slightly positive curvature to space would tend to confirm Universal Reality's prediction that the universe has a closed finite positively curved hyperspherical geometry.

Universal Reality's theory of Dark matter as a spacetime curvature produced by the predicted uneven Hubble expansion of space around the edges of galaxies and galaxy clusters should be confirmable by measurements of the strength and distribution of dark matter relative to the motion of the galactic masses that produced it.

Any detection of dimensional drift or other relativistic anomalies would tend to confirm Universal Reality's theory of a computationally based absolute spacetime background with respect to which rotation and world lines are relative. In fact confirmation of the necessity of an absolute dimensional background in relativity itself will also confirm Universal Reality.

There are likely testable consequences of Universal Reality's theory of dimensional fragments. All entangled particles should have exact instantaneous dimensional values with respect to each other at least within the limits of uncertainty that evolve in a predictable manner and may be within the range of measurement. In other words the wavefunctions of entangled particles should be coherently coupled and this coherence may be observable.

There are also many other proposals of Universal Reality that are potentially subject to experimental and perhaps theoretical falsification. The inability to falsify these would lend considerable credence to the theory. A few are of particular importance.

The METc Principle that all forms of mass and energy can be consistently modeled as different forms of spatial velocity.

A mathematical prediction of gravitational reversal in black hole to white hole transitions would lend credence to Universal Reality's theory of big bounces though it's peripheral to the main theory.

And lastly the ability to program a convincing simulation of reality on the basis of the theory of Universal Reality that correctly models the major known aspects of reality would be the best test of all and a very strong confirmation of the theory. We have already taken some initial steps in this direction that seem quite promising so far.

BIBLIOGRAPHY

Blythe, R.H. *Mumonkan*. Hokuseido Press, 1966.
Budge, E.A. Wallis, trans. *The Egyptian Book of the Dead*. Penguin, 2008.
Chaitin, Gregory. *Meta Math, the Quest for Omega*. Vintage, 2006.
Chalmers, David. *Facing Up to the Problem of Consciousness*. Journal of Consciousness Studies. 2 (3) 1995 pp. 200-219.
Chomsky, Noam. *Aspects of the Theory of Syntax*. MIT Press, 1965.
Cornford, Francis. *Plato's Cosmology: The Timaeus of Plato*. Hackett, 1997.
Eddington, Arthur Stanley. *The Nature of the Physical World*. 1928.
Evans-Wentz, W.Y., ed. *The Tibetan Book of the Dead*. Oxford Univ. Press, 1957.
Feynman, Richard. *Lectures On Physics*. Pearson, Addison, Wesley, 2006.
Goffinet, François, *A bottom-up approach to fermion masses*. Université catholique de Louvain, 2008.
Greene, Brian. *The Elegant Universe*. Norton, 1999.
Greene, Brian. *The Fabric of The Cosmos*. Vintage Books, 2005.
Halpern, Paul & Wesson, Paul. *Brave New Universe*. Joseph Henry, 2006.
Hawking, Stephen W. *A Brief History of Time*. Bantam Books. 1998.
Hofstadter, Douglas R. *Gödel, Escher, Bach*. Vintage, 1980.
Legge, James. *The Tao Te Ching of Lao Tzu (translation)*. Commodius Vicus, 2010.
Lovelock, James. *The Ages of Gaia*. Norton, 1995.
Marshak, Alexander. *The Roots of Civilization*, McGraw-Hill, 1972.
Misner, Charles W.; Thorne, Kip S.; Wheeler, Archibald. *Gravitation*. Freeman, 1973.
Musashi, Miyamoto, Harris, Victor, trans. *A Book of Five Rings*. Allison and Busby, 1974.
Owen, Edgar L. *Spacetime and Consciousness*. EdgarLOwen.info. 2007.
Owen, Edgar L. *Mind and Reality*. EdgarLOwen.info. 2009.
Owen, Edgar L. *Reality*. Amazon.com. 2013.
Penrose, Roger. *The Road to Reality*. Knopf, 2005.
Piaget, Jean, *Logic and Psychology*. Manchester University Press. 1956.
Piaget, Jean. *The Child's Conception of The World*. Littlefield, Adams & Co., 1960.
Price, Huw. *Time's Arrow and Archimedes' Point*. Oxford, 1996.

Saotome, Mitsugi. *The Principles of Aikido.* Shambala, 1989.
Schroeder, Daniel V. Purcell Simplified, http://physics.weber.edu/schroeder/mrr/MRRtalk.html, 1999.
Susskind, Leonard. *The Cosmic Landscape.* Little Brown. 2006.
Suzuki, Daisetz. *Zen Buddhism.* Doubleday Anchor Books, 1956.
Tsunemitsu, ed. *The Teachings of Buddha.* Mitutoyo, 1962.
Vilenkin, Alex. *Many Worlds in One.* Hill and Wang, 2006.
Watts, Alan W. *The Way of Zen.* Pantheon, 1957.
Wigner, Eugene. *The Unreasonable Effectiveness of Mathematics in the Natural Sciences.* John Wiley, 1960.
Wikipedia contributors. *Wikipedia, the Free Encyclopedia.* http://wikipedia.org
Wilhelm, Richard, Cary F. Baynes, trans. *The Secret of the Golden Flower.* Routledge & Kegan Paul, 1931.
Wilhelm, Richard, Cary F. Baynes, trans. *The I Ching.* Bollingen Press, 1962.
Wu, John C.H. *The Golden Age of Zen.* Pentagon Press, 2005.

Edgar L. Owen was born April 1st, 1941 and quickly realized that reality is not as it appears to be. A child prodigy, he entered the University of Tulsa aged 15 and received a B.S. with honors in science and mathematics with a minor in philosophy at 18 before completing several more years of graduate study in physics and philosophy.

In the early 60's he moved to the Haight-Ashbury in San Francisco where he hung out with notables from the Beat Generation, and conducted an intense personal study of the nature of mind and consciousness. From there he traveled to Japan where he lived for three years studying Zen and Buddhist philosophy while subsisting as a ronin English teacher.

Upon returning to the US he began a career in computer science writing numerous programs in artificial intelligence, simulations, graphics, and cellular automata while designing and managing advanced computer systems for the New York Federal Reserve Bank and AT&T. He then left the corporate world to start his own software business marketing his own CAD programs, which he ran for a number of years. Currently he owns a premier Internet gallery of fine Ancient Art and Classical Numismatics at EdgarLOwen.com.

Deeply immersed in nature since childhood, and always considering it the ultimate source of his inspiration and knowledge of reality, he has served as Chairman of his local Environmental Commission and organized several campaigns to protect the local environment and its wildlife.

Over the last several years he has worked to combine and organize the results of a lifetime of study of the various aspects of reality into a single coherent Theory of Everything. He now spends most of his time exploring the wonderful awesome mystery of reality and how it can be experienced more fully and deeply and enjoying his existence within it.

Edgar currently lives in Northern NJ in a big old house on top of a hill where he communes with nature and enjoys the company of his wild visitors including the occasional human. Edgar is currently single and looking for a younger housekeeper companion ☺. He can be contacted at Edgar@EdgarLOwen.com.

www.ingramcontent.com/pod-product-compliance
Lightning Source LLC
Chambersburg PA
CBHW081605200526
45169CB00021B/2035